THE TECTONIC PLATES ARE MOVING!

THE TECTONIC PLATES ARE MOVING!

Roy Livermore

Great Clarendon Street, Oxford, OX2 6DP,
United Kingdom

Oxford University Press is a department of the University of Oxford. It furthers the University's objective of excellence in research, scholarship, and education by publishing worldwide. Oxford is a registered trade mark of Oxford University Press in the UK and in certain other countries

First Edition published in 2018
Impression: 1

Published in the United States of America by Oxford University Press
198 Madison Avenue, New York, NY 10016, United States of America

British Library Cataloguing in Publication Data
Data available

Library of Congress Control Number: 2017953567

ISBN 978–0–19–871786–7

DOI: 10.1093/oso/9780198717867.001.0001

Printed and bound by
CPI Group (UK) Ltd, Croydon, CR0 4YY

Note on the Title of this Book

The title of this book is derived from a remark made in 2004 by John Prescott, then Deputy Prime Minister in Tony Blair's Labour Government. Referring to the impending downfall of Mr Blair, John is reported to have observed that 'the tectonic plates appear to be moving'. A year or so later, the Liberal Democrat leader, the late Charles Kennedy, referring to what he perceived as the realignment of the political landscape in Britain, informed his own party in his New Year's speech that 'What I do detect is a shifting of the political tectonic plates: a sense that change is coming'. That change came sooner than he imagined. By 7 January, he had been forced to resign by widespread opposition from his colleagues. Since then, the tectonic plates metaphor has been applied to just about every major political transition, including the UK referendum on leaving the EU and the election of Donald Trump as US President. In fact, as with most things, the politicians have got it wrong. There is no such thing as a 'tectonic plate': it is not the plates that are tectonic, but the tectonics that is plate-like.

Acknowledgements

My thanks and sympathy go to Ania Wronski at OUP, who (almost) never lost patience with me. I thank Kevin Burke, Dan McKenzie, Roger Searle, and Tony Watts for reading and commenting on individual chapters. All errors, omissions, and opinions are my own. Thanks also to Valerie and Mike for their support during difficult times.

Contents

Dedicated to the memory of my colleague
and friend, Alan Smith (1937–2017)

Introduction

> ... it is doubtful whether there will ever again be such a profusion of unexpected discoveries concentrated into so short an interval of time as there has been during the last twenty years.
>
> ARTHUR HOLMES, *PRINCIPLES OF PHYSICAL GEOLOGY* (1965)

Scientific revolutions rarely start with a bang. In 1953, a modest article, barely a page in length, appeared in the weekly science journal *Nature*, signalling the beginning of modern genetics. And just a decade after Watson and Crick's brief note, another, equally modest, contribution appeared in the 7 September 1963, issue of the same journal. Sporting the slightly arcane title 'Magnetic anomalies over ocean ridges', it heralded an even more profound advance in our understanding of the natural world.

The authors were Fred Vine, a Cambridge University research student, and his supervisor, Drummond Matthews. What they had done, in a nutshell, was to bring together the previously contentious idea that the Earth's magnetic field occasionally reverses its polarity, with the even more contentious notion that the ocean basins are created by a more-or-less continuous volcanic process known as 'sea-floor spreading', in order to explain the shapes of anomalies observed on magnetic profiles measured on ship crossings of a mid-ocean ridge. On the face of it, that doesn't sound all that exciting. Yet, in so doing, they also showed how such anomalies record not just the polarity of the Earth's magnetic field, but the entire history of the ocean basins.

Most scientists would settle for that, but it was the implications of this discovery that really shook the foundations of the scientific establishment. For, just as reversing the observed expansion of the universe took cosmologists inexorably back to the Big Bang, so rewinding sea-floor spreading closed oceans and united the continents in a single 'supercontinent', precisely as proposed—and widely dismissed—a century before. Hence, in direct opposition to the scientific consensus of the time, Vine and Matthews' conclusions required that continents and oceans must migrate constantly over the surface of the Earth—in other words, they had found the proof of continental drift. They might

The Tectonic Plates Are Moving! Roy Livermore, Oxford University Press (2018). © Roy Livermore.
DOI: 10.1093/oso/9780198717867.001.0001

well have winged into the *Eagle* pub in Cambridge, boasting 'we have discovered the secret of the Earth!'

The 'Vine–Matthews Hypothesis', as it became known, was the key that opened the door to a new era in Earth Science. There could be no doubt that a major scientific revolution had begun! Or perhaps there could. For, the following year, Manik Talwani, of the prestigious Lamont-Doherty Geological Observatory in New York, referred briefly to Vine and Matthews' hypothesis in his review of marine geophysics,[1] adding 'It should, however, be pointed out that less startling explanations are also possible for these anomalies.' Another article published in 1964, this time by Vic Vacquier and Dick Von Herzen of the Scripps Institution of Oceanography,[2] reported 14 new magnetic and bathymetric profiles across the Mid-Atlantic Ridge. Here was the perfect opportunity to put that startling explanation to the test. Yet the only mention of Vine and Mathews' paper was a passing reference to anomalies observed in the Indian Ocean—the hypothesis itself was entirely ignored and the new magnetic profiles 'smoothed' to get rid of those troublesome anomalies on the ridge flanks! Late in 1965, an article on marine magnetic anomalies by Talwani and two other Lamont heavyweights[3] appeared in the prestigious American journal *Science*, and this time Vine and Matthews' hypothesis was at least mentioned—before being dismissed with these words that must afterwards have given the trio sleepless nights: 'The flank anomalies are *not* axial anomalies at greater depths. Vine and Matthews explain the ridge magnetic anomalies by invoking the "spreading floor hypothesis" of Hess and Dietz. However, we believe that the flankward [*sic*] diminution in amplitude of the axial anomalies and the differences in character between flank anomalies and axial anomalies make their hypothesis untenable.' These comments by the eminent professors bring to mind the first, or perhaps the second, of Haldane's four phases of acceptance of a major scientific discovery:[4]

1. This is worthless nonsense.
2. This is an interesting, but perverse, point of view.
3. This is true, but quite unimportant.
4. I always said so.

[1] Talwani (1964).
[2] Vacquier and Von Herzen (1964).
[3] Talwani, Le Pichon, and Heirtzler (1965).
[4] Haldane (1963).

By 1968, the Lamont group had entered the fourth stage, and quickly published a series of papers in which they reinterpreted their magnetic profiles in terms of the Vine–Matthews Hypothesis. No mention was made of their previous opposition to Vine and Matthews' elegant explanation, or of their own complete failure to grasp its global importance. Instead, readers of these articles might well have been left with the impression that, indeed, the Lamont scientists had 'always said so'.

There are striking parallels between the discovery of 'the secret of life' by Watson and Crick in 1953 and of 'the secret of the Earth' by Vine and Matthews in 1963. Both pairs worked in Cambridge University, the former at the Cavendish Labs, the latter just up the road at the Department of Geodesy and Geophysics (now the Bullard Labs). Second, all four authors were junior scientists, little known even within their own fields: indeed, Crick and Vine were still research students at the time of publication. In each case, there was a 'third man'—Maurice Wilkins, who refused a co-authorship with Watson and Crick, and Lawrence Morley, who failed to publish an idea similar to that of Vine and Matthews. More importantly, both discoveries were of fundamental codes embedded in nature: in one case, the genetic code that records biological evolution within living cells, and in the other, the magnetic bar-code that records Earth history beneath the oceans. Both were announced in short articles in *Nature*, and both were largely ignored—at first.

But there is one major difference. Crick and Watson, together with Wilkins, eventually received a Nobel Prize for their discovery and became household names. Vine and Matthews did not. Why was this? The simple answer is that there *is* no Nobel Prize for Earth Science. Another explanation is that Earth scientists have never been as effective at communicating the really exciting aspects of their science as the biologists. Vine and Matthews did, however, receive a number of other prizes for their work, but never the public recognition they deserved.[5]

In 1966, as many Earth scientists began entering the 'I always said so' phase, Vine put most[6] of the remaining sceptics out of their misery by presenting a comprehensive exposition of the new hypothesis in the journal *Science*, demonstrating how it explained magnetic anomalies,

[5] An article in the *Daily Telegraph* by Anna Grayson, entitled 'Molehills made out of mountains', highlighting this anomaly, was published in 2003.

[6] Despite Vine's efforts, a number of geologists, particularly in the USA and Soviet Union, failed to appreciate the new theory for some years to come.

not just in the Indian Ocean, but in *all* the deep oceans, how it provided a simple method for dating the oceanic crust *anywhere* on Earth, and how it allowed the relative speed and direction of crustal motions, past and present, to be calculated. For good measure, he even provided the first polarity reversal timescale, showing the history of the geomagnetic field for the past ten million years.

Breathtaking in scope as it was, the Vine–Matthews Hypothesis turned out to be the catalyst for an even larger-scale theory—plate tectonics. During the 1960s and 1970s, the theory of the plates was erected on the foundations laid by earlier researchers, notably Arthur Holmes. It was gradually accepted as the theory that finally explained why and where earthquakes and volcanoes occur, how ocean basins form and are destroyed, why and how continents separate and collide, and how great mountain ranges are created. No previous theory had ever explained so many apparently unrelated observations. Just as profound, it was recognized that, over geological time, plate movements must inevitably lead to changes in ocean and atmospheric circulation, and hence in global climate. And, by making and breaking connections between continents, such movements had a profound effect on the evolution and extinction of living species. Plate tectonics was thus a crucial factor in creating the conditions for the evolution of advanced life, and even humans, as we will see.

PART I
FIRST GENERATION

1

Probably the Best Theory on Earth

You don't believe all this rubbish, do you, Teddy?

COMMENT BY MAURICE EWING TO SIR EDWARD BULLARD, 1966

You wouldn't think there was much mileage for an international food-and-drinks firm in sponsoring 'blue-skies' scientific research, although, in this age of green politics, there just might be. The idea is not new, however, for the Danish brewer Carlsberg has been doing just this through its Carlsberg Foundation for many years. In 1979, for example, the Foundation sponsored an expedition to Greece that discovered a new species of cabbage.[1] More relevantly, it also funded a circumnavigation of the world's oceans in 1928–30, during which a number of major submarine features were identified for the first time. One of these, a topographic high in the northern Indian Ocean, rising more than 1000 m above the surrounding sea floor, was named after the sponsors, and thus became known officially as the 'Carlsberg Ridge'. Nobody at the time could have suspected that this remote and previously unknown edifice would one day play a central role in the greatest scientific revolution of the twentieth century.

*

The Year Plate Tectonics Began

In 1962, thirty-two years after the Danish discovery, another oceanographic survey was under way over the same patch of sea floor, as part of the International Indian Ocean Expedition.[2] This time, funding came from the British taxpayer and UNESCO, and the ship was the naval

[1] *Aethionema Carlsbergii*.

[2] The International Indian Ocean Expedition was a collaborative programme of scientific cruises carried out by thirteen nations using thirty-nine ships, between 1960 and 1963. The *Owen* data were intended to contribute to reconnaissance maps as the basis for future planning.

The Tectonic Plates Are Moving! Roy Livermore, Oxford University Press (2018). © Roy Livermore.
DOI: 10.1093/oso/9780198717867.001.0001

hydrographic vessel HMS *Owen*. A number of traverses over the Carlsberg Ridge were completed using, it has to be said, somewhat crude equipment, including a home-made echo-sounder, a second-hand gravity meter, and, most significant of all, a towed magnetometer,[3] also home-made. On board was Drummond (Drum) Matthews, a thirty-year-old post-doctoral researcher from Cambridge University, whose curious Christian name was the outcome of a pact made between his father—a First World War pilot in the Royal Flying Corps—and his best friend, such that, in the unlikely event that either of them survived the war and had a son, he would be named after the other. Drum's father survived, but was badly wounded, leading to a nervous breakdown and, when Drum was eleven, suicide.

In the pre-GPS[4] days of the early 1960s, it was a laborious business to navigate a ship along the straight, parallel, tracks required for a geophysical survey, necessitating the setting up of a network of transponders, moored to the sea bed with piano wire, from which the vessel could take its bearings, while the unreliability of the equipment necessitated frequent stops for repairs. Working around the clock, apparently without technical support, Matthews managed to acquire a series of depth and magnetic profiles over the crest of the Carlsberg Ridge, the data being recorded on paper tape (it would be another twenty years before research ships were equipped with computers for data logging). Amongst those data were the profiles that would ultimately convince the world that continents really do move, and so demonstrate that the geological establishment had, for over half a century, been entirely wrong about the biggest theory in Earth Science.

On his return to Cambridge, Matthews handed over the *Owen* data to his new research student and promptly went off on his honeymoon. The student, a recent Cambridge graduate named Frederick Vine, then set to work to make sense of the new data, acquired at such personal inconvenience by Matthews and his shipmates. Vine used one of the earliest mainframe computers—the Cambridge University *EDSAC2*

[3] A magnetometer is, as you might expect, a device for measuring the magnetic field, towed on a long conducting cable because steel-hulled ships have a magnetic field of their own.

[4] It may come as a surprise, however, to note that the first navigation satellite was already in orbit at that time, and that the Transit satellite navigation system was in use by the US Navy just a couple of years later.

Figure 1.1 Magnetic anomaly profile 4A across the Carlsberg Ridge obtained by HMS *Owen*. The solid line is the observed anomaly; the dashed line is the computed anomaly assuming no reversals. [From Vine and Matthews (1963).]

machine[5]—to predict the magnetic anomalies that would be observed if two of his pet theories were correct. These were, firstly, that the rocks of the ocean floor were produced by volcanic eruptions at places like the Carlsberg Ridge, and then transported away as if on conveyor belts—an idea first mooted in 1931 by the famous English geologist Arthur Holmes and revived in the 1950s by the American marine geophysicist and ex-US Navy Rear Admiral Harry Hess. And, secondly, that, while this was happening, the north and south magnetic poles of the Earth swapped places periodically, each time reversing the direction of the geomagnetic field everywhere on Earth.

The wiggly solid line in Figure 1.1 is the fruit of one of Drummond Matthews' sleepless nights on HMS *Owen*. It is profile 4A, showing the strength of the magnetic field[6] as the ship steamed over the Carlsberg Ridge. To be a bit more precise, it shows *deviations* of the magnetic field strength from that predicted at this location—otherwise known as 'total field anomalies' or simply 'magnetic anomalies'. Vine interpreted these deviations as the effect of magnetized rocks beneath the Carlsberg Ridge, which had clearly recorded some kind of signal, a bit like the recordings we used to make on those old-fashioned compact cassettes. And, like your bootlegged 'Now That's What I Call Music 6' greatest hits compilation gathering fluff in your glove-box, it is a magnetic tape recording,[7] but with a tape made of solid rock a hundred miles long and a mile thick. The question was, how had the recording been made and what was it that had been recorded anyway (two questions)?

[5] Interestingly, Vine and Matthews included an acknowledgement to the Cambridge University Mathematical Laboratory in their paper, thanking them for permission to use *EDSAC2*. This illustrates the novel status of computers at the time.

[6] More strictly, the 'magnetic flux density', measured in 'gamma' (a now-defunct unit: the modern equivalent SI unit is the 'nanotesla').

[7] Anyone under forty may have trouble with the notion of analogue recording. I suggest Googling 'musicassette'.

In their celebrated 1963 *Nature* article,[8] Vine and Matthews explained that such recordings are produced when basaltic magma containing the iron oxide mineral, magnetite, becomes magnetized in the Earth's field as it cools and solidifies. If, as Holmes and Hess had suggested previously, the ocean floor on either flank of a mid-ocean ridge is continuously moving away from the ridge axis as if on a conveyor belt—a process dubbed 'sea-floor spreading'—then a gap is created into which magma can rise from the mantle below and form new crust. The sea floor on either side of the ridge continues to move apart, however, reopening the gap, which is soon filled by more rising magma and so on ad infinitum, creating a more-or-less continuous magnetic record on each flank.

Elegant as it was, the hypothesis outlined so far would still not be capable of reproducing the wiggles observed on profile 4A. If new crust was always magnetized in the same direction (i.e. the direction of the Earth's field at the location of the ridge), then the only variations in the anomaly recorded on a ship profile would result from differences in water depth, as shown by the dashed line in the figure. One more vital ingredient is required. For, just as a tape cassette (or, nowadays, an mp3 download) containing an hour of just one note would be unlikely to hit the top ten (but who knows?), so a uniform magnetic signal in the crustal tape recorder would contain nothing much of interest. The crustal 'music' is provided by the unstable (or rather *bistable*—i.e. equally stable in two different states) nature of the geomagnetic field, which, every so often, very conveniently flips its polarity. Since field reversals are nearly random events, the duration of periods of normal or reversed polarity can vary widely, from thousands to millions of years, resulting in a unique sequence of short and long magnetic polarity intervals, a bit like a bar-code.

Today, the magnetic field lines point towards the north geographic pole (which is actually a 'south' *magnetic* pole, but let's not worry about that now). Back in the days when a south pole *was* a south pole, that is, before about 773,000 years ago, the field lines pointed south, and before 2.61 million years ago, north again, and before 3.60 million years ago, south once more. Geomagnetists have given each of the polarity intervals defined by these reversals a name: 'Brunhes', 'Matuyama', 'Gauss', and 'Gilbert',[9] the first of these being the present interval, named after

[8] Vine and Matthews (1963).

[9] Pronounced 'broo-nez', 'mat-su-yama', 'gowce', and 'gil-bert'. Other, shorter, reversals have since been discovered within these intervals.

the French geophysicist who, in 1905, discovered polarity reversals recorded in lava flows of the Auvergne. In 1963, the idea that the entire field reverses was still highly controversial,[10] but nevertheless formed a central part of what was to become the Vine–Matthews Hypothesis. Rocks that cooled at mid-ocean ridges naturally recorded the field direction at the time, resulting in either 'normal' (i.e. like the present day) or 'reversed' (i.e. opposite) magnetization. Since sea-floor spreading was envisaged as pretty-well continuous, the bar-code signal would be faithfully recorded in the rocks as a sequence of normally and reversely magnetized strips parallel to the mid-ocean ridge, and it was these magnetized strips that produced the anomalies detected by the dodgy magnetometer towed behind HMS *Owen* (Figure 1.2). Hence the wiggly line on profile 4A must be just the latest part of a recording that extends back millions of years, a recording that makes Wagner's *Ring* seem almost succinct. In fact, if you played it through speakers (the crustal signal, not Wagner), it might sound a bit like a demented police siren, with the two notes varying randomly in length.

I said 'very conveniently' just now, because rates of seafloor spreading, typically between 30 and 150 km per million years (or millimetres per year, if you prefer), are just fast enough to allow this same pattern of reversals to be recorded with high fidelity in all the ocean basins. If reversals occurred *more* frequently, the juxtaposition of many normally and reversely magnetized rocks would tend to blur or cancel the resulting anomalies at the surface; much *less* frequently, and you would not have enough reversals in your recording to be useful.[11] Equally fortunately, if oceans were not full of water, erupted basalts would not be cooled so quickly, and would therefore flow further, overwriting the signals recorded by previous flows, as happens in Iceland.

Vine and Matthews used only a crude block model of crustal magnetization sufficient to demonstrate their hypothesis, but they noted in passing that the same hypothesis 'could explain the lineation or "grain" of magnetic anomalies observed over the eastern Pacific to the west of North America'. This was a reference to a much more extensive marine

[10] Rocks magnetized in the direction opposite to the present field had been known for some time, but were thought to result from a process of 'self-reversal'. Ironically, this was shown to have actually occurred in some cases (Nagata et al., 1956).

[11] A long interval of about forty million years without reversals occurred in the Cretaceous, causing problems for geophysicists trying to establish the history of plate motions.

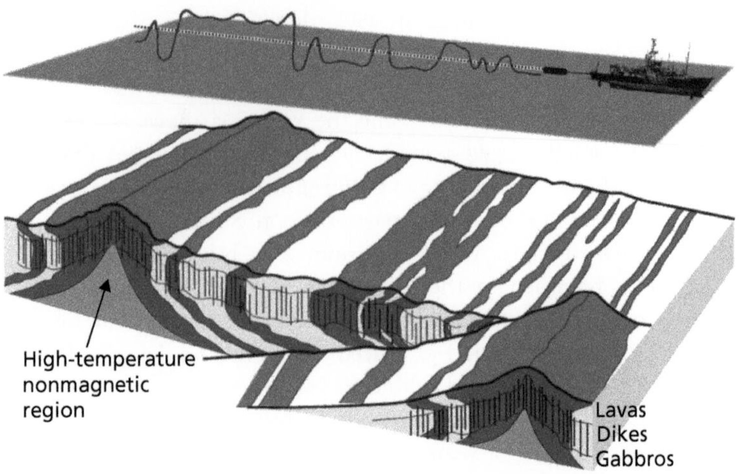

Figure 1.2 Cartoon showing how the magnetic bar-code in the oceanic crustal layers gives rise to magnetic anomalies observed at the surface. [Credit: U.S. Geological Survey.]

magnetic survey that had been carried out during the 1950s in connection with Cold War submarine operations by the US Coast and Geodetic Survey, together with the US Navy and the Scripps Institution of Oceanography. Maps had even been published in geological journals showing strange zebra stripes running more-or-less north–south over the entire area of the survey, from Baja California up to Vancouver Island (Figure 1.3). Geologists Ronald Mason and Arthur Raff realized that this pattern reflected the magnetization of basaltic crustal rocks, but failed to spot the symmetry in the pattern or the profound message it conveyed. They might later have reflected 'We had the experience but missed the meaning'.[12] Others, including Raff's boss at Scripps, were convinced that the striped pattern was just noise.[13]

Vine now collaborated with the Canadian geophysicist Tuzo Wilson (of whom more anon), using the Vine–Matthews Hypothesis to account for these magnetic stripes. Their symmetry was a strong clue to the presence of a section of active mid-ocean ridge beneath the survey area, a section that Wilson named the 'Juan de Fuca Ridge', after a sixteenth-century

[12] In fact, Ron Mason admitted much later that he could have kicked himself for not realizing the truth (see Mason, 2003).

[13] Glen (1982).

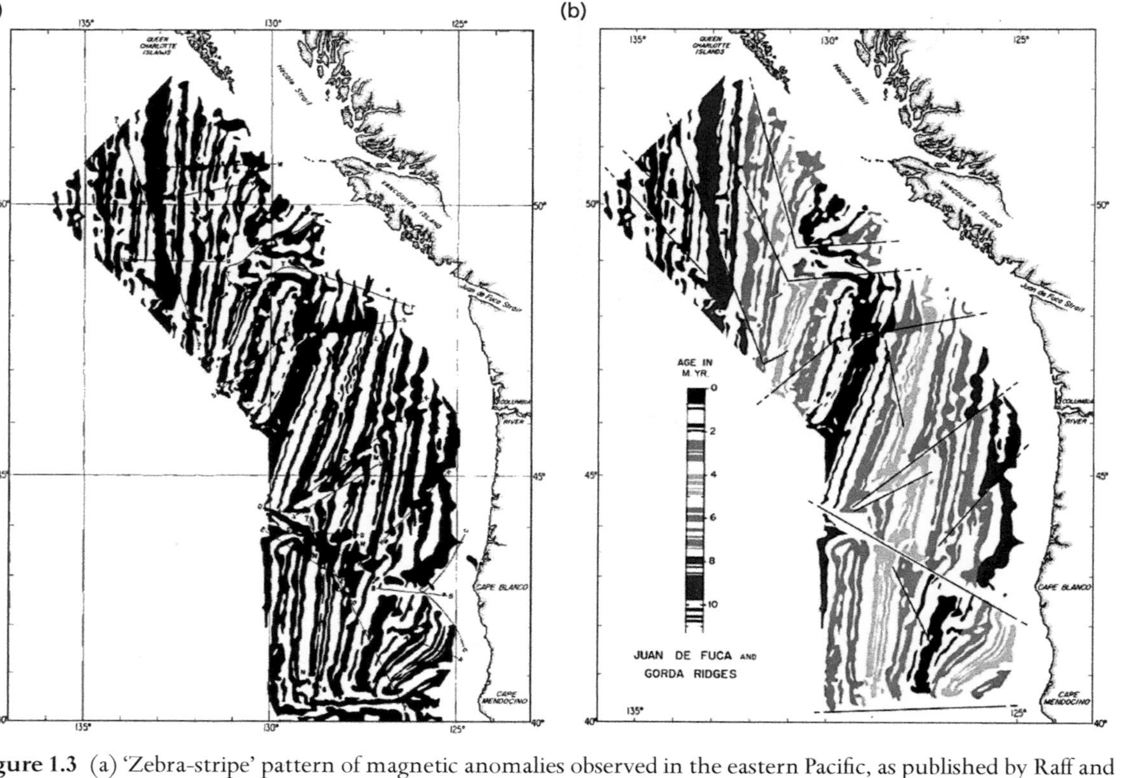

Figure 1.3 (a) 'Zebra-stripe' pattern of magnetic anomalies observed in the eastern Pacific, as published by Raff and Mason (1961). (b) The same pattern, as interpreted by Vine (1966).

Greek[14] explorer who 'discovered' the straits between Vancouver Island and the mainland.[15] Assuming a simple block model similar to Vine and Matthews' for the magnetic source rocks, Vine and Wilson demonstrated how the faster spreading rate over the Juan de Fuca Ridge resulted in high-fidelity recording of the magnetic field reversals, producing a spectacularly good fit between observed and modelled anomalies formed up to 3.5 million years ago. Furthermore, the sequence of reversals in their block model corresponded closely to that derived by independent dating of the bar-code signal in the laboratory using measurements made on magnetized rocks.[16] Their results showed beyond doubt that this area of the eastern Pacific had been formed by sea-floor spreading in the same way as the crust of the Carlsberg Ridge.

The following year, 1966, Fred Vine published a landmark article in the journal *Science* entitled 'Spreading of the Ocean Floor: New Evidence', in which he showed how the Vine–Matthews Hypothesis, in conjunction with his new polarity bar-code, reproduced fine details of magnetic anomalies recorded in *all* the major oceans, and how the new reversal timescale could be used to establish the rate of spreading at each ridge. He summed up the Vine–Matthews Hypothesis in a characteristically restrained manner: 'It is suggested that the entire history of the ocean basins, in terms of ocean-floor spreading, is contained frozen in the oceanic crust. Variations in the intensity and polarity of Earth's magnetic field are considered to be recorded in the remanent magnetism of the igneous rocks as they solidified and cooled through the Curie temperature at the crest of an oceanic ridge, and subsequently spread away from it at a steady rate. The hypothesis is supported by the extreme linearity and continuity of oceanic magnetic anomalies and their symmetry about the axes of ridges.' As a demonstration of the power of the new theory, he also speculated, in an aside, that: 'the more recent geologic history and structures of the western United States can be ascribed to the progressive westward drift of the North American continent away from the spreading Atlantic Ridge, and to the fact that the continent has overridden and partially resorbed first the trench system and, more recently, the crest of the East Pacific Rise.' In other words, much of the recent geology of the western USA and Canada could be explained by his

[14] Yes, Greek. His ship was provided by the Spanish and so his name was 'Spanishized'.

[15] Today, the Juan de Fuca Strait marks the international boundary between Canada and the United States.

[16] Cox et al. (1965).

theory as a result of the ongoing collision of the North American continent with the Juan de Fuca spreading centre discovered in the Eastern Pacific. This suggestion was subsequently pursued by a number of researchers, becoming a major topic in Earth Science.

There is no shortage of accounts of the plate tectonics revolution that followed. The young Vine was fêted worldwide, holding a lectureship at Princeton University for five years before, at the age of thirty-four, being elected a Fellow of the Royal Society and appointed Professor of Environmental Sciences at the new University of East Anglia (a post he still holds in 2018). Since then, the birth of the theory of plate tectonics has been celebrated at frequent intervals. In 1987, at a meeting to mark the silver jubilee of the theory held at Texas A&M University, Fred noted that 'Ten years ago there was a special session at the AGU to celebrate the 10th anniversary of plate tectonics, and here we are 10 years later celebrating the 25th. At that rate, I should stand quite a good sporting chance of being around for the 50th anniversary which, presumably, will be in 10 or 15 years' time.'

Incredibly, even after the main components of plate tectonics theory had been assembled, there were still some geologists, particularly in the USA and the Soviet Union, who were sceptical about the ideas of sea-floor spreading and moving continents, remaining 'fixists'. As late as 1969, for example, the geologist A. A. Meyerhoff presented the American Association of Petroleum Geologists with a long list of supposed objections to the sea-floor spreading hypothesis, before going on to say that 'sea-floor spreading and continental drift—as presently conceived—belong more properly in the realm of mythology than in the science of geology. Geologists exploring for future energy resources may find more reward in expending their energies on hypotheses which are more consistent with reality.'[17] Vine and Matthews might well have echoed their colleague, Francis Crick, by responding, 'To those of you who may be fixists, we make this prophecy: what everyone believed yesterday and you believe today, only cranks will believe tomorrow.'

Out of the Way, You Swine!

The elegance of the Vine–Matthews Hypothesis lies in its simplicity. Once you connect seafloor spreading with polarity reversals, everything fits into place like Lego bricks. But nature is a bit more complicated than

[17] Meyerhoff (1969).

Lego bricks, and such a hypothesis, revolutionary as it was, represented just the beginning of a journey towards an entirely new understanding of the Earth as a complex system. First, however, the hypothesis had to be tested by having a good look at mid-ocean ridges, to see what was really going on down there. In the mid-1960s, this was difficult, as there was no easy way to reach the ocean depths,[18] mostly exceeding 2 km. This was a job for the geophysicists.

Land-bound geologists had long been suspicious of geophysicists, particularly when the latter drew sweeping conclusions about the Earth, as it seemed, from mere numbers. Such an attitude was described as long ago as 1929 by the German scientist Alfred Wegener, who noted that many European geologists 'view the results of geophysics with a mistrust that never completely fades'. Much more recently, when the British Antarctic Survey merged its Geology and Geophysics divisions in the 1990s, an amusing cartoon appeared on the geologists' notice-board, showing a white-coated, bespectacled geophysicist with a scantily clad girl on each arm. Ahead of him, a heavy bodyguard was thrusting aside a group of puny geologists, with the words, 'out of the way, you *swine!*—a *geophysicist* is coming'.[19] This, of course, reflected the paranoia of mere geologists faced with the threat from geophysics, which had so recently solved the biggest puzzle in Earth Science. By that time, geologists elsewhere had learned to love their more rigorous colleagues, a fact well illustrated by a comment penned by a leading geochemist (who we shall meet a little later[20]): 'I gradually realized that geophysicists did not hold a monopoly on truth—or perhaps I just learned some geophysics.'

An early application of geophysics to the oceans was the measurement of heat flowing up through the ocean floor from the interior of the Earth. Just after the Second World War, a probe was developed at the Scripps Institution of Oceanography to measure this heat flow, and, in the 1950s, well before Vine and Matthews' breakthrough, their Cambridge Head of Department, Edward (Teddy) Bullard, worked with leading US marine geophysicists Arthur Maxwell and Roger Revelle, deploying the new probe in the major ocean basins. When the results for the North Atlantic were plotted, it became clear that values of heat

[18] Scientists had descended to great depths in the bathyscaphe *Trieste*, but such submersibles were not employed on a mid-ocean ridge until the 1970s.

[19] I later discovered that this was a clever adaptation of a cartoon by B. Kilban, entitled 'Out of the way, you swine! A cartoonist is coming'.

[20] Cann (1991).

flow everywhere were around 1 μcal per cm^2 (42 mW m^{-2})[21]—apart from one site. That site happened to be in the axial valley[22] of the Mid-Atlantic Ridge. Here, an exceptionally high value of 273 mW m^{-2} was recorded, more than six times the 'normal' value. Teddy speculated, in 1961, that 'It seems probable that very high values of the heat flow are characteristic of the crests of mid-ocean ridges'.[23] During the next few years, it would be realized just how profound was that remark.

Many marine geophysicists of the 1950s and 1960s had served in the military during the Second World War, criss-crossing the oceans while helping to develop new technology to detect and destroy enemy shipping, notably submarines. Their experiences led to curiosity about what lay beneath all those miles of deep ocean, and so, after the war, they beat their swords into a range of techniques with which to pursue their newly awakened interest. Sonar, magnetometers, depth charges, and sonobuoys, developed for military purposes and subsequently endowed to academic science, were adapted to become basic tools of marine geophysics. Meanwhile, with the onset of the Cold War, the military men continued with new developments to counter the Soviet threat. We have already seen how extensive magnetic surveys by the US Navy off the Pacific coast, motivated by anti-submarine warfare, produced a dataset that allowed Fred Vine and Tuzo Wilson to confirm the Vine–Matthews Hypothesis—a dataset that could never have been collected by fund-strapped academic researchers. Now, other technologies would allow geophysicists to test and develop the new theory.

The Mid-Atlantic Ridge was handy for both European and North American geophysicists hoping to understand the workings of active mid-ocean ridges. The Ridge was, by common consent, discovered in 1873,[24] when the pioneering research vessel, HMS *Challenger*,[25] made a

[21] That is, 42 milliwatts per square metre.

[22] A deep valley at depths of around 3383 m, marking the axis of the Carlsberg Ridge, had been discovered by John Wiseman of the British Museum and the expedition leader, Colonel Seymour Sewell, during the John Murray Expedition of 1933–34. It was later discovered that a similar valley marked the axis of the Mid-Atlantic Ridge.

[23] Bullard (1961).

[24] Earlier soundings, made in the North Atlantic in 1856 by Otway Berryman aboard the USS *Arctic* in connection with the laying of telegraph cables, suggested the existence of a ridge near 53°N.

[25] Soundings were made using lead lines, and were thus very time-consuming. The first successful rock dredging was also carried out during this expedition. The ship was subsequently commemorated in the names of various ships and spacecraft, as well as the deepest-known part of the ocean. An account of its exploits is given in the book *The Silent Landscape* by Richard Corfield (Corfield, 2003).

series of soundings across the South Atlantic. Measured depths decreased from around 4500 m[26] near the African and South American continents to just 2700 m in mid-Atlantic. Of course, 2700 m is still a long way to sink, but it showed, as the American naval scientist Matthew Maury had previously suggested, that the deep ocean floor was not flat, and that a major topographic feature existed right in the middle of the Atlantic Ocean. Yet the Mid-Atlantic Ridge is not a ridge in the sense that you might think of, say, a narrow line of steep hills, but more of a broad rise, occupying around a third of the total width of the Atlantic. And the axis of the Ridge is not a peak, but a narrow valley, with steep sides and a floor at depths of around 3000 m. In the Indian Ocean, the Central Indian Ridge (including the Carlsberg Ridge) has a similar form, but other mid-ocean ridges are quite different. For example, the longest mid-ocean ridge of all rises gently to a smooth, axial peak, has no axial valley, and does not even lie in the middle of an ocean, characteristics reflected in its name: the East Pacific Rise.

Throughout the 1960s, intensive exploration was carried out in an area near 45°N on the Mid-Atlantic Ridge by British and Canadian geophysicists using a technique known as 'seismic refraction' to study the structure of the oceanic crust produced by spreading. This uses artificially created seismic waves to work out the structure of the oceanic crust from the speed at which the waves, known as 'P-waves' and 'S-waves',[27] pass through it. Excitingly, this frequently involved tipping large charges of dynamite, weighing 100 kg or more, off the sterns of research ships, the subsequent explosions being recorded by hydrophones suspended from buoys or anchored on the ocean floor. This technique is ideally suited to distinguishing layers with contrasting physical properties, bounded by sharp discontinuities. Perhaps it was no surprise, then, that it revealed that the oceanic crust beneath the Mid-Atlantic Ridge was composed of layers of rocks with contrasting physical properties separated by sharp boundaries. In fact, this was a very significant result, particularly when it was realized that the oceanic crust almost everywhere is characterized by the same three layers.

[26] In those days, measurements were made in fathoms, but I have saved you the trouble of converting them.

[27] P stands for 'primary' and S for 'secondary', referring to the order in which they arrive at a seismometer. Alternatively, they can be remembered as 'push' and 'shake' waves.

The uppermost stratum—christened, imaginatively, 'Layer 1'—is a layer in which P-waves travel at velocities of 1.6–2.5 km per second. This, you may think, is pretty fast (2.5 km per second is 5600 mph, after all), but is regarded as relatively slow by geophysicists, and is characteristic of sediments deposited on the sea bed. Layer 1 is variable in thickness, and absent altogether along the axes of mid-ocean ridges. The layer beneath—you guessed it, 'Layer 2'—has higher P-wave speeds of up to 6.2 km per second, is 1–2 km thick, and corresponds, at least in its upper part, to basaltic lavas. Beneath this, 'Layer 3' is around 5 km thick, and has the highest seismic speeds within the crust, 6–7 km per second. At the base of Layer 3 is a sharp discontinuity known as the Moho[28], separating the crust from the upper mantle beneath, at which the seismic P-wave speed increases abruptly to a dizzying 8 km per second.[29] The total thickness of the layers constituting the oceanic crust turned out to be surprisingly uniform at 6–7 km in most of the places investigated, and, since it was now believed that the oceanic crust formed by sea-floor spreading, this layering must somehow have been produced at the mid-ocean ridge crest. Layer 1 was obviously a layer of sediments formed by material falling like snow from the surface layers of the ocean, but how had the layers beneath formed, and how had they preserved the magnetic field direction intact? To answer these questions, samples were needed that you could look at and analyse, starting with Layer 2.

Pillow Talk

Rock dredging is a rough-and-ready method of acquiring rock samples from the sea floor. All that's required is a large chain-link bag with steel cutters around the mouth, and a long piece of wire. Some kind of research ship is useful, too. You simply lower the dredge to the sea floor, and then edge the ship forward, so that anything on the bottom, living or not, gets caught in the bag—just like modern fishing, really. You can never be sure, of course, that whatever rocks come up from the deep were actually formed on the sea floor beneath the ship, and not, say, dropped by passing icebergs, and any bits of rock you do bring

[28] Which, as any geology student will be happy to explain, stands for 'Mohorovičić discontinuity'.

[29] To put these speeds into perspective, 8 km per second equates to 18,000 mph, compared with the speed of sound in water of roughly 1.5 km per second (3400 mph), and in air of roughly 0.34 km per second (760 mph).

up are likely to have been altered by chemical reactions with seawater over the many decades, centuries, or millennia that elapsed before you came along. But if you're lucky, you might break off a 'fresh' fragment from a basement outcrop that is not buried by sediment. Dredging on the Carlsberg Ridge in the year following the HMS *Owen* cruise[30] produced rocks suggesting that the crust beneath was surfaced with young basaltic lavas, in line with the Vine–Matthews Hypothesis. Underwater photographs of the sea floor showed pillow-shaped mounds, fragments of which were brought up in the dredge hauls. Unfortunately, these samples had been broken up and altered by seawater and by fluids circulating in the crust, making interpretation difficult.

A typical 'pillow' is a lobe of solidified basaltic lava a metre or so across. It has a thin rind of black volcanic glass surrounding an interior made up of tiny mineral crystals, almost invisible to the eye. Pillow lavas are characteristic of submarine eruptions—where they occur on land, they generally record vanished oceans. Not until 2009 did anyone manage to obtain a video recording of pillows actually forming on the deep ocean floor (for obvious reasons—the eruption temperature is close to 1000°C). A lobe of red-hot lava is extruded like toothpaste from a thin, glowing, crack in the ocean floor, and quickly turns black as its surface is chilled by the near-freezing seawater, forming that glassy rind. The extrusion soon ceases as the black outer shell solidifies, before more hot lava breaks out from fresh cracks in the newly frozen pillow, forming another lobe of incandescent lava, and so on, resulting in the characteristic hummocky form of the ocean floor.

To overcome the hit-and-miss nature of rock dredging, not to mention the expense and inconvenience of going to sea, geologists really wanted a bit of the mid-ocean ridge system on dry land, where they could hammer bits off and do what geologists do best. This was asking a lot, but Mother Nature came up trumps, providing Iceland. Here you can actually walk on the Mid-Atlantic Ridge as it spreads beneath you, but what you see is not like the sea bed, even though it does consist of basaltic lava flows. The very fact that this bit of the mid-ocean ridge system lies above sea level tells you that it is exceptional and thus its rocks may not be representative of typical spreading centres. Many Icelandic flows were erupted subaerially or beneath ice-sheets rather than under the sea, producing much ash, as we in Europe discovered in

[30] Cann and Vine (1966).

2010. And, although Iceland leaves no doubt that the Mid-Atlantic Ridge is floored mainly by young basalts, it does not tell you much about processes *beneath* the ridge axis unless you drill into the crust. This may seem a good plan and, in recent years, the Iceland Deep Drilling Project has attempted it, but the unusually thick crust that elevates this part of the Mid-Atlantic Ridge above sea level also makes it difficult to drill through to the lower crust, and a 3 km hole failed to penetrate anything other than flows, more flows, and associated feeder dykes. Slightly worryingly, drilling near the Kafla volcano was halted abruptly at just over 2 km when it hit unerupted rhyolite magma. Gaia had another trick up her sleeve, however. For there are places where whole slices of the oceanic crust, and even parts of the mantle beneath, have been conveniently dumped on land for the edification of field geologists, many in nice warm places like the Mediterranean and Middle East—ideal for a bit of fieldwork.

Snakestones and Steinmann's Trinity

As long ago as 1813, the term 'ophiolite' (or 'snake-stone'[31]) was applied to some unusually dense greenish-grey rocks with a scaly texture, found in the European Alps. The colour was due to serpentine, a silicate mineral [$(Mg,Fe)_3Si_2O_5(OH)_4$], the appearance of which was, apparently, reminiscent of a snake. The rocks, of which this was a major constituent, were thus known as 'serpentinites'. A century later, the scope of the word 'ophiolite' was widened by one Gustav Steinmann (whose name, appropriately, means 'rock man') to refer to a mysterious association of serpentinite and two other rock types, commonly found together as a 'trinity' in the Alps and elsewhere. It seemed that the three lithologies—a sedimentary rock composed of silica and containing the tiny skeletons of marine protozoa, dark-coloured basalt lavas, and the aforementioned serpentinite—occurred in a particular sequence, although it is doubtful whether Steinmann himself ever saw his complete trinity together in one place.

Serpentine is a mineral that contains water,[32] bound into its crystal structure, water that was added when serpentine replaced earlier, unstable, minerals. Those minerals were the magnesium–iron silicates

[31] From *ophis*, the Greek for 'serpent'.

[32] In fact, it is held as (OH) within the crystal lattice.

olivine and *pyroxene*, and the reason they were unstable was that they formed under high pressures, deep within the mantle, where it is hot and dry. Somehow, this mantle rock had been elevated to the surface, the confining pressure was relieved, and the minerals reacted with seawater, producing this reptilian lithology and reducing the rock's density by about 20 per cent, as well as making it much softer. Its parent would have been a rock known as 'peridotite'[33], the rock that constitutes most of the upper mantle, and thus one of the commonest rocks in the Earth. So, what all this meant was that we had an association of mantle rocks, volcanic lavas derived from melting of those same mantle rocks, and marine sediments. The rock body containing this association was now known as an ophiolite.

Six months before the publication of Vine and Matthews' *Nature* paper, a lengthy article appeared in the *Philosophical Transactions of the Royal Society*,[34] summarizing the unusual geology of southern Cyprus. The senior author was one Ian Gass, a man who would, a decade later, found the Earth Science Department at the UK's radical new Open University. The paper focused on the Troodos Massif, a curious domed rock outcrop occupying a quarter of that Mediterranean island, and comprising three very distinct formations. The first was referred to by Gass and co-author, Masson-Smith, as the 'Troodos Sheeted Dyke Complex' (the word 'dyke' has, I know, a number of meanings, but, in geology, it refers to a rock body that intruded into pre-existing rocks as a vertical sheet of hot magma, typically a metre or two in width). Dykes are pretty common, but what made these special was the lack of 'country rock'. If you should consult Wikipedia, you will be informed, somewhat unhelpfully, that 'Country rock is a subgenre of country music, formed from the fusion of rock with country'. More prosaically, country rock is also a term used by geologists to describe the older strata into which dykes are intruded, and from which lavas erupt. But in the Troodos Sheeted Dyke Complex, there was hardly any country rock—vertical or near-vertical dolerite[35] dykes made up over 90 per cent of the entire outcrop. In other words, dykes had intruded into other dykes, which had then been intruded by yet more dykes.

[33] From 'peridot', the gem form of olivine, which is a major component of the rock.

[34] Gass and Masson-Smith (1963).

[35] The name given to a medium-grained rock with the same chemical composition as basalt.

The second rock formation, immediately beneath the sheeted dykes, was known as the 'Troodos Plutonic Complex' ('plutonic' refers to rocks that originated at depth within or beneath the crust by the cooling of a magma). This was composed of dense rocks with strange names, such as gabbro, dunite, and the previously mentioned peridotite. The first of these, named after one of the less-popular destinations in Tuscany, has a chemistry similar to basalt and dolerite, but with much larger mineral crystals, suggesting slow cooling from a basaltic magma (slower cooling = larger crystals). Dunites are rocks consisting mainly of a single mineral—olivine—while peridotites include pyroxene as well. The third member of the triad was the 'Troodos Pillow Basalt Series', consisting of more than a kilometre thickness of basalt pillows *above* the sheeted dykes. We already know about pillow basalts, the toothpaste lavas squeezed out of the ocean floor, and fed by dykes in the layer below.

Gass and Masson-Smith interpreted the Troodos Complex as an 'oceanic volcanic pile' that had been thrust onto the leading edge of the African continent and uplifted without major internal deformation. It was, of course, an ophiolite, although, interestingly, they never used this term. They originally thought it was pre-Triassic (i.e. more than 250 million years old), but soon realized that it must have formed during the Cretaceous, around 100 million years ago (to most people, this seems a trifling difference, but to a geologist, it is like mistaking a mock-Tudor villa for an Elizabethan hunting lodge). Eventually, Gass realized that the geology he had described with Masson-Smith corresponded to the structure of oceanic crust, and in 1969 published a paper in *Nature*, asking 'Is the Troodos Massif of Cyprus a Fragment of Mesozoic Ocean Floor?'

Two years later, just as Fred Vine was taking up his new post at UEA, he and American geologist Eldridge Moores published an interpretation of the Troodos Complex as oceanic crust emplaced tectonically during the collision of Africa with Eurasia. That is, the ophiolite complex was shoved onto the margin of Eurasia as the intervening ocean closed—a phenomenon dubbed 'obduction'—during which it was somehow rotated by 90°. Moores and Vine also pointed to the remarkable inference that, in a 100 km zone within the Sheeted Dyke Complex, where country rock was completely absent (hooray, you may think), there must have been extension of... 100 km! This pointed to a process amounting to sea-floor spreading.

They interpreted the Pillow Lavas and upper Sheeted Dykes as equivalent to seismic Layer 2 of the oceanic crust, while the lower dykes, together with the gabbros and dunites of the Plutonic Complex, constituted Layer 3. The peridotites were, of course, assigned to the uppermost mantle beneath the Moho. Hence, apart from a few geochemical oddities, the Troodos Complex looked like a slice of oceanic crust and upper mantle that had formed at a mid-ocean ridge in some ancient ocean, and later been thrust bodily onto an adjacent continent when that ocean closed. Many similar ophiolites occur throughout the Alpine–Mediterranean region, forming a belt of relict ocean crust fragments marking the line of the lost ocean.

We humans love naming things. Wild or captive animals, machines, hurricanes, vacuum cleaners—you name it. And places—no matter how remote in space or time: faraway lands, craters on other planets, lost continents, and oceans that no longer exist (and some that never existed), we name them all. So, we have Queen Maud Land, Bradbury Crater, Lemuria, and Tethys. The last of these, Tethys, is the name given to the ocean that once separated India and Africa from Eurasia.[36] As the two southern continents migrated across the equator and eventually made contact with the margins of Eurasia, so Tethys was squeezed out of existence in a manner that we'll come back to in the next chapters. Small bits of the Tethyan ocean floor escaped oblivion somehow, and are now found in and around the crash site as ophiolite complexes such as the Troodos.[37]

There was now a tentative link between the seismic structure of the oceanic crust and identifiable rocks, which made the geologists a bit happier. But there were still problems. Ophiolites were all very well, but who could say whether they represented typical ocean crust or, like Iceland, some peculiar process associated with their unusual setting, in this case the closing of an ocean rather than its opening. And still, nobody had been able to sample lower crustal rocks from the one place for which an oceanic crustal origin could not be doubted—the oceanic crust.

A Decade of Dreams

The feature article entitled 'High Drama of Bold Thrust through Ocean Floor', which appeared in the 14 April 1961, edition of *Life* magazine, is a good example of how to sell science to the public. It was, perhaps,

[36] Dorrik Stow has written a book about Tethys, entitled *Vanished Ocean* (Stow, 2010).

[37] Later work suggests that the Troodos Massif might have formed, not as part of the original floor of Tethys, but above a subduction zone during closure of the ocean.

rather better written than the average science report, being penned by a Nobel Prize-winning author. The Nobel Prize in question was not, however, presented for science but for literature,[38] and the author was not a scientist, but the novelist John Steinbeck.

Steinbeck's curious and possibly unique status as an 'amateur oceanographer' had led him to participate in the $57 million[39] 'Project Mohole', one of the most ambitious technological and scientific projects of all time, involving (literally) cutting-edge technology, massive cost, and complex logistics. The plan was to drill right through the crust beneath the deep ocean to sample the rocks of the underlying mantle. Steinbeck described how, in March 1961, the drilling vessel—the appropriately named CUSS-1—left San Diego and 'waddled like a duck into the channel on its four gigantic diesel outboard motors'. Project Mohole succeeded in drilling five holes off the coast of Guadalupe Island, one not far short of 200 m deep, and recovering samples of rock from Layer 2: 'the first touching of a new world' as the achievement was dramatically, if hyperbolically, described in Steinbeck's article. The rock, of course, was basalt, but the really important result was that the technological challenge of deep-ocean drilling had been overcome (albeit at a cost that amounted to roughly $60,000 per foot of core recovered).

The first phase of the project was thus deemed a success and the scientists received hearty congratulations from President Kennedy. Sadly, phase two of the project was cancelled amidst recriminations in 1966, and the objective of drilling into the mantle was never achieved, so that Mohole soon became 'Nohole'. Within five years, however, a new project succeeded where Mohole had failed. This was the drilling and recovery of long sediment cores from the deep oceans within the compass of the Deep Sea Drilling Project (DSDP). Contemporaneous with the US Apollo Moon-landings, DSDP was one of the most valuable oceanographic research efforts ever attempted, and, in scientific importance, it easily surpassed Apollo.

Dangling two miles of steel pipe through a hole in the bottom of a ship in the hope of retrieving sediment samples from the sea bed may seem somewhat optimistic. Expecting that pipe, having reached the bottom, to penetrate a further mile into the sea bed, sounds, frankly,

[38] Although he did not receive his prize until the following year.

[39] $450 million at today's prices.

ridiculous. Yet an article written for *Science* in 1968 looked forward enthusiastically to what some of us would now call 'big science' in the oceans: a plan to drill into the sediments of the deep ocean floor, right through to the oceanic crust (Steinbeck's 'new world') beneath. The title of the article was 'A Decade of Dreams', referring to the desire of marine geologists to unlock the detailed history recorded in sediments deposited on the ocean floor. Noting that these deep-sea sediments contained the most complete and unaltered record of past conditions at the surface of the Earth, the author, Tjeerd Van Andel, then a professor of oceanography at Oregon State University, observed that 'investigation of the geological history of the deep ocean offers an obvious path to the study of our planet as a whole'. Prescient as these words were, Van Andel could hardly have foreseen just how much impact this project would have on geology and oceanography, and on the emerging science of environmental change, not just for a decade, but for half a century to come. A large proportion of everything we (and the Intergovernmental Panel on Climate Change) know about past conditions on the Earth derives from results obtained from deep-sea cores obtained by DSDP and its successors.

At the heart of DSDP was the *Glomar Challenger*,[40] a drilling ship built specifically for scientific work around the world and launched in March 1968. Funded by the US National Science Foundation (NSF), it was operated by a commercial firm, Global Marine (hence 'Glomar'), and coordinated by the Scripps Institution of Oceanography. Its primary claim to fame was its ability to drill in water depths of more than 7 km and recover cores almost 2 km long,[41] more than enough to penetrate right through the sediments and into the crust beneath most of the ocean basins. To do this, it was imperative that the ship did not drift away from its location above the spot where the drill string entered the sea bed (you can imagine the consequences). It achieved this using a set of thrusters in a system known as dynamic positioning, which corrected for movement relative to a set of moored buoys. In addition to carrying the drilling rig and 25 km of drill string, it had laboratory space for scientists to prepare and analyse the cores at sea, and the ability to operate for extended periods away from port. The DSDP programme was directed by a consortium known as JOIDES (the Joint Oceanographic

[40] Named after HMS *Challenger*.

[41] Even today, the record is only 2200 m, achieved in 2012.

Institutions for Deep Earth Sampling) under a series of contracts with the NSF. The eponymous institutions were at Miami, Woods Hole, Scripps, and Lamont-Doherty.

The upshot of all this was that, while other men were fooling about on the Moon, DSDP scientists were obtaining and analysing freshly recovered sediment cores to test a revolutionary new theory of the Earth. The first hole to be drilled on crust whose age had been predicted from Vine–Matthews-type magnetic anomalies was started just before Christmas 1968 at site 14, near 30°S in the South Atlantic Ocean. Using Vine's new polarity timescale, the age of the ocean crust beneath the sediments at this site was predicted to be 38–39 million years. By drilling right through the overlying sediments to the basaltic lavas beneath, it was possible to recover samples of the very earliest deposits that formed just after the crust was created at the Mid-Atlantic Ridge (assuming the Vine–Matthews Hypothesis to be correct). These sediments could be dated using the shells of tiny planktonic protozoa preserved within them. Obviously, their age must be closely related to that of the crust beneath, since sediments are constantly raining down and it takes very little time for them to accumulate on new lava flows. The date estimated by the marine geologists studying these basal sediments was 40 ± 1.5 million years. Bang on! At the next hole, the predicted basement age was 21 million years, and the estimate from drilling 24 ± 1 million years—still very close. Holes at sites 16 and 17 were both within 1 million years of the predicted ages, while the oldest sites, 19 and 20, where the predicted ages were 53 million and 70–72 million years, respectively, produced sediment ages of 49 ± 1 and 67 ± 1 million years. Considering the limited knowledge and assumptions made at the time, the agreement seemed almost magical.

This early drilling in Leg 3 was probably the most important of the entire DSDP. It established beyond doubt that the age of the Atlantic Ocean floor was just as predicted by the Vine–Matthews Hypothesis, and demonstrated that spreading had been going on at the southern Mid-Atlantic Ridge at a rate of about 40 mm per year ever since the Cretaceous. By 1983, *Glomar Challenger* had recovered more than 100 km of deep-sea cores from the world's oceans, providing one of the most important sources of data for scientists interested in the Earth's changing environment. The success of DSDP led to the construction of a new ship, the JOIDES *Resolution*, for the International Phase of Ocean Drilling (IPOD) during the 1990s (Figure 1.4), and, most recently, the Japanese-funded

Figure 1.4 The *JOIDES Resolution* drilling ship. [Photograph courtesy of the International Ocean Discovery Program *JOIDES Resolution* Science Operator (IODP JRSO).]

drilling vessel ちきゅう[42] for the latest phase of operations, planned to continue into the 2020s.

Fame at Last!

Geologists now got in on the act, and in a big way. French and American scientists hatched a plan to use manned submersibles to visit the crest of a mid-ocean ridge as part of a project with the slightly contrived title 'FAMOUS', standing for 'French–American Mid-Ocean Undersea

[42] *Chikyu*, or 'Earth'.

Study'. Submersibles would allow geologists to examine the deep-sea floor in detail, just as they might visit a rock outcrop on land (apart from the claustrophobia and the chance they might never make it back to the surface), and they could even collect rock samples—nearly one and a half tonnes of them in the end. They also carried cameras to record images of the volcanic terrain and so gather clues to the style and timing of eruptions.

Research ships generally operate in one of four modes (ignoring port-calls, overhauls, etc.). They may be in transit between port and the work area (or vice versa); they may be working 'on station', which usually involves putting gear over the side or stern of the vessel on a wire; they may be conducting 'underway' surveys, for example echo sounding or seismic profiling, as the ship steams at a constant speed; or they may be 'hove-to', waiting for better weather. For complex deployments 'on station', such as launching and recovering submersibles, involving divers and lengthy preparations at the surface, weather is a major factor, since spending days head-to-wind is not only uncomfortable, but very expensive. The French–American group therefore chose a site on the Mid-Atlantic Ridge near 37°N, WSW of the Azores, for their study, where the weather was kinder than in the area studied by the Canadians and British.

The dives, in 1973 and 1974, followed several years of reconnaissance using surface ships and aircraft to provide a comprehensive set of base maps with which to plan the operations. Once again, the US Navy provided valuable technical support by collecting bathymetric data with a new (secret) multi-beam echo sounder known as the Sonar Array Sounding System (SASS). This technology had developed from an initial idea to install multi-beam radar mapping on aircraft during the Cold War, in order to locate Soviet missile sites. The project was cancelled after the shooting down of a Lockheed U2C spy aircraft in 1962, but the idea was adapted for use with sonar instead. It became a valuable tool for rapidly mapping whole swaths of ocean floor, and subsequently found its way into civilian use on scientific research vessels, where it played an important role in the next generation of plate tectonics research (see Chapter 8). These preparatory surveys were navigated using the Navy's new NAVSAT satellite navigation system, resulting in the highest-quality charts ever produced for research purposes. A bonus was provided by the *Glomar Challenger*, which stopped off while en route to the North Atlantic, in order to drill a few holes to the west of the FAMOUS area.

The new maps showed that the Mid-Atlantic Ridge here is, like all mid-ocean ridges, segmented. That is, it consists of discrete sections of ridge, varying from 30 to 100 km in length, separated by narrow fracture zones[43] that run approximately perpendicular to the trend of the ridge. The survey was focused on a 50 km-long segment bounded by the imaginatively named 'Fracture Zone A' to the north, and, yes, 'Fracture Zone B' to the south. These fracture zones offset the ridge by about 20 km to the west and east, respectively (travelling north to south). At the ridge axis was a rift valley 30–50 km wide at a depth of 2800 m, with a narrow inner valley roughly 2 km wide along the centre-line of the ridge. This was bounded to the east and west by steep faults running parallel to the ridge axis. Interpretation of magnetic anomalies collected by ship and aircraft demonstrated that spreading occurred here at the rate of 46 mm per year—that is, 23 mm of new crust was being added to each side of the ridge each year. So, if you were to hold your breath and stand right at the ridge axis, with your bare feet on the North American side, your toenails would grow just fast enough to keep pace with Africa as it retreated towards the eastern horizon.[44]

The French made the first dives in 1973, using the bathyscaphe *Archimede*. Despite sounding like something concocted by Jules Verne, a bathyscaphe was a self-propelled submersible capable of reaching the deepest parts of the ocean,[45] so it seems strange that it took more than a decade to get one to the most important target of all—a mid-ocean ridge. *Archimede* was a large, cumbersome, contraption that used flotation tanks filled with gasoline (hexane) for buoyancy and iron shot for ballast, and was thus about as safe as the *Hindenburg*. Carrying a crew of three in a two-metre diameter steel sphere beneath the giant tanks, it

[43] A fracture zone is what it says—a linear zone (up to about 20 km wide) in which the sea floor is fractured by faults, which may or may not be active.

[44] For pedants, toenails grow at an average rate of about 20 mm per year, slightly less than the half-rate of spreading here. The oft-quoted analogy between the rate of plate separation (i.e. the full spreading rate) and the rate at which fingernails grow is accurate for slow-spreading ridges, since fingernails grow twice as fast as toenails. Spreading rates on the southern section of the East Pacific Rise may be more than three times as great as this, however.

[45] As long ago as 1960, the bathyscaphe *Trieste* reached the deepest part of the ocean in the southern Mariana Trench in the western Pacific—the Challenger Deep (named after HMS *Challenger*, the ship that made the first soundings here)—although the crew were lucky to survive and the feat was not repeated until 2012, when an egotistical Canadian film director made a somewhat futile descent in a one-man vehicle.

weighed in at 61 tons and had to be towed to the dive site, since it was far too big and heavy to bring on board. It had already been used to explore at depths of over 9000 m in the Japan and Kurile–Kamchatka trenches during the early 1960s, so the Mid-Atlantic Ridge presented few challenges, although the rugged bottom topography caused one or two hair-raising moments. *Archimede* completed seven dives in 1973, taking geologists to the floor of the axial valley, where they were able to recover rock samples using the bathyscaphe's 'telemanipulated grab'.

In the following year, a second vehicle was deployed by the French. *Cyana*, a 'diving saucer' developed from an original design used by Jacques Cousteau, was itself buoyant, and could therefore dispense with the clumsy (and dangerous) tanks. It was thus much smaller, lighter (around 9 tonnes), and far more manoeuvrable than the older bathyscaphes. Bristling with cameras and sensors, the 'yellow submarine' *Cyana* carried three personnel at a time down into the axial valley, where they examined structures between the ridge axis and the valley walls. *Cyana* and *Archimede* each made a dozen dives during 1974.

True to stereotype, the French team spent half their time diving on a 250 m-high mound situated within the axial valley that they called 'Mont de Venus'.[46] Unlike the surrounding ocean floor, this was not heavily faulted and fissured, and produced the youngest and least-altered pillow basalt samples.[47] It was thus interpreted as the site of most recent volcanism (that is, younger than 10,000 years), as you might have guessed from its location right on the axis of the ridge. Hence, despite the 4000 km width of the Atlantic, it appeared that most ocean floor had been produced in a narrow zone only about 1 km wide. This was good news, because such a narrow 'intrusion zone' was necessary to produce identifiable Vine–Matthews-type magnetic stripes.

In the same year (i.e. 1974), the Americans employed their own vehicle, which had been funded by the US Navy and constructed ten years before. It was named after another 'Vine'—Allyn Vine, the Woods Hole geophysicist behind the project—and so became known as *Alvin*. Commissioned in 1964, the vehicle used syntactic foam to obviate the requirement for large buoyancy tanks, keeping weight down to a

[46] They were, perhaps, following the advice of the bathyscaphe pioneer Jacques Cousteau: 'One should not attack the sea. You need to make love to it' (quoted in Ballard, 2000).

[47] In fact, the rocks recovered from the sea bed included some pillow fragments that turned out to be very similar to samples dredged further north back in 1947.

respectable 14 tons, and making it almost as manoeuvrable as *Cyana*. As we shall see, this was by no means the last time that Earth Scientists would thank the American military for a major technical advance, although, like a lot of really great kit developed by the US armed forces, *Alvin* had initially been assigned to classified military applications, including the recovery of an H-bomb off the coast of Spain, so it took ten years to make its first dives on a mid-ocean ridge.

Fitted with a new pressure hull to increase its maximum working depth, *Alvin* made 17 dives into the axial valley. The faulted trough looked rather like many rift valleys studied by geologists on land, where a central strip of crust subsides along sets of parallel faults. Not surprisingly, then, this was the interpretation originally placed on the axial valley: that is, it must be a strip of crust that had dropped like a keystone as the two sides separated. Here in the Atlantic, however, spreading means that the crust will gradually move away from the axis on either side, and must therefore *rise* on new faults to the elevation of the flanking highs. So, although it looks as though the axial valley has dropped, it has, in fact, formed at its present depth, and the ocean crust beneath it will, in future, *climb up* the sides of the rift on linear faults. As it does so, it will generate modest, shallow, earthquakes as observed along the Mid-Atlantic Ridge today. The axial valley is thus what scientists call a 'steady-state' feature, constantly being regenerated by spreading, but retaining the same appearance. This is just one of the many counter-intuitive aspects of sea-floor geology!

According to Robert Ballard and Tjeerd Van Andel, Project FAMOUS was designed 'to furnish data that could provide constraints on models of the kinematics and eventually the dynamics of processes at divergent plate boundaries'. In truth, most of the scientific information gleaned on these dives could have been obtained without risking the lives of the heroic submariners. Yet, despite criticism at the time that manned submersibles were merely 'toys' that did not contribute greatly to scientific understanding, the FAMOUS dives succeeded in capturing the imagination of the public in much the same way that manned Moon landings attracted headlines denied to far more significant unmanned missions launched by NASA.

Possibly the most important finding of the FAMOUS manned submersible dives was, in retrospect, something the geologists did *not* see. Having defined such a narrow zone of crustal creation, you would expect to see signs of volcanic activity, such as the odd eruption, or at least some hot springs—a group of chemists even went along to sample the latter. Nothing of the sort was observed, and despite the presumed presence of magma at around 1300°C not far beneath the sea floor, heat flow along the axis was nowhere near as high as expected. Furthermore, there were few signs of life on the ocean bed—not that you would expect much in such cold, dark conditions. The chemists went home empty-handed. Maybe they were just unlucky, or maybe the Mid-Atlantic Ridge was less active than they imagined.

Undismayed, American scientists took *Alvin* to the Pacific in 1977, where faster spreading implied a greater flow of heat, and thus a greater likelihood of hot springs. Perhaps surprisingly, in view of the fact that the longest and fastest-spreading of all mid-ocean ridges, the East Pacific Rise, runs only a few hundred kilometres to the west, they chose the Galapagos Rift, a west–east trending mid-ocean ridge just north of the equator. Signs of higher temperatures had already been detected a few hundred kilometres east of the Galapagos Islands, and so this site was chosen for the search for hot springs (or 'hydrothermal vents' to the cognoscenti). They did, indeed, find volcanic springs, although, with a water temperature of only 17°C, they were not even lukewarm. Still, compared with normal bottom temperatures of less than 1°C, this was not to be sniffed at. But what made these dives really memorable was not the geology of the ocean floor at all—it was the biology!

Aiming to find active vents and associated volcanic phenomena, the expedition this time had not thought to include any biologists. Nor had they brought any means of preserving specimens, other than some cheap Russian vodka. Nevertheless, they soon discovered beds of giant clams, each animal a foot or so in length, along with shrimps and dandelion-like organisms. Strangest of all were bright red 'tube worms', up to 50 cm long,[48] that lived in long, rigid cylinders of calcium carbonate. Their red colour, it turned out, was due to their blood, which, like ours, contains haemoglobin. Unlike deep-sea creatures elsewhere, such as the Arctic quahog clam, now believed to live for 500 years or more, these 'live fast, die young' communities were ephemeral: while a vent was active, the ecosystem made hay, but when it cooled, everything either died or moved on in search of pastures new. Being at the bottom of the ocean in what Robert Ballard refers to as the 'eternal darkness',[49] such animals could not depend

[48] Even larger specimens—up to 4 m in length—were discovered on later dives.

[49] Ballard (2000).

on photosynthesizing bacteria or plants for nutrients, so they had to derive their sustenance from somewhere else, and that could only be the hot fluids emanating from the vents. In particular, hydrogen sulphide, it seemed, was most likely to be used by bacteria living within the tissues of these creatures in a process known as 'chemosynthesis', as is the case for bacteria inhabiting hot springs on land. Maybe—a crazy thought—chemosynthesis evolved billions of years ago with the first forms of life on Earth, long before photosynthesis was stumbled upon.

These studies had shown that mid-ocean ridges were like factories producing ocean crust by means of volcanism. Studies of ophiolites and *in situ* ocean crust had given a good idea of what the product looked like in detail, but how did the production line that manufactured it operate?

Infinite Onions

Joe Cann regards himself as 'one of the old fogies of marine geology'. Just about the most eloquent communicator of Earth Science you will ever encounter, Joe has an ear for a snappy phrase or image to convey concepts that would otherwise demand whole pages of explanation, and it was Joe who came up with the notion of the 'infinite onion'.

Whilst digging in his Norwich garden back in the 1970s, it occurred to Joe that magma, formed by partial melting of mantle peridotite rock, would rise buoyantly as the crust was pulled apart at a mid-ocean ridge. He supposed that this melt rose high into the oceanic crust, where it ponded to form a magma chamber. As the crust above cracked in response to spreading, sheets of magma would rise to the surface and erupt as basaltic lavas. The lavas would be chilled rapidly by the cold seawater, while the magma in the feeder crack would cool a little more slowly, eventually becoming a dolerite dyke. The magma within the chamber would cool much more slowly, and from time to time be replenished by fresh melt entering from below. As it cooled, minerals would crystallize on the walls of the chamber, producing a rock with the same composition as basalt, but with larger crystals that were easily visible to the eye. This rock was gabbro, as had been discovered in ophiolites. Large, dense, olivine crystals in the magma might sink to the magma chamber floor as 'cumulates'.

I have heard it said that the 'onion' part of 'infinite onion' refers to the manner in which the magma chamber produces 'skins' of gabbro and

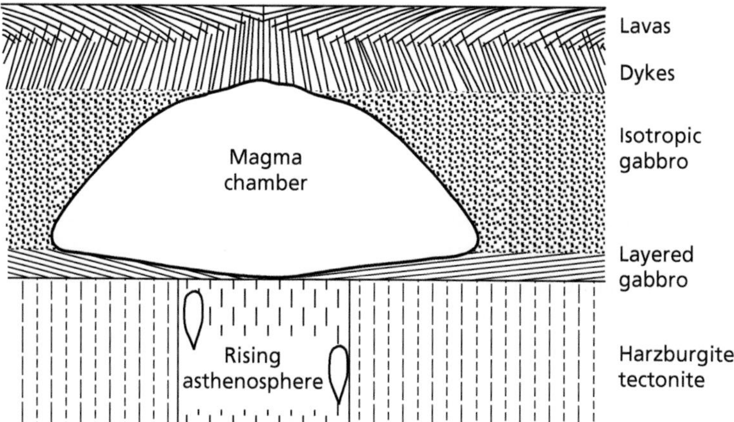

Figure 1.5 Joe Cann's 'infinite onion' model for mid-ocean ridge magma chambers. After Cann (1974).

cumulates that 'peel off' to form the lower oceanic crust. This rather sophisticated interpretation highlights the central role of the magma chamber in manufacturing oceanic crust, although the real reason for adopting the term is, by comparison, embarrassingly mundane. In fact, it was called an infinite onion because, when you slice through it (in your imagination), it looks like . . . an onion (Figure 1.5). That is, it slopes away from the ridge axis in its upper part, but towards it near the bottom, with a small spike right at the top (according to Joe).

Since sea-floor spreading is more-or-less continuous, lava flows would be supplied by feeder dykes rising from the magma chamber at frequent intervals, while mineral crystals would continue to grow slowly on the walls and floor of the chamber as the magma cooled and solidified. The result would be an upper crust composed of lava flows and feeder dykes, and a lower crust composed of gabbro with cumulates at the bottom. Beneath this would be the upper mantle that had partially melted to produce the magma that filled the chamber above, and was now what geochemists describe as 'depleted'. It was still classed as peridotite, but, having once coughed up a melt, it would, in future, be more reluctant to repeat the feat. Rocks like this were first identified in the Harz Mountains, the highest range in northern Germany (which isn't saying much), and were named after the town of Bad-Harzburg, giving us the rather cumbersome title 'harzburgite'. As a point of interest, the 'undepleted' variety of peridotite received an even worse appellation, after a place in the French Pyrenees called 'Lherz', where this type of peridotite is quite common. It thus became 'lherzolite'—if you can say it.

Like the Vine–Matthews Hypothesis itself, Cann's model possessed a certain elegance borne out of simplicity. It very neatly linked observations at mid-ocean ridges and ophiolites with the layered structure of crust derived from seismic refraction work. Its central feature—the onion—was a large, continuous, crustal magma chamber that cooled slowly to form the lower crust, and shot sheets of magma towards the surface, creating the dykes and pillow lavas of the upper crust. Encouragingly, there was evidence from the Troodos and Oman ophiolites for the existence of such large magma chambers, possibly as wide as 30 km,[50] so it looked as though Joe's elegant model was a winner.

Sadly, it was not long before things were spoiled by new observations, forcing modifications to the theory—as Joe later philosophized,[51] 'models are there to be overthrown'. The problem was that mid-ocean ridges tend to fall into two classes according to the rate of spreading. That is, 'fast-spreading' centres[52] have different characteristics to 'slow-spreading' centres. Slow-spreading centres are those like the Mid-Atlantic Ridge, Carlsberg Ridge, and Southwest Indian Ridge, where plates separate at less than about 60 mm per year. Fast-spreading centres are all the rest, with separation rates increasing to around 150 mm per year today, and perhaps over 200 mm per year in the past.[53] Slow-spreading centres, as the FAMOUS scientists showed, are comparatively rugged, with a deep valley along the ridge axis, bounded by active faults, whereas fast-spreading centres, like the East Pacific Rise, are typically smooth with subdued topography, the axis often marked by a narrow linear high. This fundamental difference is reflected in the structure of the crust beneath: whereas there was evidence[54] for a zone of low seismic wave speeds beneath fast ridges, suggesting the presence of a magma chamber, such anomalous zones were local or absent altogether beneath slow ridges. And, since a continuous (both in space and time) magma chamber was an essential, not to say, central, component of Cann's model, this represented something of a fly in the ointment. The

[50] Pallister and Hopson (1981).

[51] Cann (1991).

[52] Sea-floor spreading having been accepted, mid-ocean ridges started to be referred to as 'spreading centres'.

[53] Wilson (1996).

[54] Fowler (1978).

root of the problem, of course, is that, at slow spreading rates, magma rising into the crust has more time to cool by conduction and solidify. In that case, continued extension can only be accommodated by deformation of the solid crust, producing faults. This would explain the existence of an axial valley and all that faulting at the Mid-Atlantic Ridge.

Two other Cambridge geologists—Euan Nisbet and Mary Fowler—soon came up with a modification to Cann's model specifically for slow ridges.[55] In this, melting began around 60 km beneath the sea floor in the upper mantle, the magma segregating from the solid residue and rising to the base of the crust just 6 or 7 km beneath the sea floor. From here, magma ascended along vertical cracks, erupting at the surface to feed lava flows, or cooling and solidifying within the upper crust to form pods of gabbro. Unlike Joe's onion, melt ponded only in narrow (less than 2 km wide), more-or-less vertical, magma chambers. This model they wittily christened the 'infinite leek'[56] model.

So how might we detect whether infinite onions actually exist or whether we should go for a leek instead? Drilling would certainly establish the existence, or otherwise, of a magma chamber in the crust, but pushing a drill bit into 1300°C magma could have drawbacks. We will come back to this central problem in Chapter 8, which deals with a period in which the nature of much of the mid-ocean ridge system became familiar to scientists through concerted international programmes of exploration and mapping.

Afterthought

Through the efforts of marine geophysicists and geologists, we now had a theory that explained how continental drift could occur without flouting the laws of physics. We also had a good idea of the nature and age of the oceanic crust, as well as testable models for its production at mid-ocean ridges. The magnetic bar-code recorded in the crust beneath the oceans had revealed the secret of the Earth and provided the impetus for a complete re-think of global geology, based on the idea that the crust moves horizontally rather than just up and down.

In his momentous 1966 *Science* article, Fred Vine wrote that 'the oceanic crust is a surface expression of the upper mantle'. What this means is

[55] Nisbet and Fowler (1978).

[56] This is sometimes rendered as 'infinite leak', depending on the author's understanding of the word 'pun'.

that simply exposing the hot mantle at the cold surface of the planet is enough to produce oceanic crust. If you stripped away the oceanic crust right now, another basaltic crust would form, all by itself, like a scab. Indeed, the pioneering marine geologist Bruce Heezen[57] referred to mid-ocean ridges as 'the wound that never heals'. Sea-floor spreading is simply the mantle's response to the two flanks of a mid-ocean ridge being pulled apart. But what is it that pulls (or pushes) them apart? And, if new ocean floor is constantly being manufactured, where does the old crust go in order that the Earth does not expand like a balloon? These, and many other questions, could only be answered by a brand new theory—plate tectonics.

[57] Whose name, Richard Hazen informs us in his excellent book *The Story of Earth* (Hazen, 2012), is pronounced 'hay-zen'.

2

The Paving Stone Theory of World Tectonics

> ... there is a stage in the development of a theory when it is most attractive to study and easiest to explain, that is while it is still simple and successful and before too many details and difficulties have been uncovered.
>
> SIR EDWARD BULLARD (1969)

If your parents had saddled you with a middle name like 'Tuzo', you'd probably keep quiet about it. Yet the son of French Canadian mountaineer, Henrietta Tuzo,[1] after whom a peak in the Rocky Mountains was named, chose this appellation in preference to the forename he shared with his father, John (admittedly, John Wilson is a pretty dull name). Canada's very first graduate in geophysics in 1930, Tuzo Wilson became one of the twentieth century's leading Earth scientists, eventually outdoing his mother by having not one, but two, peaks named in his honour.[2] And, if you had to pick one name from the small army of geophysicists associated with the evolution of plate tectonics from the sea-floor spreading hypothesis, then that name would be Tuzo Wilson.

*

Sea-Floor Spreading Transformed

In 1959, four years before the publication of Vine and Matthews' momentous article, *Nature* published a brief note by the marine geophysicist Vic Vacquier (whom you may recall from Chapter 1), with the

[1] Probably derived from 'Touselle' or 'Touzelle'; see Wilson (1982).

[2] The 'Tuzo Wilson Seamounts' off British Columbia, forming part of the Kodiak–Bowie seamount chain. The original Mount Tuzo could, apparently, be seen in an engraving on the back of previous Canadian twenty-dollar bills.

The Tectonic Plates Are Moving! Roy Livermore, Oxford University Press (2018). © Roy Livermore.
DOI: 10.1093/oso/9780198717867.001.0001

catchy title *Measurement of Horizontal Displacement along Faults in the Ocean Floor*. The ocean floor in question was the same bit of the eastern Pacific that had been the target of Raff and Mason's magnetic surveys, now extended with additional profiles to the west. Vic focused on magnetic anomalies either side of a long west–east trending fracture zone known as the Pioneer Ridge Fault, named after the US Coast and Geodetic Survey ship that collected much of the data. Profiles from north and south of this feature showed almost identical patterns of anomalies, but shifted by about 138 nautical miles parallel to the fault. That is, the entire magnetic 'grain', and, by implication, the entire floor of the ocean on the northern side of the fault, had been translated 256 km to the west. And this was only one of three such faults in this part of the Pacific. What forces could possibly have transported great slices of the Earth's solid surface by hundreds of kilometres with hardly any signs of deformation (other than the narrow fault zones)? A couple of years later, Vacquier published another short note, this time with Raff, suggesting a combined offset of 1420 km on the Pioneer and Mendocino faults, again by matching the magnetic sequences north and south of the faults. Projected eastwards onto the continent, such fault movements might, they thought, explain some of the complexities of North American geology, but the faults themselves seemed to die out near the coast.

The fascinating thing about these articles is that the authors now possessed everything necessary to crack not just the theory of sea-floor spreading (the magnetic anomalies), but plate tectonics itself (the faults). Had Vic realized, as Vine and Matthews soon would, that such magnetic anomalies represent much more than merely a grain or signature in the crust, his might have been the name attached to the spreading hypothesis. And, if he had, it is likely that his interpretation of the enormous offsets on these faults would have been very different. As it was, even after Vine and Matthews published the answer to the spreading conundrum, Vacquier presented a paper to the Royal Society in 1964, in which he referred briefly to the hypothesis of the Cambridge pair, but then claimed that the origin of the north–south lineated magnetic anomaly pattern was unknown! The title of his article, *Transcurrent Faulting in the Ocean Floor*, showed that he still had not got it.

Transcurrent faults are linear, vertical, cracks along which two sections of crust move in opposite directions—like the ones Vic observed. If you stood, gazing across such a fault, and you noticed, say, that a line of fence posts or a stream (I am assuming dry land here) perpendicular to the fault had been shifted to your left on the other side, you would naturally refer to the fault as a 'sinistral transcurrent fault' (assuming you were a geologist). If to the right, then, of course, a 'dextral transcurrent fault' would be your verdict. In the case of the Pacific faults, it was clear that the mid-ocean ridge itself had been offset, in two cases to the west, and in one case to the east (when looking north). It did not take a genius to work out that it had been shifted by sinistral and dextral transcurrent faults, respectively. But it *did* take a genius to conclude that it had not, in fact, been displaced at all, that the direction of movement or 'slip' on the faults was exactly opposite to that suggested by the offset, and that here was an entirely new class of fault. That genius was Tuzo Wilson.[3]

Before the proceedings of the Royal Society meeting were published in 1965, good old Tuzo rode to the rescue with what has become a classic article,[4] explaining these enigmatic faults and the displacements along them, and providing the crucial step that led from sea-floor spreading to what was about to be called 'plate tectonics'. Central to this was the recognition that mid-ocean ridges do not get broken into segments and shifted thousands of miles by transcurrent faults as Vic and friends supposed: they are *born* segmented, and they *remain* segmented. Offset ridge segments are connected by faults along which the crust produced at each segment slides in opposite directions, both sides moving away from the ridge at which the crust was created by sea-floor spreading (see Figure 2.1). Counter-intuitively, the motion on these faults is in the reverse sense to that which would be required to offset

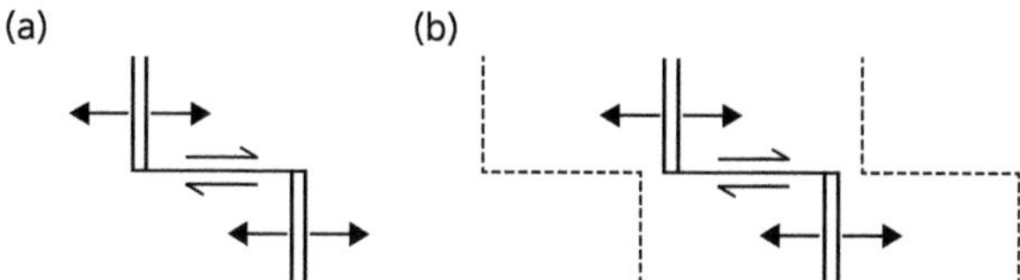

Figure 2.1 The concept of a ridge–ridge transform fault and its evolution. Note that relative motion is in the opposite sense to that of the ridge offset. [(a) and (b) are from Figures 2 and 3 ,respectively, of Wilson (1965).]

[3] Although there is a suspicion that Tuzo was given the idea by Fred Vine (see Frankel, 2012d). Another young British geologist, Alan Coode, also mentioned the idea to Tuzo after hearing him talk in February 1965 (see Coode, 2011). Coode concedes that Tuzo was far better at 'marketing' new ideas.

[4] Wilson (1965).

the ridge—that is, in the opposite direction to movement on a transcurrent fault. This can only happen, Tuzo pointed out in one of the most profound statements in the history of Earth Science, because the faults are 'connected into a continuous network of mobile belts about the Earth which divide the surface into several large rigid plates'. He had introduced the concept of a plate as distinct from the crust two years earlier, explaining 'This plate is not to be confused with the crust; it is rather to be regarded as the cooler and hence more rigid and viscous upper layer of mantle and crust resting upon the main portion of the mantle'.[5] 'Mobile belt' was a term he used to describe narrow zones of deformation that were of three fundamental types: mountain ranges (including volcanic arcs), mid-ocean ridges, and Vic's faults: in other words, plate boundaries. Tuzo christened this newly discovered class of fault a 'transform fault'.[6] It was a major breakthrough in understanding that is widely regarded as marking the birth of the global theory of plate tectonics.

A year or two later, American seismologist Lynn Sykes (♂) used recordings from a new worldwide network of high-quality seismographs to analyse ten earthquakes that occurred on faults offsetting segments of the mid-ocean ridge system, demonstrating that they were, in every case, entirely consistent with Wilson's proposal.[7]

Plates in the Round

However you look at it, 'plate tectonics' is a pretty rotten name. For a start, plates are neither flat (as in 'steel plates'), nor even plate-shaped (as in 'dinner plates'). According to that highly respected elder statesman of geology, Kevin Burke, the misnomer may have resulted from the predilection of a young geology student for rowing rather than attending lectures.[8] The student was (of course) Tuzo Wilson, who later admitted that he made an error by considering only a flat Earth when he first used the term 'plates'. A more accurate, if less concise, way of describing their three-dimensional form would be to say that plates are

[5] See Wilson (1963b).

[6] He also illustrated how similar faults might connect ridges to trenches, or trenches to other trenches.

[7] Sykes (1967).

[8] Burke (2011).

fragments of a spherical shell (a good way to get an idea of their true shape is to smash a large chocolate Easter egg). And who knows what 'tectonics' means? Actually, it means 'builder' in Greek, so you could say that it bound to be a bit suspect. In fact, it is quite a good term when used in the context of continental collisions, where new mountain ranges are constructed by folding and faulting of existing rocks, but less appropriate when applied to a major theory that involves (as we shall ultimately see) the entire planet.

It was at this stage that the baton of progress passed to—or, rather, was seized by—the physicists, who, being physicists, explained everything using maths. Two, in particular, were responsible for codifying the new theory. The first, a scientist of rare insight, was referred to by the well-known British palaeontologist Richard Fortey[9] thus: 'I was preceded in my college life by Dan McKenzie, while he was solving the problems of the construction of the world. He was a year or two ahead of me, and I was always invisible in the glare generated by his brilliance.' All geophysicists know, and many fear, Dan, whose approach to science has something in common with that of the great French mathematician Henri Poincaré: 'Accustomed to neglecting details and to looking only at mountain tops, he went from one peak to another with surprising rapidity'.[10]

In 1963, just as the September issue of *Nature* containing Vine and Matthews' article appeared, Dan became a graduate student of Teddy Bullard in the same Cambridge University department, and was thus exposed immediately to their revolutionary hypothesis, as well as Bullard's ideas about the fit of the continents around the Atlantic. Despite this stroke of luck, and the presence in the Department of Tuzo Wilson and Harry Hess, he remained unconvinced and did not join the revolution, choosing to complete his thesis on mantle convection instead. It was not that Dan was especially stupid (as he later remarked[11]), but rather that 'it simply was not obvious to us that what Fred and Drum were doing was so important'.

By 1967, he was convinced, and, as a postdoctoral fellow in the earthquake seismology laboratory at the California Institute of Technology, Dan took some physics courses to develop his skills in analytical methods,

[9] In his book *The Earth: An Intimate History* (Fortey, 2004).

[10] Belliver (1956).

[11] McKenzie (2003).

despite the annoying showmanship of the lecturer, one Dr R. P. Feynman. He later obtained a postdoctoral position at Scripps, where he and ex-Cambridge colleague and fellow convert Bob Parker wrote a short article introducing the fundamental principles of plate tectonics on a sphere, or, to put it another way, they wrapped Wilson's rigid plates around a globe. In this much-cited paper,[12] they used the North Pacific as a kind of laboratory for testing the new theory, and made frequent mention of the 'plates'. They did not, however, use the term 'plate tectonics' explicitly, preferring the rather more evocative, if wordy, 'paving stone theory of world tectonics'. Other authors later referred to the 'new global tectonics', a title doomed from the outset, but, as we know, everyone finally settled for 'plate tectonics'. Personally, I think it's a shame that McKenzie and Parker's suggestion did not catch on.

Influenced by Teddy Bullard, Dan and Bob explained how the motions of plates could be described as rotations on a sphere using a theorem from the eighteenth-century Swiss mathematician Leonhard Euler. Fortunately for us, the theorem is fairly simple. Imagine the outline of a continent on a globe. Then imagine the same outline shifted to a different location, in different latitudes, with a different orientation. Now, what is the minimum number of movements needed to shift the outline from its original position to the new one? Unless you are a mathematician or have an unhealthy interest in geometry (in which case you might want to skip the next two paragraphs), you may well say 'two'. That is, one to translate the outline to the desired position, and a second to orient it properly.

The correct answer, however, is 'one'. Both the 'translation' and the 'orientation' are, in fact, rotations about some axis that passes through the centre of the Earth[13] (you may need a minute to appreciate this). And Euler's theorem says that any two such rotations can be contracted into a single rotation—that is, you can move any shape from anywhere on a sphere to anywhere else, in any orientation, with a single rotation (you may need another minute to convince yourself of this—Euler took a lot longer). So, you could represent the displacement of one plate (let's call it—I don't know—'plate B') with respect to another ('plate A', say) by sticking a drawing pin into a globe to define a pole of

[12] McKenzie and Parker (1967).

[13] Just like the Earth's spin axis, which passes from the north pole to the south pole via the centre of the planet, but with a different orientation.

rotation (i.e. where the rotation axis cuts the surface), and then connecting it with a bit of string to plate B. To keep it simple, let's assume plate B is moving eastward with respect to plate A (see Figure 2.2a). If you try this in your imagination, you will see that, for west–east relative motion, the drawing pin has to be stuck in the north (or south) geographic pole. The speed of plate B relative to plate A can now be represented by the rate at which you rotate plate B on the end of the string, or, in maths-speak, the rate at which the angle subtended at the drawing pin—sorry, pole of rotation—increases: what we might call the 'angular rate of rotation'. Also, and perhaps surprisingly, you will find that the actual speed of separation of the plates at any point on the boundary between them varies with distance from the pole of rotation. As you move away from the drawing pin, the speed increases until you are 90° from it (i.e. at the equator for west–east plate motion), then decreases again. At the drawing pin, the speed of separation actually decreases to zero. Furthermore, if the rotation pole lies anywhere other than the geographic pole, the *direction* of relative plate motion will also change with location (see Figure 2.2b).

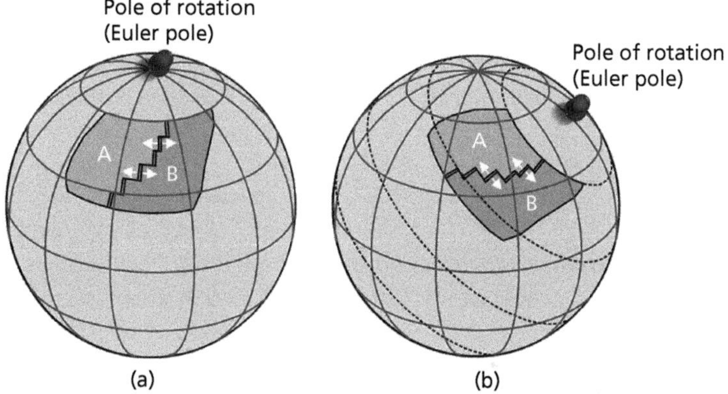

Figure 2.2 Rotation of fictitious plates on a sphere. Plate B rotates anticlockwise about the Euler pole relative to A. White arrows show the resulting plate motions at a mid-ocean ridge. *a.* if the Euler pole should coincide with the geographic pole, then plate motion would be parallel to lines of latitude; *b.* if the Euler pole is elsewhere, then both rate and direction of relative motion change with distance from the Euler pole. Dotted lines denote small circles (plus 'equatorial' great circle) drawn around the Euler pole. Note that the rotation is concerned only with the *relative* motion between the two plates.

Mathematically, a plate's motion is specified by the longitude and latitude of its pole of rotation, together with the angular rate of rotation about that pole in degrees per million years. This motion could be 'absolute motion' (i.e. with respect to the solid body of the Earth) or 'relative motion' (i.e. with respect to another plate). This may all sound somewhat dry to non-mathematicians, yet this approach provided geophysicists and their computers with the power to move, not just mountains, but entire continents and the plates that bore them.[14] If you could establish the location of the pole of rotation for each pair of plates, together with the associated angular rate of rotation about that pole, then you would have a complete mathematical description of all the plate motions on Earth, which you could use to calculate the motion at any point you chose. Of course, you would need to measure the speed and direction of motion at many points first of all, in order to determine those pole positions and angular rates, but this was already being done by mapping magnetic anomalies and recording earthquakes.

As Fred Vine had shown, the rate of spreading could be obtained by modelling the magnetic barcode over mid-ocean ridges. Knowing the age of any two conjugate magnetic stripes on opposite flanks of a ridge, and the distance between them, it was child's play to work out the average rate of spreading between them. This, in turn, would give the angular rate of rotation, once the Euler pole position had been located. Transform faults were very useful here, because they had to be parallel to the direction of motion between the two plates involved, and were thus, geometrically speaking, segments of small circles centred on the pole of rotation. If you were so minded, you could draw great circles perpendicular to several transforms, and these should (in a perfect world) all cross at the pole of rotation (Figure 2.3). By analysing the first motions of earthquakes on ridges and transforms, as Lynn Sykes had done, it was also possible to derive the direction of relative motion from what he called the 'slip vector', simply the horizontal direction in which slip had occurred on the fault that generated the quake. If you drew imaginary small circles of 'latitude' centred on wherever the Euler pole turned out to be, then, as McKenzie and Parker pointed out,

[14] Some years later, in the 1970s, Tuzo Wilson lamented that, had he known of Euler's theorem, he could have 'nailed the theory of plate tectonics'. See Dewey (2015).

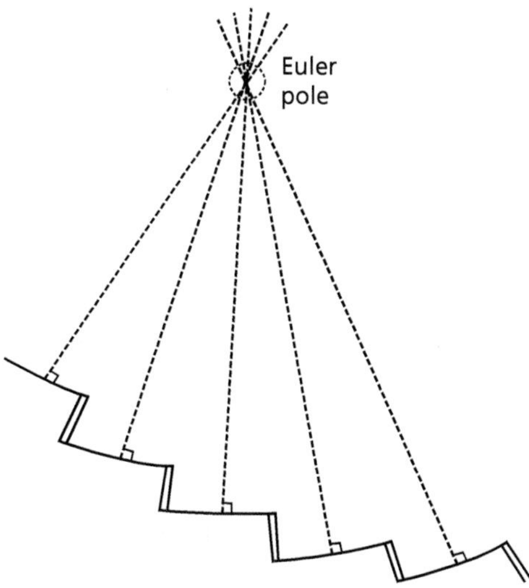

Figure 2.3 Great circles drawn perpendicular to transforms converge at the Euler pole.

'If the paving stone theory applies, all slip vectors must be parallel to the latitudes which can be drawn with respect to this pole.' Slip vectors could thus be used in reverse to help define the location of the pole of rotation.

Applying all this to the North Pacific, an area representing roughly a quarter of the Earth's surface, McKenzie and Parker were able to explain its character and gross geology with uncanny accuracy. Using earthquake slip vectors from aftershocks of the 1964 Alaska earthquake together with the trend of the San Andreas Fault, they estimated the location of the Pacific–North America Euler pole to be somewhere north of the Great Lakes (Figure 2.4). They then drew predicted vectors, based on this pole, along the west coast of the USA and Canada, showing how the direction of relative plate motion would change as you proceeded from south to north. Baja California, being part of the Pacific plate, moves north-westward relative to the rest of Mexico, neatly explaining the opening of the Gulf of California by sea-floor spreading. West of the San Andreas and Queen Charlotte transform faults, the Pacific coastal strip was predicted to be moving northward parallel to the overall trends of the faults—exactly what you would expect. But between these two faults, something queer was

Figure 2.4 Euler pole and vectors for rotation of the Pacific plate with respect to North America. [From McKenzie and Parker (1967), Figure 4.]

going on. The Cascadia margin of North America (from Vancouver Island to northern California, including the cities of Seattle, Portland, and Eureka) had a quite different character, which could not be explained by the movement of just two plates. An additional oceanic plate, the 'Juan de Fuca' plate, identified previously on the basis of magnetic anomalies by Tuzo Wilson, was required. As an illustration of the power of this new, quantitative, approach. Dan and Bob demonstrated that, in order for the plate boundaries to be stable, the Juan de Fuca plate had to be underthrusting the North American continent in the queer region north of the Mendocino Fault, which would explain the presence of active volcanoes such as Mount Meager, Mount

Baker, Mount St Helens, Mount Rainier, Mount Hood, and Mount Shasta, and the frequent occurrence of thrust-type earthquakes in this region.

North of the Queen Charlotte Fault, up in the Gulf of Alaska, Pacific plate vectors were directed *onshore*, implying direct convergence with North America—again, just what you would expect from the large earthquakes recorded there. It was clear that the character of a plate boundary (i.e. constructive, conservative, or destructive) is determined not simply by the relative plate motion vector at any location, but by the combination of this vector with the orientation or 'strike' of the plate boundary. Thus, an abrupt change in strike could give rise to a dramatic change in tectonic style. For example, where the orientation of a plate boundary changes from parallel to oblique with respect to the vector, the boundary might switch from a conservative to a convergent type. This is exactly what happens at the sharp bend in the Pacific coast of Alaska.

Having explained the tectonics of a quarter of the Earth's surface so satisfactorily, McKenzie and Parker concluded that it seemed reasonable to expect that the same principles would also apply to the remaining three-quarters. The paving stone theory thus provided quantitative support for the daring speculation made by Vine in his 1966 masterwork, that North America had overridden the East Pacific Rise. The Juan de Fuca Ridge is thus the surviving relic of the northern section of the East Pacific Rise that collided with a volcanic arc, and the Juan de Fuca plate is a surviving relic of a huge plate that once filled the eastern Pacific, a plate dubbed the 'Farallon' plate after a group of islands off the Californian coast. The zebra-stripe magnetic anomalies on the flanks of the Juan de Fuca Ridge that confounded Raff and Mason thus record the history of the collision, which continues today. The 'survivors' are therefore doomed.

The second physicist, in case you're wondering, was Jason Morgan. While Dan McKenzie was dazzling everyone in Cambridge and elsewhere, Jason, working alongside Fred Vine at Princeton University in New Jersey, also concentrated on the geometry of the boundaries between the plates, and the relationship of these boundaries to plate motions. The impressive title of Jason's 1968 *magnum opus*, 'Rises, Trenches, Great Faults and Crustal Blocks',[15] referred to the three types of plate

[15] Morgan (1968).

boundary established by Tuzo Wilson (i.e. rises = mid-ocean ridges, trenches = deep-sea troughs associated with volcanic arcs, great faults = transform faults). In this paper, Jason, like Dan and Bob, rectified Tuzo's omission by 'extending the transform fault concept to a spherical surface'. He did not use the word 'plate' at all, referring to 'crustal blocks' instead, thereby introducing a *faux pas* of his own, though he corrected himself by concluding that 'The required strength cannot be in the crust alone', and favouring instead 'a strong tectosphere,[16] perhaps 100 km thick'. In other words, a plate must consist mostly of rigid *mantle* rock surmounted by oceanic or continental crust.

Like Dan and Bob, Jason applied the rotation pole idea to the plates, estimating pole locations for several pairs of 'crustal blocks' by drawing great circles perpendicular to transform faults. He also demonstrated how, by adding together rotations for several pairs of plates, you could estimate the motion between two plates that were converging, and thus where no sea-floor record existed. So, for example, if you wanted to know how fast Africa was approaching Europe, you could find the rotations for Africa to North America and also for North America to Europe from magnetic anomalies and transform faults on the Mid-Atlantic Ridge, and then simply 'add' the two rotations (using matrix algebra) to get the resultant Africa-to-Europe Euler vector, assuming 'closure'. Soon after this, McKenzie and Morgan combined forces to look at how plate boundaries might evolve with time, including the junctions where three plates meet.[17] They claimed, somewhat optimistically, that 'problems of present day tectonics reduce to determining the plate boundaries and relative rotation vectors of all plates on the Earth's surface'.

Surprisingly, it has never been possible to say exactly how many plates there are. 'About a dozen' was, and still is, a common approximation, although one estimate in 2002[18] put the total at 52, which by 2016[19] had multiplied to 159. One reason for doubt about the existence of such large and obvious things as plates is that they are not always large or obvious. Some plates, or perhaps we should call them 'platelets', are too small for geophysicists to be bothered with, while relative motion

16 Another name for 'lithosphere' or, more simply, 'plate'.
17 McKenzie and Morgan (1969).
18 Bird (2003).
19 Harrison (2016).

between others, such as India and Australia, is so slow that no-one can agree whether a boundary exists at all.

Forward to the Past

McKenzie and Parker noted in 1967 that plate theory, as originally formulated, applied primarily to present-day motions, and that the past evolution of the plates was then only poorly understood. But, by 1970, Dan, now back in Cambridge, had rectified this by working out the history of East Africa and Arabia based on plate tectonics principles. The Carlsberg Ridge, which had so recently and so unexpectedly risen to prominence, continued westward into the Gulf of Aden, where it was offset by many small transforms. It did not, however, die out, but continued as far as Djibouti, where it intersected the southernmost Red Sea. This was also the place where the East African Rift terminated, making it a 'triple junction', such as McKenzie and Morgan had previously described. The Rift, of course, ran through the African continent and, unlike the other two branches, had not (yet) developed into a narrow ocean. Nevertheless, the East African Rift, the Red Sea, and the Gulf of Aden, were all identified as plate boundaries, defining three plates that were dubbed the 'Nubian', 'Somalian', and 'Arabian' plates, the first two forming modern Africa.

Using magnetic anomalies and transform faults from the Gulf of Aden and Red Sea, McKenzie and colleagues calculated rotations for the three plates, using the 'closure' argument to calculate the rate and direction of extension on the East African Rift following the separation of Arabia around 15 million years ago. Rates of spreading calculated for the Gulf of Aden and Red Sea were about 20 mm per year, lower than those on the Mid-Atlantic Ridge or East Pacific Rise, and opening of the Rift was slower still, while it was clear that the Carlsberg Ridge had propagated into the African continent from the east, producing a new ocean basin that eventually connected with the similarly youthful Red Sea, thus separating Arabia as an independent plate.

The following year, 1971, Dan, together with John Sclater at Scripps, was able to publish a comprehensive history of the Indian Ocean. In a magnificent *tour de force*, the authors compiled data for the entire ocean basin, including magnetic anomalies, bathymetry, heat-flow measurements and earthquake epicentres. McKenzie and Sclater's interpretation of these data led to calculations of best-fitting plate rotations for each of seven time-slices going back to the Late Cretaceous, from which they

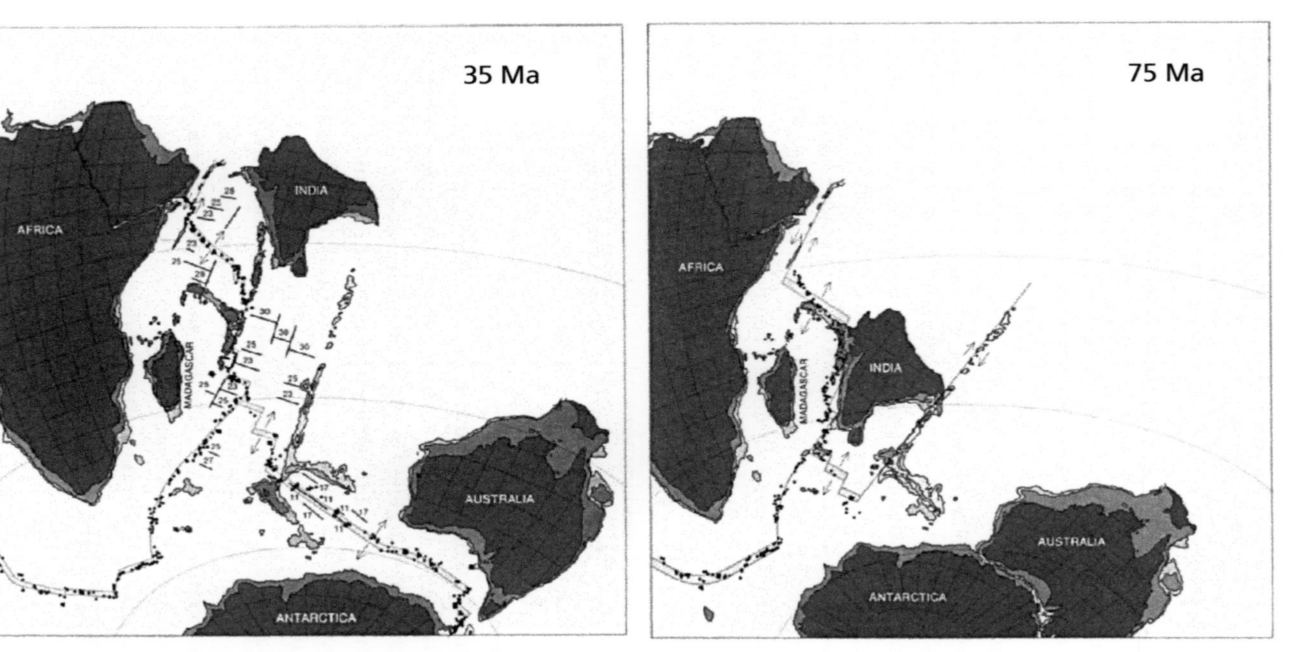

Figure 2.5 Reconstructions of the Indian Ocean by McKenzie and Sclater. [From McKenzie and Sclater (1973).]

plotted a series of maps showing the stages of growth at each of these times (see Figure 2.5). To their surprise, these reconstructions implied that India must have been moving northward at speeds of around 170 km per million years prior to 55 million years ago, much faster than any other continent-carrying plate moves today. Its abrupt deceleration 55 million years ago appeared to reflect the onset of the mother of all collisions with Eurasia. At 91 pages, not including fold-out plate reconstructions, McKenzie and Sclater's article was exceptional both in quantity and quality, earning Royal Society Fellowships for its authors and setting a standard for marine geophysicists working on the history of the plates elsewhere. Thus, scarcely eight years after Vine and Matthews published their initial interpretations of a few short magnetic profiles over the Carlsberg Ridge, the theory spawned by their work permitted McKenzie and Sclater to unravel a detailed history of the entire Indian Ocean basin for the past 75 million years. Ironically, Dan was 'very disappointed' because, in his opinion, nothing new had been discovered. He therefore moved on to investigate the forces that move plates.

Pipes to the Deep Mantle

When you fly from Europe to North America, it is obvious that you are moving westward since, in the eight to ten hours that you are in the air, those two continents stay where they are (for all practical purposes), providing you with a reference frame in which your motion—speed and direction—is easy to measure. But what happens when this reference frame itself moves? Where do we find another 'fixed' frame to which we can refer the movement of continents over millions of years? Since all the plates are moving, we have to abandon the idea of using any single plate as our reference,[20] and look for an alternative. One possibility is to take the mantle far beneath the plates as our 'absolute' reference frame, assuming that any motion there is very much slower than that of the plates. But how do you relate the motion of a plate to the deep interior? Once again, the answer lay in the ocean, and once again, it involved you-know-who.

The Hawaiian Islands had been a mystery to geologists for many years prior to the sixties revolution. What on Earth was a huge volcanic

[20] Although the tardiest plates, Antarctica and Africa, have sometimes been used in this way (e.g. Burke and Wilson, 1972).

structure like Mauna Kea—rising 10 km from the sea bed—doing in the middle of the Pacific Ocean? And what was the explanation for the existence of a tail of islands, seamounts, and atolls running for 3600 km to the WNW of the Big Island as far as Kimmei seamount? This was not even the end of the story, for sea-bed mapping revealed that a chain of seamounts and guyots[21] continued beyond Kimmei with a much more northerly trend, forming a pronounced elbow at the junction. Named (like Kimmei) after Japanese 'mikados', the Emperor Seamounts extended as far north as Meiji at 53°N, close to the Kurile–Kamchatka Trench. In total, the combined Hawaii—Emperor chain consisted of more than 100 volcanoes, with a total length close to 6,000 km.

In 1960, geologists at the United States Geological Survey (USGS)[22] described how the eight principal Hawaiian Islands became increasingly dissected by erosion WNW of the Big Island, eventually giving way to a string of tiny islets and submerged seamounts. That they represented a sequence of subsiding extinct volcanoes had been suggested much earlier by Charles Darwin and others, but most geologists persisted in believing that their origin was associated with a giant fault or crack, through which magma leaked to the surface. In 1963, Tuzo Wilson dismissed this idea on the grounds that no amount of mapping of the sea bed around the islands had revealed any such fault. Instead, he proposed,[23] the volcanic islands were created by movement over a fixed or very slowly moving magma source deep in the mantle. He imagined that a shallow horizontal 'jet stream' carried the ocean floor towards the WNW, and that 'lava' generated in the middle of a deep mantle convection cell rose up through the moving upper layer to erupt at the surface. Although he got the 'jet stream' part wrong, the essence of his idea was that a moving surface layer above a fixed heat source within the mantle could give rise to a sequence of volcanoes that became progressively older as they were carried away from that source. Arthur Holmes put the whole thing much more succinctly[24] thus:
'The two extreme possibilities [...] are both equally exciting:

[21] Erosion during their subaerial phase causes islands to be planed-off so that, when they eventually sink, they become flat-topped seamounts or 'guyots', as Hess had previously christened them.

[22] Eaton and Murata (1960).

[23] In a paper rejected by one leading US journal of geophysics, and eventually published as Wilson (1963).

[24] In the new edition of his fine volume, *Principles of Physical Geology* (Holmes, 1965).

(a) The sub-crustal magma source slowly migrated linearly from the end of the chain with atolls, where it began millions of years ago, to the end with active volcanoes, where it is still operating;
(b) The magma source remained in its present position while the overlying crust migrated in the opposite direction, i.e. towards N. Japan in the Hawaiian case.'

He concluded: 'Neither of these unfamiliar processes can be ruled out, but crustal migration is probably the dominant one.'

Either the concept was so revolutionary that it stunned the world of Earth Science, or it was so ridiculous that it was simply ignored, but the fact is that nobody took much notice of these ideas for nearly eight years. In 1970, Bob Dietz and John Holden, working at the forerunner of the US National Oceanic and Atmospheric Administration (NOAA) in Miami, adopted Tuzo's idea in order to provide 'absolute' coordinates for a series of maps showing past reconstructions of the plates, assuming that volcanic centres or 'hot spots' in the South Atlantic and Indian oceans were fixed. Then, a year later, Jason Morgan recast Wilson's suggestion as the 'fixed-hotspots' hypothesis. It was not, he hypothesized, mid-ocean ridges that were stuck above rising convection currents, as was generally believed, but the hotspots.[25] Hawaii marked the surface expression of one of ten or so 'pipes to the deep mantle', as he described them. Within each pipe, perhaps 150 km in diameter, hot rocks rose at around two metres per year, eventually beginning to melt beneath the Pacific plate as it passed over them. Throwing caution to the wind, he suggested that such pipes, or 'mantle plumes', were 'manifestations of convection in the lower mantle which provides the motive force for continental drift'. The motive force argument turned out to be a step too far,[26] but his hypothetical mantle plumes sowed the seeds of a controversy that would lead to bloodshed in the next century.

You are probably wondering why continuous motion of a plate over a 'fixed' hotspot should result in a chain of volcanoes rather than a single

[25] Within a year, the volcanic centres had acquired a hyphen. 'Hot-spots', Jason declared in an article published in the *Bulletin of the AAPG*, 'are surface expressions of deep mantle plumes' (Morgan, 1972). The mysterious appearance and disappearance of punctuation has even affected Hawaii, which, in recent years, seems to have acquired an apostrophe or, rather, an 'okina': Hawai'i.

[26] Although it has been resurrected in a slightly different form recently, as we shall eventually see.

volcanic ridge—in other words, why are the Hawaiian Islands islands? Jason had an answer for this, too. Magma, rising from the plume, would become trapped beneath the moving plate, forming a reservoir from which vents to the surface formed, allowing the melt to ascend and erupt. Once formed, a vent system could effectively drain the trapped magma over a region with a diameter of roughly 100 km, so that eruptions would continue even after the volcano had been transported away from the hotspot. Eventually, a new vent system would form, building a new island, while the old one used up its remaining melt and then died.[27]

Whatever their origin and punctuation, as long as their locations were fixed with respect to the deep mantle, hotspots provided the sought-for reference frame within which the absolute motions of the plates could be established. As Kevin Burke and Tuzo Wilson later put it,[28] 'The hotspots and their volcanic trails are milestones that mark the passage of the plates'. Radiometric dating of rock samples from milestones in the Pacific showed that they increased in age towards the WNW, while young active volcanoes were found only at the eastern end of each chain, as Arthur Holmes had predicted. The distances between the islands and seamounts along a chain thus allowed estimates to be made of the speed at which the Pacific plate had moved over each hotspot, while the trends of the various chains represented 'flow lines' of absolute motion. Jason showed that the Hawaiian chain, together with the Tuamotu and Austral chains in the South Pacific, could be fitted quite well by rotating the Pacific plate by 34° about a pole at 67°N, 73°W (see Figure 2.6). Remarkably, the Tuamotu and Austral chains also had bends marking a change to more northerly trends in the older seamounts. These older trends could, Jason demonstrated, be fitted simultaneously with the Emperor seamount trend by a separate rotation of the Pacific plate. It looked as though something dramatic had occurred about 45 million years ago, causing the giant Pacific plate to change its motion very abruptly, from the NNW trend recorded by the Emperor Seamounts to the WNW trend of the Hawaiian chain. Like the mantle plume idea, this was a conclusion that has kept geophysicists at each others' throats ever since.

In theory, as long as the motion of one plate in the fixed-hotspots frame was known, that of all other plates could be found using their

[27] An alternative explanation was later advanced by Skilbeck and Whitehead (1978).
[28] Burke and Wilson (1976).

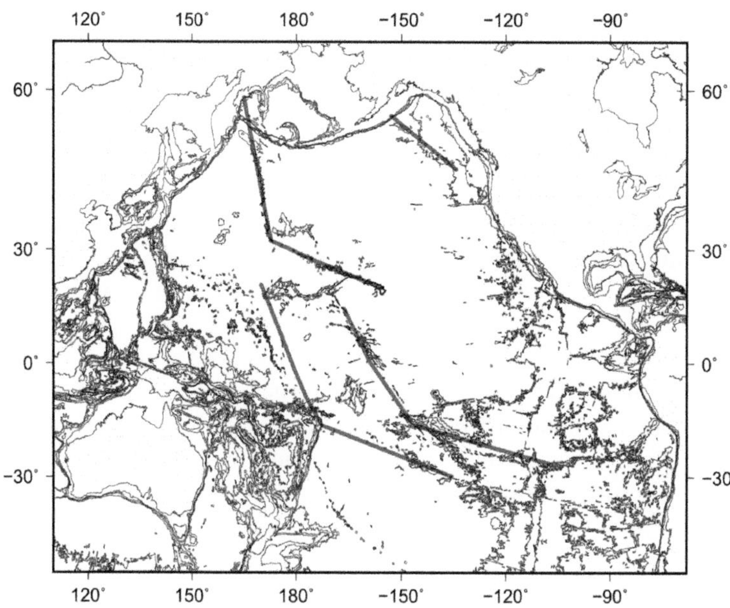

Figure 2.6 Trajectories of island/seamount chains predicted by Pacific plate motion derived from dating of the Hawaiian chain. [Redrawn after Morgan (1972).]

relative motions. Jason had thus established a method for calculating the absolute motion of any plate, as long as its motion could be linked to the Pacific plate by a series of relative plate rotations. Using his rotations, plates could be reconstructed back to about 45 million years ago, the age of Kimmei seamount that marked the western end of the Hawaiian trend. Jason's fixed-hotspot reference frame thus allowed the absolute velocities of the plates to be estimated for the first time. The Pacific plate was, despite its huge size, the Usain Bolt of plates. Significantly, it was almost entirely oceanic, carrying no continents apart from the slivers of continental crust that made up the western seaboard of the USA and Canada, and it was surrounded on three sides by deep ocean trenches. Its maximum speeds exceeded 110 mm per year in a band across the central Pacific region, and it was migrating towards the north-west, as recorded by the numerous island/seamount chains. Other oceanic plates also moved at a fair lick, the Nazca plate motoring eastward towards South America at a rate of nearly 40 mm per year, while the neighbouring Cocos plate headed north-east towards Central America just as eagerly. Plates carrying continents, such as Eurasia, Africa, and Antarctica, tended to progress in a more dignified manner at less than half the rates observed in the Pacific. In fact, parts of these plates hardly appeared to be moving at all. On a human timescale, of course, all these velocities are tiny: North America, for example, drifts about the length of one's body in a lifetime, while the Pacific plate moves nearly as fast as a coconut palm grows on one of its atolls.[29] Geologically, however, plate motions are blindingly fast, the Pacific plate covering over 1000 km in just ten million years.

Hotspots, it appeared, were unevenly distributed about the globe. The Pacific Ocean seemed to suffer from volcanic acne, while the African plate, including the eastern Atlantic and western Indian Ocean, likewise had more than its fair share. They came in a variety of forms, some large and long-lived, like Hawaii, others rather puny and ephemeral, such as Samoa or Tahiti. Some were isolated far from plate boundaries, while others lurked beneath mid-ocean ridges—notably under Iceland—or had migrated away from a mid-ocean ridge, like Tristan. Some island and seamount chains terminated in vast stacks of lava flows that, apparently, marked the first arrival of plumes at the surface, while others, like the Emperor Seamounts, simply petered out. As if all this were not controversial enough, Jason went on to suggest that his pipes to the deep mantle might play an important role in the break-up of continents, generating huge volumes of lava that flooded the landscape with numerous basalt flows, now stacked up on the rifted edges of today's continents, examples being the Deccan Traps in India and the Paraña basalts of South America. This suggestion was slightly discomfiting, for, as Jason pointed out, it suggested a strong likelihood that a postulated plume beneath Yellowstone heralded a future split in the North American continent that would make the Civil War look like a tea party.

A landmark was reached in 1972, when Clement Chase at the University of Minnesota published the result of an 'inversion' of seafloor data for eight major plates to obtain rotation poles and angular rotation rates describing 'instantaneous' (i.e. present-day) relative plate motions. Two years later, Bernard Minster and colleagues at the California Institute of Technology in Pasadena published a similar model, using the trends of 20 island and seamount chains on several plates to place the motion of each plate in a fixed-hotspots (i.e. 'absolute')

[29] Dietz (1977).

reference frame. The goodness of their fits persuaded Minster et al. that the hotspots did, indeed, provide a valid absolute reference frame, at least for the past 10 million years. The subject of global plate models will be taken up in more detail in Chapter 8.

Fit or Bust

If I told you that Sir Edward Bullard's great-grandfather started out with a single goat, you might be anticipating a heart-warming rags-to-riches story. The goat, or rather the Goat, however, was a public house in nineteenth-century Norwich, and its owner, Robert Bullard, soon went on to found his own brewery in that fine city. By the time young Teddy came along in 1907, the family were prosperous pillars of the Establishment. His grandfather, Sir Harry Bullard, served three times as Mayor of Norwich and was elected as the Member of Parliament for the city, subsequently establishing his aristocratic credentials by being expelled from the House for bribery. The Anchor Brewery, for many years a city landmark with its great brick chimney bearing the name 'BULLARDS' in large white capitals, survived for a century until it was brought to its knees by a deeply unpleasant 1960s phenomenon known as 'Watney's Red Barrel', after which it remained, for some years, as a ruin. I realize this is getting away from the point, but perhaps not as far as you might think, for it was profits from Bullard's brewery that allowed Edward Crisp Bullard, or Teddy as we know him, his privileged academic life as the great-grandson of the founder. The Bullards, it seems, were to Norwich what the Ewings were to Dallas (with apologies to any highbrows). And, by a remarkably contrived twist of fate, Teddy Bullard's name is often linked with an American contemporary as the joint founders of the science of marine geophysics, no less. The name of his transatlantic counterpart? Yes, Ewing!

One of Teddy's many claims to fame is known as the 'Bullard fit'. Nothing to do with his health, this refers to an important piece of work done in Cambridge in the mid-1960s, with graduate student Jim Everett and research assistant Alan Smith, in which they used Euler's theorem, before it became widely appreciated in Earth Science, to reconstruct the continents around the Atlantic that had, for centuries, flattered to deceive observers with the apparent fit of their coastlines. By so doing, the trio hoped to demonstrate mathematically (or at least statistically) the goodness of the fit and dismiss, finally, one of the specious objections to drift advanced by Sir Harold Jeffreys, Plumian Professor of Astronomy and Experimental Philosophy in Cambridge. Jeffreys was a highly respected physicist and author of the definitive textbook on the Earth, entitled *The Earth*. His views were thus very influential and played an important part in maintaining a fixist view of geology through much of the twentieth century. By 1965, when Bullard, Everett, and Smith published their work,[30] Jeffreys had been succeeded in the Plumian chair by the equally misguided Fred Hoyle,[31] but maintained, Canute-like, his determined opposition to the tide of new discoveries until his very last breath nearly a quarter of a century later. In fitting the continental edges, Teddy ignored the arguments concerning possible driving forces raised by Jeffreys and his ilk, and concentrated simply on the evidence that movement had taken place.

Here, we should make a clear distinction between Euler's theorem applied to the *motion* of the plates, and Euler's theorem applied to *reconstructions* of the plates. This may seem a trivial, even semantic, difference, but appreciating it is vital for a proper understanding of plate tectonics. On the face of it, we can simply replace the *angular rate* of rotation about the Euler pole with an *angle* of rotation, and the Euler vector becomes a finite rotation that shifts a plate from here to there on the globe. But, while this produces a reconstruction of two plates (or continents in case of the Bullard fit), it does not say much about how the plates got from there to here. This subtle point was illustrated by Jason Morgan in his 1968 article, in which he plotted small circles around the Bullard reconstruction pole for South America and Africa, showing that their trends were quite different to those of fracture zones in the equatorial Atlantic. The reason for the difference was that the 'instantaneous' pole had changed its position during the 140 million years of Atlantic opening, each time altering the trends of fracture zones produced during the subsequent interval of spreading.

But, to get back to the continents, the important thing about the Bullard fit of North America, Europe, South America, and Africa is the fact that they matched, not the coastlines, but the continental edges at

[30] The work had been presented earlier the same year at the same Royal Society symposium addressed on the subject of transcurrent faults by the unfortunate Vic Vacquier.

[31] Famous for such non-starters as the panspermia hypothesis and the idea that plate tectonics results from expansion of the Earth caused by a change in the gravitational constant of the universe, known as 'big G'.

the 500 fathom isobath.[32] This depth occurred some way down the continental slope into the abyss—a much better approximation to the place where continental crust ends and oceanic crust begins than ephemeral coastlines. Indeed, it is not widely appreciated that a quarter of the continent of Eurasia is water-covered, while nearly a third of North America is likewise flooded. The EDSAC2 computer, the very same machine used previously by Fred Vine to compute magnetic anomalies, was set to work to find the longitude and latitude of the rotation poles, plus the angles of rotation, that gave the closest match between the chosen contours on opposite sides of the Atlantic.[33] It came up with a fit that was much too good to be dismissed by the fixists (Figure 2.7). As Teddy put it, 'either the fit is due to chance similarities, on a par with the similarity of the coast of Italy to a boot, or the continents

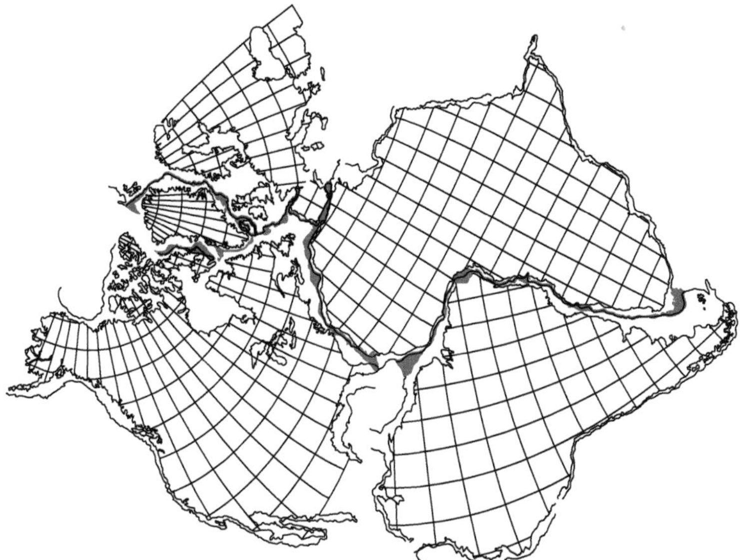

Figure 2.7 Figure from Bullard et al. (1965), showing the fit obtained by minimizing misfits at the 500-fathom isobaths.

[32] A fathom was an old depth measure equal to around 1.8 m or six feet, so this would be 914 m in today's money.

[33] For those concerned with such things, this was achieved in a least-squares sense. In other words, a fit was found that minimized the squared misfits between the two contours.

were once united and have separated with the formation of the Atlantic Ocean'. For South America to Africa, the average misfit was less than 100 km, not bad when you consider the width of the Atlantic and the age (roughly 140 million years) of the two margins, showing beyond doubt that, apart from a couple of overlaps where sediments had been dumped by rivers *after* continental separation, the fit was remarkable.[34]

The same fitting technique was applied later to the southern continents (plus India) by Alan Smith and Tony Hallam.[35] Here, the coastline fits were less obvious, and a variety of possible solutions to the puzzle had been suggested previously, particularly concerning the location of Madagascar against east Africa. Like the Bullard fit, Smith and Hallam's reconstruction has stood the test of time, and modern reconstructions differ only in detail. The fact that such good fits could be made tells us something important about the continents. That is, they must be pretty strong, for, had they deformed to any significant extent since they rifted millions of years ago, their shapes would have changed and the fit would be degraded. Their apparent strength derives from the fact that they are transported on the backs of plates that are even stronger and thus resistant to deformation. 'Continental drift' was therefore a poor name for this phenomenon, and began to fall out of favour in the late 1960s.

The Origin

To a biologist, mention of *The Origin* can mean only one thing: Charles Darwin's *On the Origin of Species*, a nineteenth-century book that continues to have a profound influence on evolutionary biology even in the twenty-first century. But there is another *Origin*, one that contains ideas just as revolutionary, just as profound, and just as heretical. *The Origin of Continents and Oceans* by the German polar explorer, Alfred Wegener, first published in 1912, presented a startlingly wide range of evidence in support of his theory of Kontinentalverschiebungen (or 'continental drift', as it was later translated), involving the dispersal of the present continents from a single parental 'supercontinent', beginning around 250 million years ago. Like Darwin's bestseller, Wegener's monograph was published in several editions, each containing significant amendments and

[34] A commentary on the Bullard, Everett, and Smith article and its role in foreshadowing the arrival of plate tectonics was published by John Dewey (see Dewey, 2015).

[35] Smith and Hallam (1970).

additions. In it, Wegener referred, as did Jason Morgan fifty years later, to the movement of 'continental blocks'. These constituted what he called the 'sialsphere', in contrast to the 'sima' of which the ocean floors were thought to be composed, the portmanteaus 'sial' and 'sima' denoting the dominance of the elements silicon and aluminium (in continental-type crust) and of silicon and magnesium (in oceanic-type crust).

Although anonymous in the first edition of *The Origin*, Wegener's primordial continent had, by the 1920 edition, acquired the name of 'Pangæa',[36] from the Greek, meaning 'all land'. Pangæa consisted of two subassemblies, a northern supercontinent, known as Laurasia, comprising North America, Greenland, and Eurasia, and a southern supercontinent, known as Gondwanaland,[37] comprising South America, Africa, Madagascar, India, Australia, and Antarctica (Figure 2.8). The term 'supercontinent' is thus ambiguous, as pointed out by Kevin Burke.[38] Is it a very large continent, or is it *all* the continents? To avoid confusion, I suggest we define 'supercontinent' simply as a large continent comprising two or more present-day continents. Hence, Gondwana and Laurasia are supercontinents. A conglomeration of all (or nearly all[39]) the continental crust on Earth, such as Pangea,[40] I will refer to as a 'pangea', rather in the way that the generic term 'hoover' was derived from 'Hoover'.

According to Wegener, Pangæa had always existed, or at least had been around for a very long time—until 200 to 300 million years ago, when, for some reason, it decided to fragment into what we know as the continents. He had no explanation for why it self-destructed so suddenly, just as he had no viable forces to drive the continents apart. As a result, English translations of Wegener's books fuelled a controversy that burned until the 1960s, a controversy that became highly personal,

[36] As an interesting historical aside, it seems that Wegener himself was not too fussed about the name, which, having appeared in the second edition of the book as 'Pangäa', was translated into 'Pangæa' in the English translation of the third, before disappearing again in the fourth, in which there is reference only to the 'great Ur-Continent'.

[37] Which we will, from now on, refer to simply as 'Gondwana'.

[38] Burke (2011).

[39] In fact, even during the lifetime of Pangea, there were other continental areas, such as the north and south China blocks, wandering across the Tethys Ocean, so it seems unlikely that there was ever a continent comprising 100% of all continental crust.

[40] In modern usage, the diphthong is commonly dispensed with.

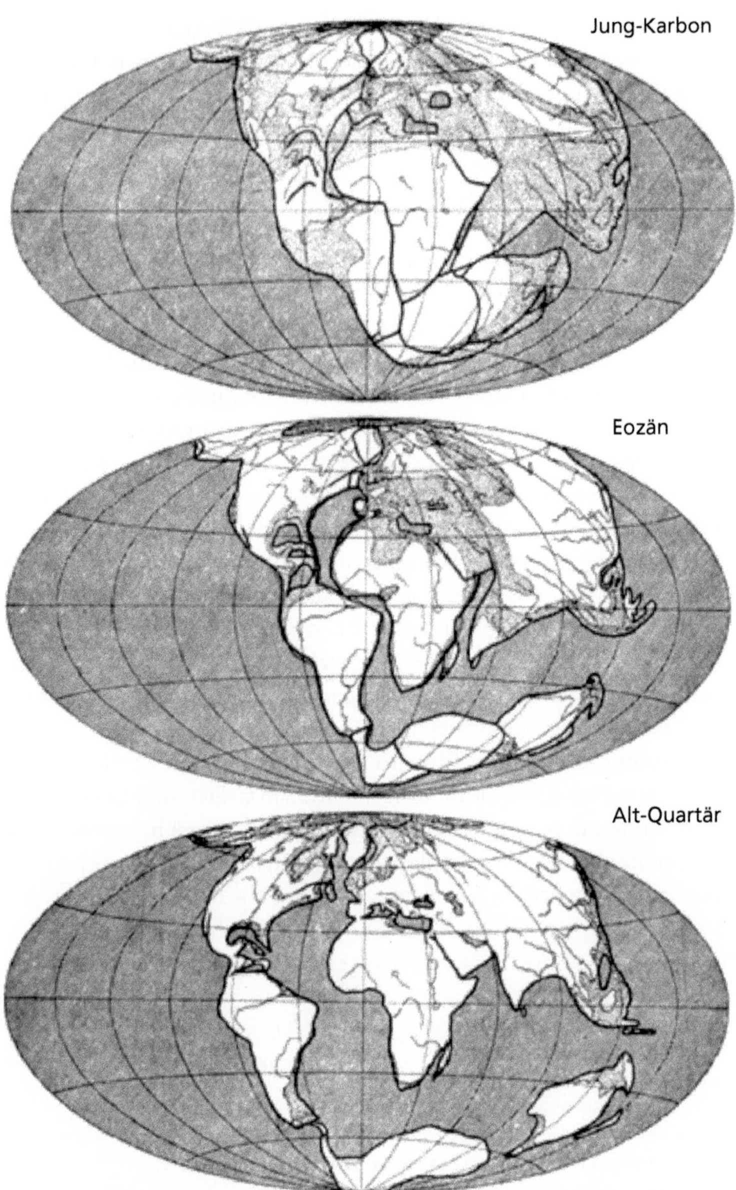

Figure 2.8 Alfred Wegener's reconstructions of the continents in the Late Carboniferous, Eocene, and Quaternary. Note that the position of Africa is fixed in each map. [From Wegener (1929).]

even vitriolic, with Alfred's brainchild theory being referred to as 'foot-loose' and 'unscientific', and Wegener himself dismissed as a 'meteorologist'. Yet Wegener was an explorer of some note. Living in an age when your grandmother can make it to the pole on a pogo stick, it is easy for us to forget that, in the early part of the last century, visiting high latitudes entailed great hardship and a significant probability that your trip might not end happily. This was, of course, the 'heroic' age of polar exploration, when brave men risked their lives without maps, landmarks, decent food, or Gore-Tex. Like Scott's *Terra Nova* expedition, Wegener's final journey to the polar ice was *einfach*, and, like Scott, he remains entombed in the ice to this day. In recent decades, he has shed his slightly crackpot image to emerge as one of the twentieth-century's great heroes of science, whereas Scott's reputation, sad to say, seems to have been heading in the opposite direction. Fittingly, Germany has honoured her far-sighted pioneer by naming her national polar research institution *Das Alfred-Wegener-Institut*.

A very long book could be written on the controversy surrounding Wegener's theory of continental drift. Indeed, just such a work of scholarship, Henry Frankel's *The Continental Drift Controversy*,[41] was published a few years ago, taking up four volumes and over 2000 pages. It is a fascinating story, containing lessons for us all when it comes to rejecting (or, indeed, accepting) scientific arguments, and the curious reader is referred to Frankel's tomes for many hours of enlightenment on this topic. Here, subject to somewhat more environmentally friendly space restrictions, I offer just one or two of the more amusing episodes from the long and difficult gestation experienced by Wegener's theory of moving continents.

Made in the USA

Back in 1857, a beaming, white-bearded, James Hall rose to make his presidential speech to the American Association of Petroleum Geologists in Montreal, using the occasion to present a wonderful new orogenic theory of his own devising. Sadly, James' conjecture was nonsense and many of his audience left the meeting shaking their heads. Later, the Secretary of the Society, the eminent physicist Joseph Henry, wrote Hall a friendly letter begging him to be cautious in promoting his new

[41] Frankel (2012a–d).

ideas.[42] James would not hear of it and proceeded to promulgate his hypothesis, leading geologists up one of the greatest blind alleys in the history of science.

James had been impressed by the remarkable sequence of rocks outcropping within the Appalachian Mountains, where a thickness of 7.5 miles of sediments had been deposited. Even more surprising than their thickness was the nature of these sediments: they had clearly been deposited under water, yet they were all of a type suggesting *shallow* water deposition. How was it possible to pile up a sequence of sediments with a thickness greater than the greatest depths in the deep ocean, entirely in water shallower than 1000 m? James' answer was that the rocks had been deposited in a shallow marine trough within the continent, and, as the sediments accumulated, the crust beneath began to sag,[43] thereby making room for more deposition at shallow depth. As they subsided, the sediments became folded and deformed, as observed in the rocks mapped in the Appalachians. James' hypothesis was all the more remarkable in view of the fact that no contemporary examples of such a phenomenon were known anywhere on Earth, but as a theory of mountain building, it had one very large and very obvious shortcoming: it did not include the building of any mountains.

The omission was highlighted in 1873 by another prominent American geologist, James Dana. Naming Hall's sediment-filled trough a 'geosynclinal', Dana added the missing component to the theory, which thus became the 'geosynclinal theory of mountain building'. At some stage, lateral pressure brought about a 'crisis' in which the great thickness of rocks was folded and uplifted into a mountain chain. The cause of the pressure, according to Dana, was the cooling and shrinking of the Earth, widely accepted by geologists at the time as a viable theory of planetary evolution. By the mid-twentieth century, the geosynclinal theory had been elaborated into what was then regarded as 'probably one of the greatest unifying principles in geologic science',[44] but has since become as discredited as the shrinking Earth theory itself. The various components of the theory were given confusing and clumsy

[42] Probably, because of justified doubts about James' theory, his address was not published until 1882.

[43] This was, in effect, an early appreciation of what became known as the theory of isostasy, presented by George Airy in 1855.

[44] Knopf (1948).

names by Hans Stille at the University of Göttingen in 1941.[45] German is, of course, renowned for agglutination, so that we may have what Mark Twain called 'the awful German language' to thank for ugly compound nouns such as 'eugeosyncline' and 'miogeosyncline'. Meanwhile, at Columbia University in New York, another prominent American geologist, Marshall Kay, attempted to reconcile the wide range of conflicting views that had been advanced since the work of Hall and Dana, adopting Stille's extravagant terminology and elaborating it further with an even more confusing plethora of terms, including 'zeugogeosyncline' and 'epieugeosyncline', for which he was rightly upbraided by colleagues and even, on occasion, subjected to verbal abuse. Nevertheless, while working in the type example in the Appalachians, Marshall made an important distinction between geosynclines with abundant volcanism and those without, foreshadowing the modern concept of 'active' and 'passive' continental margins.

Much later, in 1964, just as the plate tectonics revolution was getting under way in Britain, Marshall invited a promising young English geologist to accompany him on fieldwork in Newfoundland. The younger man had previously conducted fieldwork in western Ireland and, as soon as he set foot in Newfoundland, became aware of the 'extraordinary correspondence' between rocks in the two locations. This was John Dewey, who, being a geologist (and despite having recently been appointed to a Cambridge lectureship and thus being aware of Fred Vine and Drummond Matthews' *Nature* paper), still regarded the oceans as 'an irrelevant mystery', and thus failed to make a connection between sea-floor spreading and mountain building at the time. Fortunately, later that year, while he was contemplating some arcane feature in structural geology at his desk in the Sedgwick Museum back in Cambridge, in walked a grinning Canadian visiting scientist clutching a large sheet of coloured paper. The Canadian was Tuzo Wilson, and the coloured paper, in which he had made several parallel cuts, was his famous demonstration of transform faults. 'This moment transformed my life', John recalled later,[46] not without irony. He quickly became a convert to, and a prominent ambassador for, plate tectonics, subsequently collaborating with Marshall Kay, who had by then recognized that the new theory of the plates rendered his controversial nomenclature

[45] Stille (1941).
[46] Dewey (2003).

redundant. Together, they sketched geological structures of the Maritime Appalachians of Canada and the British Caledonides on Teddy Bullard's 1965 reconstruction of the Atlantic-bordering continents,[47] demonstrating the remarkable continuity of the rocks.

Dewey, now working with colleagues Jack Bird, in New York, and Kevin Burke, in Toronto, soon put mountain belts into the context of plate tectonics.[48] When a continent ruptures, rift basins are formed in which thick sediments and volcanic rocks accumulate. Eventually, a mid-ocean ridge forms and the two passive margins migrate away from it as sea-floor spreading gets underway. Passive margins could thus be fitted together like the pieces of a broken plate (which, of course, is exactly what they are), as was done by Teddy Bullard. If a trench (subduction zone) were later to develop near such a continental margin, a volcanic arc like the Andean margin of South America would result, and the ocean might begin to close. Alternatively, subduction might begin within the ocean basin itself, forming an island arc like the Tonga arc, which is separated from Eurasia by the Lau Basin. Wandering continents might eventually collide with each other or with island arcs. In the case of a continent–continent collision, the narrow trench zone would be replaced by a wide zone of deformation and major uplift, dominated by thrusting (as in the Himalayas). Sediments previously deposited on the continental margins and on the ocean floor would be deformed and thickened, creating high mountain belts. A continent–island arc collision (as in the Caribbean or Taiwan) would produce somewhat smaller mountains (like the Andes), a shorter-lived orogeny, and probably a reversal in the direction of subduction. Not satisfied with the remarkable scope of his model, John went even further, suggesting that 'cycles of oceanic expansion and contraction related to the growth, movement, and destruction of lithosphere plates are the fundamental cause of all major global tectonic features from at least the late Pre-Cambrian to the present day'.

Following a brief recrudescence of the redundant geosynclinal terminology,[49] the last rights on the concept were read in 1970 by

[47] This, of course, was what had led Tuzo Wilson to ask his famous question in 1966.

[48] Dewey and Bird (1970); Dewey and Burke (1973).

[49] Dietz and Holden (1966). In fact, outmoded terms like 'miogeocline' and 'geosyncline' may still be found in the recent literature.

American geologist, Peter Coney, who concluded that 'the concept of a geotectonic cycle is obsolete as a model of mountain system evolution, since definition of each of the so-called phases is obscure, relationship of any one phase to another is in doubt, and there is considerable evidence that the entire cycle can not be attributed to any single internal deterministic mechanism'.[50] Work in the Himalayas and elsewhere had shown that 'drifting continents rather than preparatory geosynclines produce mountains'. Thus, the mysterious world of geosynclines could be left behind and the first major geological concept 'made in America'[51] laid to rest.

Don't Schoot the Messenger!

When it comes to ridiculous names, the Dutch take some beating. Probably the silliest name of any geologist (or any other scientist for that matter) was that borne by one Willem Anton Joseph Maria van Waterschoot van der Gracht, a moniker that would no doubt have appealed to Groucho Marx. Van Waterschoot van der Gracht (let's just call him 'Willem' from now on) was a founder member of the American Association of Petroleum Geologists (AAPG) and a prominent supporter of drift during the early twentieth century, a combination that made him, if not unique, then something of a rarity in North America at the time.

In 1926, the AAPG decided to convene a symposium on continental drift in conjunction with its November gathering in New York—a symposium subsequently described as 'notorious'. Wegener could not attend, but sent a couple of papers detailing aspects of his theory to be read at the meeting. In his absence, Willem gave the introductory speech and subsequently drafted a 75-page 'Introduction'[52] to the proceedings volume, published two years later. If the meeting was intended to open minds to the revolutionary new theory, then it was a spectacular failure, with a 'fixist' majority of contributors taking turns to revile Wegener and his big idea. One palaeontologist remarked with disdain—and what can now be seen as prophetic irony—that 'It is not, as he

[50] Coney (1970).

[51] Dott (1979).

[52] As well as 29 pages of 'Remarks regarding the papers offered by the other contributors to the symposium'.

[Wegener] says [...] that the palaeontologists need a geophysicist to show them the road on which they should travel' (this, of course, was exactly what they *did* need). According to Sam Carey, one of the prime movers behind drift in the 1950s, the outcome of the symposium 'was that American geologists almost universally rejected continental drift'[53] and went on believing that the Earth was shrinking like a dried prune.[54] Perversely, Carey himself became convinced of the verity of drift.

Nevertheless, in his lengthy introduction, Willem not only advocated the theory of moving continents, he went further, asserting that: 'It would seem as if in the older Paleozoic,[55] before the Caledonian diastrophism, America might have moved westward faster than Eurasia, opening a Paleozoic Atlantic geosyncline, which was partly closed again during the Caledonian diastrophism, on account of the westward drift of Eurasia, which then again became more rapid than that of North America. Thus the former, overtaking the latter, closed the old Atlantic again and folded the Caledonian chains.' By 'Caledonian diastrophism', he meant the building of the Caledonian mountains of Scotland, Greenland and Canada, 400 to 500 million years ago; by 'geosyncline', he meant a deep ocean full of sediment; and by 'old Atlantic', he meant 'old Atlantic'. This last was a breathtaking concept, implying that the present Atlantic had been born just where an earlier ocean had closed.[56] In terms of plate tectonics, which of course nobody in 1928 knew anything about, this implied that plates must have been transporting continents around the globe for at least 500 million years, more than twice the length of time recorded by Vine and Matthews' bar-codes.

So, did the Atlantic close and then re-open? In 1966, *Nature* published a paper with precisely this title, placing the question squarely in the context of plate motions. I'm sure you don't need me to tell you the name of the questioner and author of that article. Yes, dear old Tuzo again.[57] Fortunately, the modern Atlantic had opened along a slightly different line to the one along which the 'proto-Atlantic' had closed,

[53] Carey (1988).

[54] Ridiculous as this idea may sound, contraction theory has recently been applied to the planet Mercury (Byrne, 2014).

[55] The interval from the beginning of the Cambrian to the end of the Permian, approximately 541 to 252 million years ago.

[56] The idea of a proto-Atlantic actually goes back as far as 1924, and the work of Swiss geologist Emile Argand.

[57] Wilson (1966).

leaving identical rocks on opposite sides of the Pond, while fossils from quite different faunal realms lay next-door to one another on either shore. The answer to the conundrum in the article's title was therefore 'yes', implying that plate tectonics was a general feature of the Earth's history. The proto-Atlantic referred to by Wilson was later rechristened *Iapetus*[58] by Cambridge geologists Brian Harland and Rodney Gayer in 1972, after the father of Atlas (for whom the Atlantic was named) in Greek mythology. Echoing Willem, Tuzo wrote in 1968, 'If continental drift has been going on for an appreciable part of geological time, at such rapid rates as recent work suggests, it means that a succession of ocean basins may have been born, grown, diminished, and closed again.'

Here was evidence that Wegener's Pangæa had not simply sat around for billions of years before exploding in slow motion during the Mesozoic. Plates must have been cavorting around the globe for at least 400 million years, and probably much longer, so that who-knows how many pangeas might have been assembled and fragmented previously. Confirming Willem's suggestion that the Caledonian-Appalachian mountains represented the site of a proto-Atlantic, John Dewey estimated[59] in 1970 that the oceans undergo complete turnover every 250 million years, and, not being a man renowned for diffidence, boldly suggested that plate tectonics may have been operating for as long as 3 billion years. Yet, with no sea floor older than 180 million years left *in situ*, reconstructions of plate configurations prior to Pangea would have to be based on other data, if such could be found.

Maggot

The plate tectonics revolution had its share of heroes, some of whom we have met, but many others who, in a fair world, would have received greater recognition for their contributions. There is one hero, however, whose name you will not find in any textbook or history of science. Mr Margetts. Not '*Dr* Margetts' nor yet '*Professor* Margetts', just—Mr Margetts. I cannot even add a Christian name because I never knew it. Indeed, to me and the rest of his pupils, he was only rarely known even as Mr Margetts, being more commonly referred to (when out of earshot) by the soubriquet 'Maggot'.

58 Pronounced 'ee-ap-it-us'.

59 Dewey and Horsfield (1970).

Maggot was a geography teacher at West Hatch School in Essex. Back in 1968, he taught his subject for the General Certificate of Education at 'Ordinary' and 'Advanced' levels. As a very ex-pupil, I can vouch for the fact that his lessons were a cut above the average, and far more entertaining. Admittedly, Maggot was somewhat eccentric, choosing to march the three miles to and from school each day, scorning the bus that ran the entire way, and possessing what can only be described as a military, if not actually sadistic, attitude towards his pupils. But, in those days, there were many such teachers, most of whom had served in the war or else done National Service. No, what made him stand out was what he taught.

As we entered the classroom for our first advanced lesson in physical geography, we each had a well-worn textbook thrust into our hands, which we were ordered to cover with brown paper by the next lesson. The book turned out to be the first edition of Holmes' *Principles of Physical Geology*. Written during the Second World War, it was, apart from the second edition mentioned earlier, the last of the comprehensive geological texts covering every aspect of the subject between two (now sadly defaced or missing) covers. Dull as that may sound, many geologists owe their careers to the spark provided by Holmes' great book and, even now, as you flip through it, you feel inspired to visit faraway places to inspect curious rock outcrops in stony deserts inhabited by white-robed Arabs with their camels. But what caught our imaginations then was the final chapter on 'Continental Drift'. Here, Holmes reproduced what he described as Wegener's 'strange, but now familiar, maps', showing the fragmentation of Pangæa into today's continents. Referring to the separation of North America and Greenland from Europe, Holmes informed us that 'The Atlantic is the immense gap left astern, filled up to the appropriate level by sima from below'.

Maggot was most definitely a mobilist, presenting the ideas of Wegener, du Toit,[60] and, of course, Holmes, as established fact. But he went further. He introduced the topic of 'PLATE TECTONICS', chalking the words in large capitals at the top of his blackboard and underlining them emphatically. He proceeded to explain some of the ideas we have already looked at, including sea-floor spreading and even magnetic anomalies, at a level appropriate for reluctant 16-year-olds. Here was a modest geography

[60] Another giant of the drift theory, whose famous book *Our Wandering Continents* (du Toit, 1937) is well worth a library ticket.

teacher in an obscure state secondary school, teaching a subject so new that it had not yet appeared in any textbook other than Holmes', and so controversial that half the geologists in the USA still refused to accept it. Most likely, he had read of the insights of Vine, Wilson, McKenzie, et al. in *Nature* or *Scientific American*, and, unlike so many professional geologists, quickly recognized the sea change. He would have been pleased, and not a little amazed, to learn that his insight led to one of his own students conducting plate tectonics research in the deep oceans.

A recurring theme in physical geography at the time was the 'life-cycle'. The essence of this was the use of the phases of a person's life as a metaphor for the apparently endless cycles of nature. Best-known was the temperate landscape cycle of W. M. Davis, which anyone over forty will recall having had drummed into them. Variants existed for arid and glacial climates, which, like the Davis model, generally involved the stages of 'youth', 'maturity', and 'senility'. Following the fashion for once, Tuzo Wilson likewise expressed the global variation in ocean basins in terms of a life-cycle,[61] but one with six rather than three stages (Table 2.1), each with its characteristic features and tectonics. These were 'embryonic', 'young', 'mature', 'declining', 'terminal', and 'relic'. For each stage, Tuzo presented a set of examples from the present-day globe, selecting one as the 'type' example in each case. Hence, the East African Rift and the Red Sea provide the type examples of the first two categories, while the Atlantic and Pacific represent the mature and declining stages, respectively. The terminal stage is typified by the Mediterranean, while the Indus Line in the Himalayas illustrates the relic scar or suture. As mentioned in Chapter 1, the scars of such vanished oceans are frequently festooned with ophiolite complexes—slices of ocean crust and upper mantle thrust onto the margins of the closing ocean.

This life-cycle was later christened the 'Wilson Cycle'[62] and soon became a central concept in Earth Science. Of course, not all oceans pass through every stage: as with life itself, the cycle might be cut short by some blind hand, such as that which caused the death of the mid-ocean ridge in the Labrador Sea around 34 million years ago, thereby restoring Greenland to the North American plate whence it had rifted 60 million years earlier. Still, the Wilson Cycle provided a useful idealized structure

[61] Wilson (1968); the concept was developed further in the textbook *Physics and Geology* by Jacobs, Russell and Wilson (1973).

[62] Dewey and Burke (1974).

Table 2.1 The Wilson Cycle.

Stage	Example	Motions	Sediments	Igneous rocks
1. Embryonic	East African rift valleys	Uplift	Negligible	Tholeiitic flood basalts, alkali basalt centres
2. Young	Red Sea and Gulf of Aden	Uplift and spreading	Small shelves, evaporites	Tholeiitic sea floor, basaltic islands
3. Mature	Atlantic Ocean	Spreading	Great shelves (miogeosynclinal type)	Tholeiitic ocean floor, alkali basalt islands
4. Declining	Western Pacific Ocean	Compression	Island arcs (eugeosynclinal type)	Andesitic volcanics, granodiorite–gneiss plutonics
5. Terminal	Mediterranean Sea	Compression and uplift	Evaporites, red beds, clastic wedges	Andesitic volcanics, granodiorite–gneiss plutonics
6. Relic scar (geosuture)	Indus line, Himalayas	Compression and uplift	Red beds	Negligible

From Wilson (1968).

into which a myriad of observations could be fitted, and perhaps said something about what might be called 'the pulse of the Earth',[63] to which we will return to in Chapter 7.

Afterthought

On reflection, perhaps 'paving stone theory' wasn't such a good title. Paving stones tend to be not just flat, but regular in shape and size, which plates manifestly are not. Maybe 'crazy paving theory' would have been better? Anyway, it's too late now, but whether you prefer plates or paving stones, the important thing, as noted by Tuzo, is that they are *rigid*—rigid spherical caps that tend to keep their shape and don't deform easily under stress.

The first generation of plate tectonics provided a neat, self-consistent, theory of the movements of these spherical caps. Forged in the furnace of mid-ocean ridges, plates gradually lost much of their heat to the oceans as they migrated away, carrying with them a magnetic record of their history imprinted in the ocean crust. Since they were rigid, their motions could be described quite simply by means of three numbers that were easily calculated from sea-floor data. And if that wasn't enough to make geophysicists happy, a set of hotspots had been thoughtfully located beneath the plates to record their absolute motions, too.

For the first time, there was a global kinematic structure within which to begin making sense of traditional geology. The challenge was summed up neatly by Ian Gass and his colleagues at the Open University, in their introduction to one of the first textbooks[64] to incorporate plate tectonics: 'The success of the plate tectonics theory is not only that it explains the geophysical evidence, but that it also presents a framework within which geological data, painstakingly accumulated by land-bound geologists over the past two centuries, can be fitted.' Tuzo Wilson referred to those same geologists a little more cruelly:[65] 'They have been like sailors so occupied in examining the decks of their ships that they have failed to look further and thus notice that their ships were under way.'

[63] This was also the title of an important book by the influential Dutch geologist, J. H. F. Umbgrove (Umbgrove, 1947).

[64] Gass, Smith, and Wilson (1972).

[65] Wilson (1968).

3

Poles Apart

> Later discoveries have clearly shown that the conclusions based on palaeomagnetic measurements were correct, and their failure to convince the wider community of geologists and geophysicists remains to me the most interesting part of the history of plate tectonics.
>
> DAN MCKENZIE (2003)

In just a few years, the magnetic bar-code secreted beneath the world's oceans had provided detailed intelligence on the motions of the plates. When combined with other data from the sea floor, this allowed geophysicists to reconstruct the history of entire ocean basins following the rifting of Pangea. Some folk, however, are never happy, and 'glass-half-empty' types might well have complained that, impressive as all this was, it accounted for less than 200 million of the 4600 million years of Earth history, that is, just 4 per cent. What about that other 96 per cent? Did plate tectonics operate through part or all of this long history and, in any case, how could you ever know, since the evidence had all been shredded by the closure of earlier oceans?

There was hope. The same process that had so conveniently sequestered the recent history of the plates in the sea floor had also been at work throughout much of earlier geological time, recording the story in rocks onshore. By comparison with the high-definition picture of plate motions offered by bar-codes and fracture zones, this recording was monochrome, fuzzy, and incomplete. Yet, by the mid-1950s, it had already provided conclusive evidence that continents were truly mobile. Curiously, hardly anyone noticed.

*

French Letter

As with so much of the story of plate tectonics, the roots of paleomagnetism can be traced to a time long before the sixties revolution. And here I

The Tectonic Plates Are Moving! Roy Livermore, Oxford University Press (2018). © Roy Livermore.
DOI: 10.1093/oso/9780198717867.001.0001

don't just mean a few decades, but many centuries. In fact, the scientific study of the Earth's field could be said to have begun in 1269 with a letter written in Latin by Petrus Peregrinus ('Peter the Pilgrim') of Maricourt in northern France, to his *amicorum intimus*, Syger de Foucaucourt.[1] Exactly why anyone should occupy themselves by writing a thirty-page epistle on the subject of magnetism to an old chum is anyone's guess: one can only speculate that Petrus was a man with few friends.

In his missive, Petrus described how a compass would find the poles of the Earth and how, by forming a sphere from lodestone (lodestone is a naturally occurring magnetized chunk of the iron oxide mineral, magnetite), one could reproduce the directions shown by a compass at any point on the Earth. There was something of a delay before final publication of Petrus' *Epistola*—nearly 300 years, in fact—but it eventually became a classic, representing one of the earliest examples of scientific writing. What Petrus did not know, however, was that the geomagnetic field has other mysterious properties that provide clues to its origin, yet have confounded attempts to explain its behaviour, right up to the present day.

The magnetic compass, of course, played an important role in the explorations of Cristóbal Colón in 1492, and later of Vasco de Gama, Ferdinand Magellan, and other notables. On Cristóbal's first voyage of discovery, so the story goes, concern was raised by his crew when, in mid-Atlantic, their compass ceased to agree with sightings of the Pole Star. About twenty years later, the Nuremberg sundial maker Georg Hartmann made measurements in Rome showing that the magnetic field there was directed about 6° to the east of true north. Later, he measured the difference in Nuremberg too, concluding that there the field pointed 10° east of north. The difference between magnetic north and true north is referred to as the 'variation' or 'declination'. Corrections for such magnetic variations were later applied by compass makers for use in specific areas, so that the notion of a variable difference between magnetic north and True North, depending on location, became widely known.

Not much else happened until the 1570s, when a London instrument maker and retired mariner, Robert Norman, noticed that the north-seeking end of a compass needle not only sought north, but insisted on

[1] Even stranger, the 'Letter on the Magnet' (*Epistola de magnete*), was possibly written by Petrus, an engineer in the French army, from the trenches in Italy.

dipping downward. To make the needle horizontal, he needed to attach a small weight to the south-seeking end or else trim off the tip of the north-seeking end to balance. He devised a simple experiment to investigate this phenomenon further.[2] By setting up a sort of vertical compass that he called a 'dip circle', in which a magnetized needle was pivoted about a horizontal axis, he showed that, in London at least, the magnetic field tended to dip down at an angle of about 70°. In 1581, he published his famous book, *Newe Attractive, shewing the Nature, Properties and manifold Vertues of the Lodestone, with the declination of the Needle, touched therewith, under the Plain of the Horizon*, a title that speaks for itself.

In 1600, some 87 years before the publication of the *Philosophiæ Naturalis Principia Mathematica* of Isaac Newton, another memorable book appeared, with the only slightly less memorable title, *De Magnete, Magneticisque Corporibus, et de Magno Magnete Tellure*, often referred to simply as *De Magnete*. This, the world's first truly scientific monograph,[3] was written by the world's first geophysicist,[4] William Gilbert. William was an Elizabethan Essex man, who rose from modest beginnings to become physician to the Queen[5] and President of the Royal College of Physicians. Although somewhat less exalted than Isaac's bestseller, his great tome is highly impressive, setting out everything that was then known about the phenomenon of magnetism (and electricity, for that matter). Noting that '*magnus magnes ipse est globus terrestis*' (the terrestrial globe is itself a great magnet), William described the direction of the Earth's field at any given location in terms of two angles, the 'variation' with respect to true north (as measured by Georg Hartmann),[6] and the downward dip

[2] The experiment was significant enough to be included in a book entitled *Great Scientific Experiments: Twenty Experiments that Changed our View of the World* (Harré,1981).

[3] William's approach to what was then called 'natural philosophie', and now known as 'science', was, in the twentieth century, the subject of a heated debate between folk with nothing better to do. The question of how revolutionary was his experimental approach to the testing of ideas was rehearsed in politically inspired academic debates by so-called 'philosophers of science', in the same way that the sequence of events leading to the plate tectonics revolution has been contorted into daft theories by those same idle folk.

[4] The publication of Gilbert's great book is commonly regarded as the beginning of geophysics.

[5] His appointment was in 1601, the year following publication of *De Magnete*. The honour was short-lived, for he died just two years later of bubonic plague.

[6] The fact, revealed by early compasses, that the field lines tend to point towards the poles, had been realized more than a thousand years earlier by the Chinese.

with respect to the horizontal (as discovered by Robert Norman and now known as the 'inclination'), which he called, confusingly, the 'declination'. In the southern hemisphere, the direction of the field pointed upward (negative inclination), while in the north, it pointed downward (positive inclination), becoming steeper at higher latitudes. Using a spherically carved lodestone or 'terrella', William mapped out what he supposed to be the changes in these angles over the surface of the Earth, from which he concluded that the main field configuration closely resembled that of a giant bar magnet. Hence the obvious explanation was that the Earth must be magnetized much like a piece of iron.

Thirty years after the publication of William Gilbert's great work, Edmund Gunter and Henry Gellibrand, successive professors of astronomy at Gresham College in London,[7] discovered that the magnetic variation changed not only from place to place, but also with time. Measurements made previously in 1580 by the comptroller of the Queen's Navy, William Borough, at Limehouse, in what is now the east end of London, were repeated by Edmund and Henry in Deptford (about three miles away), demonstrating a systematic change in field direction that became known as the 'secular variation'. Whereas old William had measured a declination of more than 11° fifty years earlier, the pair now found their compass pointing only about 5° east of north. If such changes were usual, then naval charts drawn up on the basis of regional measurements of magnetic variation had a limited shelf-life. It has been customary to credit this discovery solely to Henry, while Edmund's contribution has tended to be forgotten.[8]

In 1683, Edmond Halley, the great astronomer, made yet another surprising discovery. He observed that the part of the magnetic field that was *not* like that of a dipole—in other words, the noise—tended to drift slowly westward with respect to the Earth's surface at about 0.2° per year on average.[9] These changes (the secular variation and westward drift) were remarkably rapid by geological standards (of course,

[7] Appointed in 1619 and 1626, respectively.

[8] On the other hand, old Henry has been defamed publicly by the English biographer Gordon Goodwin, who described him libellously as 'a plodding industrious mathematician, without a spark of genius' (Malin and Bullard, 1981).

[9] Note that the non-dipole field is greater in some places than others, and almost absent over the Pacific Ocean, while the rate of westward drift also varies, in places even becoming an eastward drift (Merrill, 2010). Today, it represents up to 30 per cent of the observed field.

just about any change is rapid by geological standards), pointing to movement of whatever was causing the field on a timescale far too short for it to be solid material.[10] In 1692,[11] despite having only the flimsiest of data, Edmond gave the broad structure of the planet's interior and an explanation of the westward drift of the geomagnetic field, coming up with this inspired insight: 'So than the external parts of the globe may well be reckoned as the shell, and the internal as a nucleus or inner globe included within ours, with a fluid medium between. Which having the same common centre and axis of diurnal rotation, may turn about with our Earth each 24 hours; only this outer sphere having its turbinating motion some small matter either swifter or slower than the internal ball.'

If, as William Gilbert had claimed, the Earth was a giant magnet, the field at the surface would be similar to that of a fictitious bar magnet (a dipole) located at its centre. Since the same field configuration is produced by a uniformly magnetized sphere, it could be that the entire Earth was permanently magnetized, from the surface to the core. However, it was difficult in this case to see how the surface field could change rapidly enough to produce the secular variation and westward drift. Moreover, a quick back-of-an-envelope calculation showed that the temperature inside the Earth must greatly exceed that at which permanent magnetization could be retained by mantle rocks. This temperature[12] is around 600°C, and is reached just 25 km beneath the surface, while the temperature in the core exceeds 4000°C. Besides, the intensity of magnetization required to produce the surface field greatly exceeds that of any known rock and, furthermore, there would be no reason why the magnetic poles of a uniformly magnetized Earth should coincide with the geographic poles unless the magnetization was somehow caused by rotation. The bar magnet analogy had therefore to be abandoned, and physicists needed to think more deeply for an explanation.

[10] In his 'Lecture on Experimental Physics' in 1871, James Clerk Maxwell, one of the greatest physicists of all time, described such measurements poetically as a record of the 'pulsations and flutterings' of the 'never-resting heart of the earth', and the secular variation as 'that slow but mighty working which warns us that we must not suppose that the inner history of our planet is ended'.

[11] Halley (1692).

[12] Known as the Curie temperature after the great French physicist Pierre Curie.

Core Science

To save time and paper, we now jump more than two centuries to 1939,[13] and shift the scene somewhat jarringly from seventeenth-century England to the California Institute of Technology in Pasadena, just as war was breaking out across the Atlantic. Walter Maurice Elsasser, a Jewish refugee from Nazi Europe, was in the process of publishing an article in the American Physical Society journal, *Physical Review*, entitled 'On the Origin of the Earth's Magnetic Field'. This was a subject in which Walter had invested much thought, concluding that, since alternative theories involving uniform magnetization of the planet had been discounted, as had 'fundamental' theories in which magnetic effects were produced as a direct consequence of rotation, there remained only one viable mechanism for field generation, and that was 'thermoelectric currents in the metallic interior of the earth'. Motions within the 'central part of the earth' (i.e. the core) were subject to the influence of the planet's rotation, the upshot being (following pages of rather tedious mathematics) an external field very like that of a dipole or bar magnet.

Now, you might think that the discovery of a vast ocean of molten metal within our planet, with a volume 100 times that of the Pacific, Atlantic, and Indian oceans combined, would warrant a Nobel Prize (assuming that geophysicists were eligible for Nobel Prizes, which they are not). Surprisingly, it is difficult to pin down the discoverer, but credit must go to dear old Harold Jeffreys, who, it seems, established this startling fact in the 1920s and 1930s on the basis of the Earth's tides and the propagation of seismic waves through the interior of the planet. The most convincing evidence came from seismic body waves—the push and shake waves radiating from large earthquakes. The speed of P-waves increases gradually throughout the mantle, from about 8 km s^{-1} at the top, to around 11 km s^{-1} at a depth of 2900 km, where it drops back suddenly to around 8 km s^{-1} again. At the same depth, S-waves exhibit an even more spectacular drop, from around 7.5 km s^{-1} to zero, leading Harold to venture, rather coyly: 'There seems to be no reason to deny that the earth's metallic core is truly fluid'. The discovery in

[13] Since this is not a history of geomagnetism, we reluctantly have to forego the exciting developments that occurred in the interim, including the work of great men like Johan Carl Friedrich Gauss. Happily, this history has been documented elsewhere, and the reader is referred to Merrill (2010) and references therein.

1936, by Danish seismologist Inge Lehmann,[14] of a solid iron region the size of the Moon at the very centre of the Earth showed that the core was not completely molten. Nevertheless, the region surrounding the inner core, with a volume seven times that of the Moon, had to be liquid.[15]

Anyway, Walter began by noting the obvious, but highly significant, observation that, since the geomagnetic field is more or less aligned with the rotation axis of the Earth, it must be related in some fundamental way to the rotation of the planet. This could not be a direct effect, since rotation had only a tiny influence at the atomic level at which magnetic effects are generated. Hence, the role of the Earth's rotation must be indirect, and the geomagnetic field was thus 'a rather accidental consequence of fluid motions which take place in the interior of the earth'. Walter went on to construct a theory that might explain how a magnetic field with the characteristics observed at the surface could be generated from the movement of hot metal in the outer core. Somehow, he suggested, the electric currents within the molten iron core were configured in a way that emulated a self-sustaining dynamo. This was a reference to the Irish physicist, Joseph Larmor, who, twenty years previously, had made the suggestion[16] that a magnetic field could be generated by such a mechanism. Using the example of Michael Faraday's 'disc dynamo', in which a copper disc, rotating in a magnetic field, generates an electric current, Joseph noted that the induced current would, in turn, produce a magnetic field of its own. If that field could be aligned with the original 'starting' field, thereby reinforcing it, then you might have a dynamo that could maintain the magnetic field indefinitely. He demonstrated this using a rotating conducting disc, with a wire connected by a brush at the rim (Figure 3.1). When the rotating disc is placed in a starting magnetic field, a current flows through it and down the wire to the spindle. In passing through the coil, the current generates a magnetic field that reinforces the original field, which then drives the disc dynamo, and so on. Of course, you don't get something for nothing, and, as anyone who has used a bicycle dynamo knows, energy has to be continually invested in turning the

[14] Referred to in the literature, quaintly, as 'Miss Lehmann'.

[15] This newly established structure of the Earth, derived from the study of earthquake waves, was summarized in 1939, in a book edited by Beno Gutenberg.

[16] Larmor (1919).

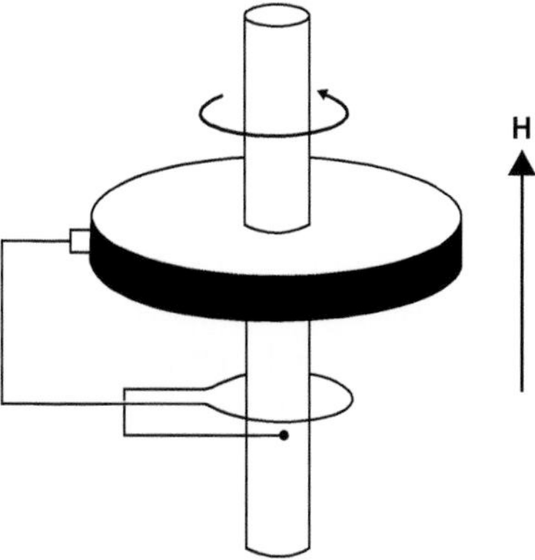

Figure 3.1 Larmor's disc dynamo. The vertical arrow shows the direction of the magnetic field **H**. Rotation of the disc in the magnetic field produces a current that flows into the attached wire, which is wound round the rod in such a way as to reinforce the initial field.

spindle, but, as long as rotation can be maintained, the magnetic field will persist. Joseph thought that such a dynamo might be operating inside the Earth to generate the observed field.[17] Motion of a conducting medium would somehow interact with the magnetic field to produce electric currents that, in turn, generated the magnetic field observed at the surface. He also proposed that rapid changes in the magnetic field—that is, the secular variation and westward drift—could be accounted for simply by changes in the 'internal conducting channels' in which the electric currents flowed, noting that this 'would require fluidity and residual circulation in deepseated [*sic*] regions'. At the time, Joseph had little idea of exactly where this fluidity might be located, nor of what currents might be responsible for generating the field.

Here we get into a branch of physics that is best avoided, the very name of which is guaranteed to make you choke on your corn flakes—magnetohydrodynamics.[18] The problem, in a nutshell, was to find a pattern of current flow within the core that could produce and maintain an external magnetic field like that observed here at the surface.

[17] Faraday himself had realized that a conducting liquid moving through a magnetic field would serve to generate an electric current, and had tried to detect this in a famous experiment in which he rigged a wire across the Thames at Waterloo Bridge. The experiment failed, owing to electrical 'noise', but the principle was sound.

[18] An ugly term attributed to the Nobel Prize-winning physicist Hannes Alfvén.

This turned out to be a problem that would keep geophysicists occupied for the next eighty years. Yet, although it was impossible to provide mathematical rigour, Walter developed the first theoretical model for dynamo action generated by a conducting fluid within the Earth. For this, three conditions had to be fulfilled. Firstly, the dimensions of the fluid core had to be large, because otherwise the magnetic field generated by its motion would decay too fast for important feedback mechanisms necessary for dynamo action to take effect. Secondly, there must be rotation of the body in order to provide asymmetry in motion, since symmetrical motion is unable to produce dynamo action (it would take too long to explain this even if I understood it). This asymmetry would then be provided by the Coriolis force. And thirdly, there needed to be a source of energy to drive the dynamo. In Walter's model, this was the Earth's internal heat, derived from radioactive decay.

Walter pointed out that a metallic fluid under the conditions obtaining within the core would have a low viscosity comparable to that of water, so that temperature variations within it would lead to thermal convection, such as occurs in the atmosphere or a saucepan of tomato soup. Hot liquid iron would rise from the bottom of the outer core towards the core–mantle boundary 2200 km above, where it would lose some of its heat (although still at around 4000°C) and sink. As it rose and fell, the hot liquid would, like rising or sinking air, be affected by the rotation of the planet (i.e. the Coriolis force), causing it to corkscrew, twisting the electric field lines embedded within it, and generating a magnetic field that would extend outside the core. Very likely, Walter surmised, the necessary heat for convection was provided by radioactive decay of unstable elements forming impurities within the molten metal.

Shortly after the war, during which he moved to Columbia University in New York, Walter published another article,[19] on which he had been moonlighting while employed at the National Defense Research Council in New York. Giving his address as the 'Department of War Research' at Columbia, he set out in this tripartite blockbuster his updated views on field generation. His analysis led to the conclusion that the necessary amplification of the field required flow parallel to lines of magnetic longitude, or 'poloidal flow', in a pattern symmetrical about the equator. The secular variation was a good guide to the pattern and rate of flow in the outer core. Walter showed that the field

[19] Elsasser (1946).

inside the core had to be very different to that observed here at the surface. For a start, it must be much stronger, and, secondly, it would be much more complex.

A couple of years later, Teddy Bullard, now working unhappily in Toronto, got in on the act. Whereas Walter had discussed field generation 'with great generality and elegance', Teddy attempted to offer a more specific model. Like Walter, he dismissed alternative theories for field generation, such as tidal friction and precession, concluding that fluid motions within the core provided the only realistic solution. He also agreed that thermal convection was likely to be the crucial process, shifting material vertically to regions with higher or lower angular momentum. A heat supply much less than produced by radioactive decay within surface rocks, and about the same as that found in meteorites, would suffice to raise the thermal gradient in the core sufficiently for convection to commence. As with moving air masses at the surface, this would lead to a Coriolis effect that would cause the rising or sinking iron to interact with the existing field to generate electric currents that, in turn, would produce a strong magnetic field within the core. By this time, there was already good evidence that the Earth's field existed in two states: the present or 'normal polarity' state, in which the lines of force point downward in the northern hemisphere, and 'reversed polarity' in which they previously pointed up.

Reversals

In the spring of 1906, the director of the Observatoire du Puy de Dôme gave a talk in which he explained how he had measured the direction of magnetization of clay sediments in a road cutting in the Massif Central. The clay had been baked by the heat of overlying lava flows[20] erupted more than 6000 years ago from three volcanoes. The director, Antoine Joseph Bernard Brunhes (Bernard for short), a bespectacled, goatee-bearded French geologist, was intrigued to know whether this natural 'firing' had produced a strong, stable, magnetization, such as had previously been discovered in ancient pottery.[21] He had also chiselled out two samples of the overlying lava and measured their magnetizations, too. His results showed very similar directions in both the lavas and the

[20] Volcanic rocks generally have much stronger fossil magnetism than sedimentary rocks, which is thus easier to measure.

[21] Laj et al. (1996).

clays beneath, which he explained by suggesting that the magnetization of both must have been acquired during eruption and remained the same ever since.[22]

Even more surprising, rocks of Miocene age (between 5 and 23 million years old) near Saint-Flour gave directions roughly 180° opposed to those elsewhere, implying 'that at a certain moment of the Miocene epoch, in the neighborhood of Saint-Flour, the north pole was directed upward: it was the south pole which was the closest to central France'. This was not, Bernard opined, because France had suddenly switched hemispheres, but (only slightly less incredibly) because the Earth's field had reversed its polarity. He published his findings that same year. By any standard, this was a breathtaking scientific discovery that might well be considered as startling as finding that the gravity field once acted upward.[23] It was therefore ignored by the geological community.

Despite confirmation that volcanic rocks frequently exhibit reversed polarity, no further progress was made until Motonari Matuyama, geology professor at Kyoto Imperial University, made a surprising discovery as he was surveying magnetic anomalies over volcanic provinces in Japan, Korea, and Manchuria. While investigating the cause of reduced local magnetic anomalies, he sampled some associated basalt lavas and measured their magnetization, finding, like Brunhes, that many were reversely magnetized. Having reasonably good knowledge of the age of the lavas, Motonari was able to determine that the reversed samples were mostly early Pleistocene in age (older than 770,000 years), while all the younger samples had normal polarity. Even older (Miocene) samples had normal polarity once again. He proposed that the Earth's field had entered a period of reversed polarity around 2 million years ago, flipping back to its present polarity around 773,000 years ago.[24] Geologists continued to pay little or no attention.

Fatal Attraction

'I hope you rot in hell.' Unlikely words to hear from a judge in a British court of law. But this was the cheery send-off proffered by Judge William

[22] Bernard's measurements were repeated using modern methods in 2002, with results indistinguishable from his. See Laj et al. (2002).

[23] This was noted by Allan Cox in Cox (1973).

[24] This period of reversed polarity is now known as the 'Matuyama epoch'[24], while the subsequent normal period, including the present day, is known as the 'Brunhes'.

Mudd in the San Diego Superior Court in 1996, as he sentenced prize-winning kick-boxer Paul Cain to twenty-five years for murder. His victim, a 73-year-old British scientist, had been in town visiting colleagues at the Scripps Institution of Oceanography, when he was robbed and strangled by Cain in his hotel bedroom. Such an undignified end was in sharp contrast to the esteem in which Keith Runcorn, once described as the 'theoretical visiting professor of physics in Newcastle',[25] was held, particularly for his pivotal role in developing the branch of science that became known as 'paleomagnetism'.

From 1950 to 1956, Runcorn was Assistant Director of Research at the Department of Geodesy and Geophysics in Cambridge University. In some ways, Keith could be regarded as the José Mourinho of geophysics: a flawed genius with a gift for establishing the composition, if not always the tactics, of a winning team without kicking a ball himself.[26] So, although he may not always have backed the right horse when it came to continental drift, or used the correct methods in his research, his influence permitted others to make critical contributions to the subject, and it is his name, above all others, that springs to mind when considering the role of paleomagnetism in the birth of plate tectonics.

The story, however, began a few years earlier, in Manchester, where Patrick Blackett was Langworthy Professor of Physics. Patrick, described as a 'tall, dark and handsome scientist with a taste for natty grey suiting',[27] was an ex-First World War naval officer turned physicist. In the Second World War, he played an important role in connection with the development of the atomic bomb, a weapon he later despised. Not a man to let a complete lack of formal academic qualifications inhibit him, Patrick earned himself a Nobel Prize for Physics in 1948, and was eventually elected President of the Royal Society in 1965. He was also an influential figure in British politics during the 1960s, serving as scientific adviser to Harold Wilson's socialist government, reflecting his own political principles, for which he had previously been reviled.

Notwithstanding that seismic studies had confirmed the existence of a molten metallic core, and that Walter Elsasser had all but solved the problem without reinventing the laws of physics, Patrick, one of

25 Obituary by Raymond Hide (1996).

26 In fact, that's a bit unfair on Keith, who continued to pursue research on important questions such as the source of the geomagnetic field and polar wandering.

27 *News Review* feature, 29 May 1947. See Nye (1999).

Britain's leading physicists and, like Teddy Bullard, an ex-student of Ernest Rutherford, came up with an alternative theory. Fooled by an apparent relationship between the magnetic moments of the Earth, the Sun, and the star 78 Virginis, and their respective angular momenta (that is, their magnetic fields seemed to depend on their rotation), Patrick revived a suggestion by his Manchester predecessor Arthur Schuster that, despite the Sun being at far too high a temperature to be magnetizable, it might nevertheless sport a small magnetic field by virtue of its rotation alone.[28] This led Schuster to ask 'is every rotating mass a magnet?' If so, this would amount to an entirely new law of nature. Despite Schuster's own description of the idea as 'extravagant', Patrick was undismayed, and in May 1947 presented his ideas to the Royal Society, publishing them in the same month in a *Nature* article entitled 'The Magnetic Field of Massive Rotating Bodies'. Here, he discussed the core-source ideas of Walter Elsasser and others, and noted previous unsuccessful attempts to prove a connection between rotation and magnetization, including one using a 10 cm copper sphere rotating at 12,000 rpm. In the latter case, Patrick attributed the failure to an inability to measure the tiny fields involved. Teddy Bullard had reservations about the theory and suggested that magnetic field measurements in deep mines might resolve this question, since Patrick's theory implied that the field should vary with depth in a manner quite different to that predicted if the source lay in the core. A former student of Teddy's at Cambridge, Keith Runcorn, by then a junior lecturer at Manchester, made such measurements in the coal mines of Lancashire and Yorkshire, eventually reporting in 1951[29] that the variations in horizontal and vertical field components were much closer to the predictions of core-source models.

Patrick, however, was not convinced. To test his theory in the lab, he needed two things: a dense mass and a sensitive device to measure the tiny magnetic field produced by its rotation. He solved the first problem by procuring a range of metal cylinders, including one of pure gold, 10 cm in diameter, weighing more than 10 kg. Patrick's position of influence within the British Establishment following his distinguished war service undoubtedly helped in obtaining the not inconsiderable funding

[28] Others, including Albert Einstein, had previously suggested some kind of relationship between rotation and the generation of a magnetic field.

[29] Runcorn et al. (1951).

required for this.[30] The idea was that the cylindrical mass would rotate with the Earth and produce a small, but measurable, magnetic field. Unfortunately, a suitably sensitive device with which to measure the field did not exist, so Patrick set about constructing the world's most sensitive magnetometer. In essence, this was very simple: two magnets of equal magnetic moment suspended on a fibre, one above another, such that their moments cancelled, creating 'astatic' conditions (the device was thus known as an 'astatic magnetometer'). Any magnetic item (hopefully including the gold cylinder) placed near the lower magnet caused a twisting of the fibre, which could be measured very accurately using a mirror to deflect a light beam. In order to detect the tiny fields expected, Patrick arranged things for maximum sensitivity, requiring more than half an hour to make a successful measurement. The device was housed in a wooden hut in a corner of a field at Jodrell Bank Observatory near Manchester. The Earth's field was backed off with three giant Helmholtz coils, creating a field-free space in which measurements could be made.

Today, we are used to overselling in science—in geomagnetism, for example, publicity is often sought by suggesting that the decline in intensity of the geomagnetic field over the past few centuries is a sign of an impending polarity reversal that will likely leave us all roasted by cosmic rays. Yet, even in the 1940s, the practice was well known, and Patrick Blackett himself was a leading culprit. Using his many connections, he was able to generate widespread interest in his theory, such as a *News Review*[31] article in May 1947 under the modest headline, 'Newton, Einstein—and now Blackett', describing his plan to test his fundamental theory using a spinning metal ball—an experiment that he never performed. This was a crucial period for research on the source of geomagnetism and, as it turned out, for Patrick personally. In 1948, he received the Nobel Prize for his earlier work with cloud chambers and the discovery of elementary particles: it seemed that his place in Westminster Abbey was already reserved. His prospects were not helped, however, when he became involved in politics with the publication of a book critical of the Allied bombing strategy during the recent war and opposed to nuclear weapons, leading the FBI to perceive

[30] The gold was, apparently, borrowed from the Bank of England. What became of it afterwards is anyone's guess.

[31] See Nye (1999).

him as a threat thereafter. Fate was turning against him with regard to his fundamental theory, too. Between 1947 and 1949, he received communications from several theoretical physicists, all raising objections that threatened to scupper his hypothesis. New stars were being discovered with varying magnetic fields, some of which actually reversed their polarity, while a paleomagnetic study of dolerite dykes of Tertiary age in northern England[32] showed unequivocally that these rocks had reversed polarity, which, according to Patrick's theory, could only occur by reversing the direction of rotation of the planet or star concerned. Worse still, new measurements showed that the magnetic moment of the Sun was much less than assumed by Patrick in developing his new 'law', which, in any case, was not derived from first principles, but was 'a kind of intuitive guess'.[33] Blackett, the brilliant experimentalist, had strayed unwisely into the unfamiliar waters of theoretical physics and immediately struck a mine.

In a remarkable example of scientific objectivity, Patrick then spent two years, beginning in 1949,[34] making meticulous measurements with his magnetometer using a variety of masses, eventually proving beyond any doubt that his theory was indeed wrong. Given the deep mine measurements of Keith Runcorn and the recent progress of Walter Elsasser and Teddy Bullard on a likely source within the core, it came as no surprise to anyone but Patrick that the predicted magnetization could not be detected. Finally, in 1952, he published a paper admitting defeat[35]—but in the most positive terms. Whilst he made no bones about what he now described as the 'Schuster–Wilson' theory being incorrect, and confessing that the use of the gold cylinder was a mistake, he devoted most of this 61-page article to a description of his astatic magnetometer, coming to the highly auspicious conclusion that 'When the magnetometer was completed it was found to be very suitable for the measurement of the remanent magnetism of weakly magnetised specimens, in particular certain sedimentary rocks'. This idea had occurred to Patrick even before he commenced his ill-fated experiments, when, back in 1948, he had come across a paper written by Ellis Johnson, Thomas Murphy, and Oscar Torreson, entitled 'Pre-history of

[32] Bruckshaw and Robertson (1949).
[33] Benfield (1950).
[34] See Hore (2002).
[35] Blackett (1952).

the Earth's Magnetic Field'.[36] The trio worked in the Department of Terrestrial Magnetism at the Carnegie Institution in Washington, where they had measured the weak magnetizations preserved in layered glacial sediments known as 'varves', collected in New England, and also in sediment cores from the Pacific Ocean collected during the US Antarctic Expedition of 1947/48. From this work, they concluded that the geomagnetic field had remained essentially constant for the past million years, and expressed a desire to extend their investigation back as far as a billion years before present. Most likely, Patrick was attracted by the abstract of their paper, in which they stated that their results were 'consistent with the "fundamental" theory proposed by Schuster, Babcock, and Blackett', whilst covering themselves with the proviso that these same results did not, however, 'provide positive evidence to support this theory'.

In many ways, work at the Carnegie Institution could be said to represent the birth of modern paleomagnetism. The group was well equipped to carry out the crucial studies that could prove continental drift, having begun a decade earlier by developing a magnetometer sensitive enough to measure the weak polarization in sediments. However, it all went badly wrong for them, largely as a result of the failure of US scientists to see the potential in their work. In 1948, the group travelled more than 7000 miles across the USA with their truck-mounted magnetometer, sampling sedimentary rocks in eight states. Following this reconnaissance study, they concluded that the Earth's magnetic field had remained more-or-less constant in strength and direction for at least 400 million years, and convinced themselves that polar wandering, magnetic reversals, and continental drift were all fiction. In fact, their samples had been overprinted by magnetizations produced by the present-day field, obscuring or erasing the original remanence. John Graham, a leading member of the group, submitted his PhD thesis in 1949, describing important new tests for the stability of magnetization in sedimentary rocks, publishing simultaneously in the prestigious *Journal of Geophysical Research*. Yet, while the latter has become one of the classic papers in paleomagnetism, his thesis was very nearly failed by his sceptical referees.[37] Sadly, paleomagnetic research then died a slow death at Carnegie during the very period in which, on the other side of

36 Johnson et al. (1948).
37 Doell (1971).

the Atlantic, it was making its greatest contributions of all. John Graham, who might easily have become one of the great names in twentieth-century science, ended up sniping at the conclusions of others.[38]

Mad Rise

In 1950, Keith Runcorn returned to Cambridge as Assistant Director of Research at Madingley Rise (the Department of Geodesy and Geophysics), following the departure of Teddy Bullard, who generally referred to the labs, with some justification, as 'Mad Rise'. Keith immediately began to assemble a formidable array of geophysical talent in the pursuit of the new science of paleomagnetism. And it was these protégés that now carried out the really crucial work.

Paleomagnetic research had been initiated in the Department the previous autumn by a young Dutch research student, who had come to Cambridge to measure the magnetization of rock samples collected in Iceland. This was Jan Hospers, now credited with having been the first to demonstrate that the direction of magnetization of rocks of similar age from the same location, sampled sufficiently densely to average out short-term noise, resembles that of a fictitious dipole located at the Earth's centre and aligned with the rotation axis. Having completed his MSc in Utrecht, Jan relocated, for purely financial reasons, to Cambridge on a Royal Dutch Shell scholarship, to do his PhD. Following the collection of Icelandic lava samples, his former professor, Rein van Bemmelen, suggested that Jan might make an interesting project by measuring their intensity of magnetization for use as a tool in correlating rocks of similar age. The idea was that variations in field strength might have caused equivalent variations in magnetization intensity in the rocks. Matching these variations in different parts of Iceland could, perhaps, provide a means of identifying the relative ages of lava flows.

Arriving in Cambridge with his boxes of rock samples, Jan got his first surprise. There was no magnetometer in Cambridge with which to measure their magnetization—something, you may say, he should have checked before setting out. Luckily for him, after Keith Runcorn arrived at the beginning of 1950, Jan was taken under his wing and Keith then fixed it for Jan to decamp to Imperial College in London, where a new 'spinner' magnetometer had been constructed, and make his

[38] See comments by Irving in Frankel (2015b).

measurements there.[39] Jan quickly realized that his original objective—to provide a new and reliable means of age correlation from magnetic intensity—was not feasible, but to his credit he did not throw in the towel. He had noticed that around half his specimens were magnetized in a direction opposite to the present Earth's field, but, in order to demonstrate this to sceptical scientists, he needed a statistical method to represent the dispersion of his results (a bit like a standard deviation, but on a sphere). Unfortunately, such a method, like the Cambridge magnetometer, did not exist, so, in a wonderful example of lateral thinking, he approached a geneticist.

In fact, this was not as illogical as it sounds, since the geneticist in question was Ronald Fisher, sometime Professor of Genetics at Cambridge University, recently nominated by Richard Dawkins as the greatest biologist since Darwin.[40] Ronald also had more than a passing interest in statistics, and had, a few years earlier, sketched out a method of dealing with directions on a sphere. Predictably, Jan was introduced by Keith Runcorn, who regularly had breakfast with Fisher in college. Fisher must have been a very accommodating sort of person, for he not only developed a new branch of statistics for the young paleomagician, but even went as far as calculating the statistics for Jan's Icelandic data himself. Ronald published his method in 1953, and 'Fisher statistics', as it became known, quickly formed the basis of paleomagnetic (and many other) data analyses worldwide. If this seems like something of a detail in the story, then it is worth noting the view of another major figure (who we shall shortly meet[41]) that 'It is my belief that the early lead established by Runcorn's Cambridge group was due, in considerable measure, to their early access to and exploitation of Fisher's statistics'.

Jan returned to Iceland in 1951 to collect more samples, including lavas and more weakly magnetized sedimentary rocks. Armed with Fisher's statistics, he was able to demonstrate that the geomagnetic field recorded by his rocks, when averaged over a few thousand years, was pretty well identical to that of a simple bar magnet, regardless of polarity. Moreover, the virtual bar magnet would have to be situated close to the centre of the Earth and aligned along the rotation axis. In science-speak, he had established the 'geocentric axial dipole hypothesis' that

39 Frankel (1987).

40 See https://edge.org/conversation/armand_marie_leroi-who-is-the-greatest-biologist-of-all-time/.

41 Irving (1988).

provided the foundation for the success that paleomagnetism was about to enjoy. Jan also established a simple, yet powerful, method of comparing the directions of remanence measured in rocks from different locations by calculating the position on the surface of the Earth of the pole corresponding to the measured remanence direction. Using the declination to determine the direction of the pole, and the inclination to derive its colatitude (the angular distance to the pole), Jan thus calculated a 'paleopole' for each sampling site. For rocks of Recent age, this pole was very much closer to the geographic pole than the present geomagnetic poles, indicating that much of the short-term 'noise' (i.e. secular variation) had been averaged out by the paleomagnetic recording process. His findings were soon confirmed by colleagues working on magnetized rocks from North America, Australia, and South America. Although Jan himself failed to appreciate the veracity of the drift theory, his methods proved instrumental in proving that same theory, just a few years later, by allowing results from different continents to be compared graphically.

Meanwhile, Keith Runcorn had decided to set up his own paleomagnetic laboratory in Cambridge, and began to think about ways to acquire suitable rock samples. Knowing little or nothing of field geology, he asked a lecturer in the Geology Department if he would like to cooperate. Unfortunately, this colleague knew very little about paleomagnetism. According to Keith, the lecturer then 'went away on some sort of field trip and they brought back a lot of little chippings from some formation which, you know, they said was a pretty important formation—it was the greywackes of Wales. After I saw this, I knew that I wouldn't get anywhere unless I went and did it myself.'[42]

Ken and Ted's Excellent Adventure

On a summer's day in 1951, Ted Irving, a young Natural Sciences graduate, waited outside the Black Bull public house in Kendal in the English Lake District. Ted had been recruited as a research assistant by Keith Runcorn to help him gather rock samples for paleomagnetic analysis in support of his objective of investigating secular variation in the distant geological past. When Keith arrived, the pair had lunch in the pub before setting out on their poorly planned mission. Lacking maps and

[42] Frankel (2012b).

any transport of their own, they took the local bus to Mealy Gill, where they collected eleven mudstone blocks. Afterwards, they caught the train to Glasgow, where they nipped into the University library to read up on Scottish geology, before proceeding by train and bus to Loch Torridon, hoping to sample the famous Torridonian sandstone of Precambrian age (about 600 million years old). Having removed large slabs of rock oriented with a compass, they continued their search for suitable sedimentary rocks, eventually reaching Wick, in the far north of Scotland. On returning, Ted took the samples to Jodrell Bank, where Runcorn had arranged for him to use Patrick Blackett's astatic magnetometer to measure remanence directions. Only one rock type, the fine-grained Torridonian red sandstones, gave consistent results, and these directions were highly oblique to the present field. Keith, realizing that his objective to sample the ancient secular variation had been overambitious, lost interest, but Ted decided to return to resample the red sandstone and investigate its remanence further.

By the following year, 1952, Runcorn had obtained funding for a new lab, allowing a second Natural Sciences graduate, Ken Creer, to join his 'burgeoning geomagnetism group'[43] at Mad Rise, where both Ken and Ted became PhD students. Unlike Ted, Ken had majored in physics rather than geology, and so brought a valuable technical facility to the team. The two research students 'became excited at the prospect that the work we were about to begin might, in some yet obscure way, contribute to this unresolved debate [concerning continental drift]'. During the next few years, the pair would carry out what they described as 'the first successful physical test of Wegener's Theory of continental drift'.[44] But, in order to do so, they needed to establish their laboratory and develop routines for the measurement of samples and analysis of results. Ken began constructing a new magnetometer based on Patrick Blackett's astatic design, but optimized for paleomagnetic work, allowing rapid measurement of numerous samples. By January 1953, it had been built and installed in a wooden hut in an old gravel pit near Mad Rise. They were now ready to begin the task of measuring the remanence of thousands of rock samples from all over the UK and elsewhere.

Magnanimously, after disproving his own theory, Patrick Blackett had allowed Keith Runcorn and his newly assembled team in Cambridge

[43] Creer and Irving (2012).
[44] Creer and Irving (2012).

access to the Jodrell Bank magnetometer to measure the remanence of their sedimentary rock samples. At the same time, however, Patrick assembled a rival paleomagnetism group led by John Clegg, a former member of the radio astronomy group at Manchester. This group also contributed mightily to the cause of drift, but their contribution has never, I think, been fully appreciated. Having read Arthur Holmes' great book,[45] in which the existing evidence for continental drift was presented, Patrick appreciated, more than did Keith Runcorn, the possibilities of this research. By 1952, a year earlier than Ken Creer, John Clegg had constructed a modified version of the astatic magnetometer specifically for the measurement of remanence in rock samples. In 1953, the group, including Mary Almond and Peter Stubbs, moved to Imperial College, London, and the following year published an article[46] in the *Philosophical Magazine* that had a major (although still not decisive) effect on opinion concerning the value of paleomagnetic studies.

Meanwhile, Ted was able to show that the fine-grained Torridonian sediments did indeed preserve ancient field directions and that these could be grouped into an upper part, in which directions were either down towards the southwest or up towards the northeast, and a lower part, in which the directions were downward but in a direction very different from the upper part. Ted had discovered a major time gap in sedimentation of more than 200 million years, a gap that was invisible to geological study,[47] thereby proving the effectiveness of paleomagnetism. Unfortunately, these sediments contained few fossils, making dating uncertain, but their magnetization had to be older than Triassic (more than 250 million years). After visiting the Torridonian outcrops on five occasions and carrying out extensive tests, Ted concluded that the rocks were very likely Precambrian in age (around 600 million years old). Some samples, however, gave directions close to the present dipole field, while others lay between the two. Ted drew the obvious conclusion that the ancient magnetization had been overprinted by the present-day field. By sticking to red sediments that had been cemented into a rigid mass, he was able to derive a stable direction that indicated a pole far from the present geographical pole. Either the poles had wandered or the continents had drifted (or maybe both).

[45] Holmes (1944).
[46] Clegg et al. (1954).
[47] This was later confirmed using radiometric dating by Turnbull et al. (1996).

Ken then visited as many potential sampling sites with well-dated rock outcrops as he could in an effort to extend the analysis back in time as far as possible. Most of his early forays produced negative results owing to weak magnetization or scattered directions. Eventually, he sampled volcanic rocks from the strongly magnetized Exeter Traps, which preserved a reversed magnetization in a direction more than 90° from the present geomagnetic field. Moreover, their dips (inclinations) were shallow, indicating a low latitude at the time of eruption in the early Permian (300 million years ago). Meanwhile, Ted sampled other red sediments, similar in appearance to the Torridonian sandstones, in Devon and elsewhere, that also seemed to retain reliable ancient field directions.

Jan Hospers' main conclusions had been that (i) the Earth's field reverses occasionally, so that normal and reversed directions could be combined to give the mean field direction, and (ii) by averaging measured directions, a paleopole could be calculated for each rock formation, approximating the geographic pole at the time of magnetization. Accepting these, Ken recognized that Jan's technique of plotting paleopoles on a polar projection could provide a simple means of displaying and comparing results from rock formations of different ages. If, as the Carnegie guys believed, the positions of the continents were fixed and the geomagnetic field had always been stable in its present configuration—that is, a dipole field aligned with the rotation axis—then all paleopoles from rocks with a stable magnetization would plot close to the present north or south pole, no matter what the age of the rock. Crucially, Ken and Ted had performed field and laboratory tests on their samples to demonstrate that the magnetizations they were measuring were original and not later overprints. By plotting the mean paleopole for each age of rock measured, Ken constructed a path[48] showing that, while the youngest rocks gave paleopoles close to the present geographic poles, those from rocks older than a few million years migrated away from the present poles with age, winding through the Pacific Ocean back to a point on the equator during the late Precambrian. A generalized polar path derived from climatic indicators in sedimentary rocks was also shown. The agreement was far from perfect, but, given the uncertainties in the latter, the similarity in shape

[48] Ken got the idea for plotting continuous polar paths from an article by Beno Gutenberg (1939) in which he discussed similar paths derived from geological data.

was at least encouraging.[49] The new polar wander path was published in a landmark article[50] in the summer of 1954, authored by Ken, Ted, and Keith Runcorn (Figure 3.2). There they noted that, since only paleopoles from Britain were included, it was not yet possible to distinguish between continental drift and polar wandering as explanations of the path observed.

Ken presented his results to the British Association for the Advancement of Science meeting in Oxford in September of the same

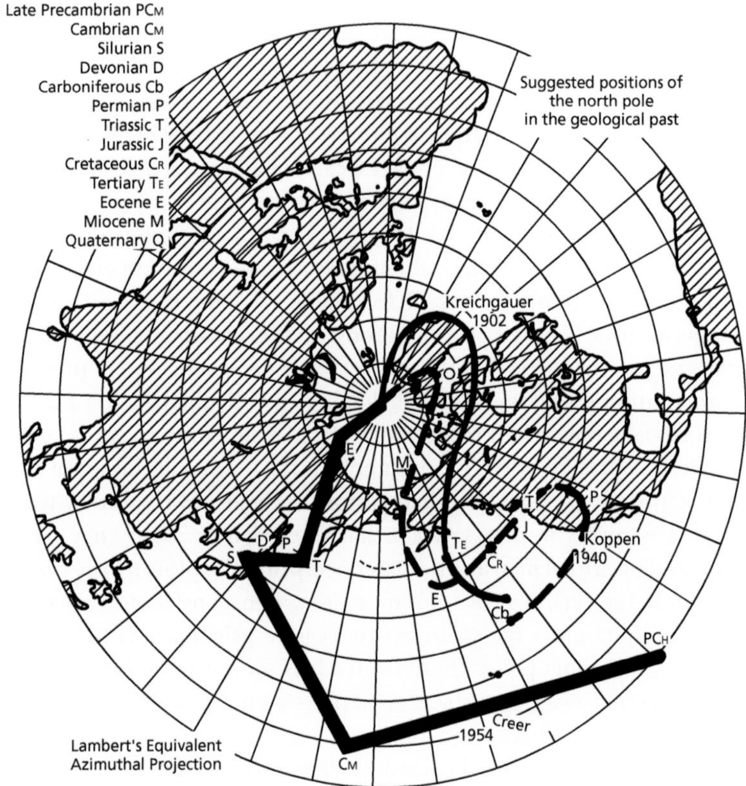

Figure 3.2 Ken Creer's initial polar wander path.

[49] It turned out subsequently that much of the discrepancy resulted from incorrect assumptions built into the climatically derived path.

[50] Creer et al. (1954).

year. To their credit, the press quickly picked up the story, which was then published in *The Times* and the *Manchester Guardian*, journalists having appreciated the message that Ken's results were suggesting that the continents may have drifted independently over the Earth's surface. A report in *Time* Magazine[51] included a redrafted plot of Ken's apparent polar wander path, which, owing to the inertia in academic publishing, actually became the first version to appear in print.[52] Ken had, by then, added an even earlier Precambrian paleopole located in the Western USA, leading to the slightly misleading headline 'Arizona Arctic'.

Ken's results could be interpreted in two ways. Firstly, the rotation axis (together with the north and south poles) could have migrated (somewhat erratically) relative to the entire crust of the Earth, something that became known as 'true polar wander'. Another way of describing this is to say that the Earth had tumbled bodily relative to the axis of rotation. Secondly, the continents may have drifted independently relative to the poles, as suggested by Wegener and his followers.[53] To test which hypothesis was correct, it was necessary to assemble comparable apparent polar wander paths for other continents. In an historic revision of his groundbreaking apparent polar wander path, Ken used Fisher's statistics to calculate a mean paleopole and associated 95 per cent errors from results published by John Graham for the Silurian (about 420 million years old) Rose Hill Formation in the Appalachians.[54] The calculated paleopole plotted well away from the polar wander path for Britain, suggesting that North America had a different path of its own, and that the North American continent probably lay around 2000 km closer to Europe at that time. He later added 95 per cent confidence ellipses around all the older paleopoles to demonstrate the statistical uncertainties involved. This version he included in his PhD thesis, but did not publish until 1957.

Getting the Drift

In November 1954, Ted, unable to secure a post in Cambridge, went to work at the Australian National University (ANU) in Canberra, where

[51] See *Time* Magazine, 27 September 1954.
[52] Frankel (2012b).
[53] Of course, it could also have been a bit of both.
[54] Graham (1949).

he began work on constructing an Australian polar wander path. Keith, frustrated at losing trained staff and having no prospect of setting up a permanent geophysics group, left Cambridge in January 1956 to take the chair of physics at Newcastle University,[55] where he remained for the next 32 years. Despite his instrumental role, Keith was still not a drifter at the time. His motives in recruiting Ken and Ted as research students had nothing to do with proving or disproving drift, but rather with an attempt to study secular variation in ancient rocks, to see if the geomagnetic field had behaved in the past as it does today. Keith's epiphany came only in July 1956, after he had left Cambridge for good. In March, he had published a paper in which he compared paleomagnetic results from Britain with his new results from North America, claiming that 'The agreement is sufficiently close to give strong support to the hypothesis of polar wandering'. That is, he believed that the two continents shared a common polar wander path and therefore true polar wander was the best explanation. Six months later, he published two short papers, in the most obscure places, finally accepting drift between the two continents.[56] It seems that the arguments of his students, Ken and Ted, had convinced him of the reality of mobilism: Runcorn had become a convert to drift and, like all proselytes, soon gained a reputation as an enthusiastic evangelist for the theory of moving continents.

One of the (by now) ex-Cambridge group's failings was in marketing their achievements. You'd have thought that they had accumulated more than enough solid evidence to claim a few pages in *Nature* or *Science*, and thereby secure recognition for one of the major breakthroughs of the century. Yet, somehow, they failed miserably at the final hurdle, publishing instead in such august publications as *Geofisica pura e applicata* and *Geologie en Mijnbouw*. Runcorn had, it is true, published his American results the previous year in *Nature*, but said nothing about the big question—continental drift. In a masterclass of bad publishing strategy, Ted submitted a paper presenting the crucial step, the comparison of polar wander paths from different continents (a paper in which he did not even use the words 'continental drift', preferring the slightly less inflammatory 'relative movement of the continents'), to the *Journal of the Geological Society of Australia*, where it was promptly rejected. Advised by his new boss at ANU, he lowered his sights still

[55] Then still a college of Durham University.

[56] Runcorn (1956a, b), Frankel (2012b).

further and submitted a revised version to *Geofisica pura e applicata*, which was less fussy and accepted the article immediately.[57] This contribution has only recently been recognized as 'seminal', and was republished in 2014 as a classic paper in Earth Science.[58]

Nevertheless, Ted's conclusions were clear. The polar wander paths of Europe (constructed by combining the British results of Ken, Keith, and himself) and North America (constructed from Runcorn's recently published results) were similar in form, but diverged as one stepped back in time (Figure 3.3), suggesting that the two continents had been much closer in the past, just as predicted by Alfred Wegener. Ted made one final tactical error at the end of his paper, where he noted that his results depended on the assumption of an axial dipole magnetic field in

Figure 3.3 Polar wander paths from Irving (1956). The bold line joins paleopoles from North America; the faint line joins paleopoles from Europe. Poles 8 and 9 are from Australia and India, respectively.

57 Irving (1956).
58 Frankel (2014).

the geological past, before handing his critics a loaded revolver by saying 'If this assumption is abandoned the positions of the pole given in this paper cease to have any significance'.

In the following year, 1957, the trio finally managed to publish a major article[59] in a reputable journal (the *Philosophical Transactions of the Royal Society*). Polar wandering still reared its ugly head, but they noted that 'There is also evidence for a relative movement of 1000 miles between America and Europe since Triassic times'. Any lingering doubts about the paleomagnetic method should have been dispelled by a letter to *Nature* in November,[60] in which results from volcanic rocks and red sandstones of Triassic age (approximately 220 million years ago), collected from widely scattered locations in North America, and measured at labs in Newcastle, Cambridge, and Canberra, were presented. Despite the differences in sampling location and rock type, the positions of the Triassic north pole calculated from measurements at each site agreed remarkably well. The authors included a pole of similar age derived from the English Keuper Marls for comparison, pointing out that the discrepancy between the North American and British results could be explained by continental drift and the opening of the Atlantic Ocean.

In 1958, Ted published a review in the *Geophysical Journal of the Royal Astronomical Society*, in which he again set out the overwhelming evidence. In this summary, he noted the discrepancies between paleopoles older than 20 million years from different continents, concluding, rather nervously, that 'explanations other than relative movements between continents (continental drift) are not promising'. Finally, in December 1959, Patrick Blackett, together with John Clegg and Peter Stubbs, submitted an article to the *Proceedings of the Royal Society of London* in which they presented a compilation of paleomagnetic data available at the time.[61] Interestingly, they eschewed Fisher statistics and polar wander paths, but argued nevertheless that the 'most plausible' explanation of their combined paleomagnetic data was continental drift. Perhaps chastened by Patrick's previous experience with big theories, they ventured to speculate, only slightly less cautiously than Ted, that 'On the tentative assumption that the rock magnetic results can be explained

[59] Creer et al. (1957).
[60] Du Bois et al. (1957).
[61] Blackett et al. (1960).

by continental drift, it is possible to estimate the ancient latitude and the orientation relative to the earth's rotational axis, of each continent'. From such estimates, they concluded that, for the past 200 million years, Europe, North America, India and Australia had all been moving steadily northward at rates between 22 and 89 mm per year. Despite their timidity, the data were now overwhelmingly in support of continental drift throughout the past 700 million years. Paleomagnetism had arrived: the case for drift was, to all intents, proven.

The following year, American paleomagicians Allan Cox and Richard Doell of the US Geological Survey wrote an exhausting, if not exhaustive, 'Review of Paleomagnetism', some 124 pages in length.[62] Recent developments in the fledgling science had probably been as spectacular as in any branch of contemporary research, with the possible exception of Watson and Crick's activities in Cambridge. The dragon of fixism had been slayed and the mobilists had won the day with their exciting proof that continents really had moved vast distances. Yet none of that excitement was apparent in Cox and Doell's review. In their summary, they noted four rather dull general conclusions:

1. The Earth's average magnetic field, throughout Oligocene to Recent time, has very closely approximated that due to a dipole at the centre of the Earth oriented parallel to the present axis of rotation.
2. Paleomagnetic results for the Mesozoic and early Tertiary might be explained more plausibly by a relatively rapidly changing magnetic field, with or without wandering of the rotational pole, than by large-scale continental drift.
3. The Carboniferous and especially the Permian magnetic fields were relatively very 'steady' and were vastly different from the present configuration of the field.
4. The Precambrian magnetic field was different from the present field configuration and, considering the time spanned, was remarkably consistent for all continents.

They were mistaken on all four counts. Ted was incensed and wrote a forthright letter to Cox, suggesting that he rename his review 'paleomagnetism apologetics' and observing that 'For, I suppose, a total of about 12 years or so in the subject, you really have very little to show for it, and this

[62] Cox and Doell (1960).

retarded stunted growth [the review] is not going to bear fruit'.[63] Nevertheless, it was a further five years before the Americans came down off their high fence.

Indian Summer

In truth, it should all have been over by 1956. In Wegener's Pangæa reconstructions,[64] India was shown as part of Gondwana, lying adjacent to Antarctica and Australia. This was based on independent geological evidence of a widespread Permian glaciation compiled by Alfred and his father-in-law, Wladimir Köppen, implying that India had migrated from high southern latitudes to middle northern latitudes, crossing the equator en route. India could thus provide the acid test of Wegener's 'unscientific' theory. Ronald Fisher enters the story here for a second time. According to Ted Irving, Ronald was embarking for India in 1951, just as Ted thought about this test, and generously arranged to visit the Director of the Geological Survey of India in support of Ted's request for oriented rock samples. It appears that Fisher was also motivated by a desire to get one over on Harold Jeffreys, with whom he had frequent disagreements.

Only six samples turned out to have stable magnetization, all of them from the Deccan Traps, a vast series of flood basalts covering half of western India. The Traps consist of numerous horizontal lava flows erupted around 66 million years ago, stacked one upon another, with a total thickness of over a mile. The flows are undisturbed and largely unaltered, so that their original magnetizations, and the ancient field directions they represent, are faithfully preserved. Ted's results showed, firstly, that all stable magnetizations were of reversed polarity and, secondly, that the inclinations were steeply upward, corresponding to high southern latitudes. Here was the scientific equivalent of a winning lottery ticket. Fame, a Fellowship of the Royal Society, possibly a knighthood, beckoned. Unaccountably, Ted never got round to publishing these results at the time, although he included them at the back of his PhD thesis, which was failed by the examiners in 1954.[65] Ted explained later that results such as these, and those of other young

[63] Frankel (2012b).
[64] Wegener (1920).
[65] The results were eventually published as an appendix in Ted's Italian job of 1956.

researchers in the group, were touted at international conferences by Keith Runcorn, who then delayed publication until whole sets of themed articles could be published together, by which time their contents were no longer novel.

A few years afterwards, John Clegg and colleagues at Imperial College in London conducted a much more extensive paleomagnetic study of the Deccan Traps, collecting half a ton of rock samples, from which they calculated a paleopole at 28°N 85°W, far from the British and North American end-Cretaceous mean pole at 76°N 130°E. Their results, published in a respectable journal in 1956,[66] were internally consistent and close to Ted's. The evidence was clear—India had migrated from latitude 34°S at the time of eruption of the Deccan Traps, drifting more than 5000 km northward before running into Asia. Bearing in mind the older glacial sediments, it must have travelled at least twice as far since the Permian (250 million years ago). Somehow, even this spectacular and convincing evidence—confirmed in later publications of the Imperial group—was insufficient to convince a majority of geologists. Somehow, they clung to their misguided fixist view of the Earth until the advent of Vine and Matthews' hypothesis in the 1960s, and, in many cases, for years beyond.[67] Naturally, old Harold Jeffreys continued to man the barricades, objecting to the new paleomagnetic results on the basis of his own crude experiences half a century before, and ignoring the advances in sampling, cleaning, measuring, and testing of rock samples that had taken place during the 1950s. In the fourth edition of his textbook *The Earth*, published in 1959, he acknowledged that recent work had demonstrated directions of rock magnetization in India and Australia very different from those measured for the northern continents, noting, somewhat condescendingly, that 'it has been asserted that they prove continental drift'. Predictably, he then went on to offer vague and unconvincing opposition based on the observation that the ocean floor was rough rather than smooth. By this time, however (as Ted noted many years later), 'His prejudiced statements were treated as comic relief'.[68]

[66] Clegg et al. (1956).

[67] Such inertia is a rich subject for debate among historians of science. See e.g. Oreskes (1999).

[68] Frankel (2012b).

Earlier Reconstructions

Paleomagnetism had shown that the continents had moved independently across the surface of the planet since the break-up of Wegener's Pangæa, beginning some time after the Permian (250 million years ago). But it had shown more than that. The establishment of apparent polar wander paths for the continents provided strong evidence that continental drift had been a feature of the Earth's geological history since at least late Precambrian times (more than 600 million years ago). If, as Wegener had suggested, Pangæa had existed throughout geological history until its demise in the Jurassic and Cretaceous, then every continent would have been rigidly attached to every other continent prior to break-up, and their earlier polar wander paths would be identical. In 1958, Ted, now happily ensconced in his new billet in the antipodes, published a paper with his research student Ron Green, in the *Geophysical Journal*, presenting a new polar wander path for Australia.[69] Like those of the northern continents, it suggested relative movement—continental drift—since the Carboniferous (i.e. during the past 300 million years), but they also presented paleopoles from earlier Australian formations as old as Proterozoic (600 million years). These showed clearly that the differences persisted throughout geological history, prompting them to conclude 'it is necessary to suppose that relative movements also occurred prior to the Carboniferous'. Pangæa had therefore not existed for eons, but had been assembled from pre-existing continental fragments during the Paleozoic.

The idea was picked up in 1968 by Tuzo Wilson, and incorporated in his ocean cycle (described in Chapter 2), leading to an explosion of ideas concerning the geography of the Earth during the Paleozoic and earlier times. Perhaps there had been earlier supercontinents, created and disrupted in a grand 'supercontinent cycle' comparable to Tuzo's ocean cycle? However, the problem in reconstructing these past geographies lay with the nature of the data. Lacking any *in situ* ocean floor magnetic anomalies or fracture zones, geologists could not reconstruct the continents uniquely in their former juxtapositions. Their latitudes and orientations could be determined by paleomagnetic study of the natural remanence in rocks, but their relative longitudes could only be assessed on the basis of geological evidence. This was a problem to be addressed in the second generation of plate tectonics theory.

[69] Irving and Green (1968).

By now, various laboratory and field tests had been developed to check whether the remanent magnetization measured in rock samples was 'original'—that is, frozen in at the time of deposition or eruption—or whether it had been acquired during some later event, such as burial and reheating during mountain-building. Dating by means of fossils (for sedimentary rocks) or radiometric methods (for igneous rocks) then allowed the sampling site and, by extension, the continent of which it formed part, to be placed in its correct orientation and paleolatitude at the time of magnetization. For the first time, there was a quantitative method of reconstructing the past geography of the planet, but corroborating evidence was required to convince the sceptics.

Climate of Opinion

Alfred Wegener's continental reconstructions had originally been drawn with Africa fixed in its present location. But, in 1924, Alfred and his father-in-law Wladimir Köppen, knowing nothing of paleomagnetism, superimposed paleolatitudes on these maps, which they had estimated from climate-sensitive sedimentary rocks, such as coals, evaporite salts, and desert sandstones.[70] Wladimir, a pioneer of climatology, had previously established that the recent climate could be classified into bands of latitude according to mean temperature, rainfall and seasonality. This 'zonal' pattern was disrupted by perturbations resulting from the distribution of land and sea, yet latitude remained the dominant control on climate. Assuming that this had always been the case, Köppen and his son-in-law applied this 'zonal principle' to the geological record, using the distribution of dated climate-sensitive rocks to determine the ancient latitude bands in which they had formed. Incorporating these ideas into Alfred's reconstructions, they produced the first crude set of paleoclimatic maps, which were subsequently reproduced in the 1929 edition of Alfred's classic work *Die Entstehung der Kontinente und Ozeane*.

Perhaps the most valuable climate indicators were the sediments dumped by large glaciers, which, as far as anybody knew, only existed in high latitudes.[71] 'Till' or 'boulder clay' is a glacial sediment consisting of unsorted rock fragments of all sizes, contained in a muddy matrix. In

[70] Köppen and Wegener (1924).

[71] Here, we are referring to extensive ice sheets at sea level rather than valley glaciers at high altitude.

Britain, it forms a component of the 'drift' that inconveniently blankets the 'solid' geology of much of the island. Ancient, indurated, rocks with these same characteristics were interpreted, naturally, as fossil tills, hence the name, 'tillite'. The occurrence of tillites is thus a strong indicator of past glaciations, particularly where they rest on a foundation of hard rock, scratched and polished by the movement of glaciers. The first discovery of such a combination was in the north-west of Scotland, where geologist James Thomson of Kilmarnock found a metamorphic (i.e. altered by pressure and heat) mudstone at Port Askaig on Islay, containing fragments of granite, of which there are no outcrops on the island.[72] He guessed that the granite inclusions must have been transported from the north by ice, as evidenced by striated quartzites, and identified the rock formation as a relict glacial deposit of great age.

As early as 1912, Alfred Wegener had noted that 'One of the strongest proofs of these ideas is to be found in Permian glaciations'. The 'ideas', of course, were his notions of continental drift and, specifically, the existence of the great southern supercontinent of Gondwana.[73] He was referring to the discovery of tillites of Carboniferous to Permian age (around 300 million years) in Australia, Africa, India, and South America. These Gondwana tillites were all, so far as could be determined at the time, of similar age, confirming that a widespread glaciation had been experienced by the supercontinent during the latest Carboniferous and early Permian, presumably as it lay at the south pole. Apparently, places that now are redolent of warmth were icy wastes 300 million years ago. Alfred's argument was that, when plotted on the *present-day* map of the world, these tillites suggested a Carboniferous–Permian glaciation of vast extent, assuming the intervening oceans were also covered by ice. As he noted, 'With a glaciation of this magnitude no part of the Earth's surface would have been ice free. With such a south polar location, the north pole would fall in Mexico where no trace of Permian glaciation is found. The South American glacial outcrops would lie on the equator.'[74] On the other hand, when the continents were reconstructed in their Pangæa configuration, the extent of the glaciation was reduced to

[72] Thomson (1871).

[73] Wegener (1912) referred to 'Gondwana-Land'. Since then, there has been ongoing debate about whether 'Gondwana' or 'Gondwanaland' is the correct term. In the interests of economy, we use the former here.

[74] Translation by Roland Von Huene (2002).

something like that of the most recent ice age in Laurentia (essentially, North America and Greenland), and could therefore be explained by ice extending from a pole located in southern Africa.

A major part of Ted Irving's classic 1956 'Italian' article was a comparison of the newly compiled polar wander paths with climatic evidence. Coal seams, fossil sand dunes and coral reefs, red sediments, and salt beds were all evidence of warm climates (hence low latitudes), while tillites pointed to much colder climates (hence high latitudes). Ted found that the paleomagnetic latitudes for Britain and the USA were in good agreement with such sediments, whereas the latter were clearly at odds with their present latitudes. The one exception was the ancient Torridonian formation sampled so diligently during that summer of 1951, and several times subsequently. Here, the rocks gave magnetic inclinations suggesting low latitudes, despite the occurrence of glacial sediments of similar age nearby. Ted pointed out that paleoclimatic evidence could only be used in a gross way to confirm or refute the latitudes assigned by means of paleomagnetism, since other factors, such as ocean currents, were also involved (think of the effect of the North Atlantic Drift on the climate of north-west Europe). Nevertheless, taking into account all the uncertainties, his overall conclusion was that the comparison provided support for continental drift. Once again, Patrick Blackett's Imperial group published the results of a more complete study in a more reputable journal.[75] Comparing the paleolatitudes suggested by coral reefs, salt deposits, and glacial sediments with those derived from the magnetic inclination of contemporaneous rocks (using the relation that the tangent of the angle of inclination is twice that of the latitude), Patrick also concluded that general agreement existed between latitudes derived from paleomagnetism and those based on paleoclimatic evidence, tentatively supporting drift of the continents.

By 1960, Keith Runcorn had become the guru of the mobilists. He used his considerable influence to organize a series of international meetings in Newcastle at which renowned experts were invited to present evidence from their own specialism for and against drifting continents, subsequently publishing the contributions in volumes of collected papers. A meeting on paleoclimatology had already been held in 1959, resulting in the publication of an influential volume entitled *Descriptive*

[75] Blackett et al. (1961).

Palaeoclimatology, edited by Keith's colleague, Alan Nairn. Keith now planned, with the distinguished American geologist, Walter Bucher, a much bigger gathering in Newcastle, bringing together paleoclimatologists and paleontologists to discuss the evidence for ancient climates in the light of continental drift. Walter attempted to obtain funding for the conference from the US National Science Foundation, but, predictably, they did not consider the subject worth discussing. Even more remarkably, when Walter approached the future 'father of plate tectonics' for help in 1960, Tuzo Wilson refused on the grounds that continental drift was an 'entirely discredited theory'.[76] The meeting was eventually funded by the NATO Advanced Study Institute.

Following this meeting, another volume edited by Alan Nairn was published in 1964, entitled *Problems in Palaeoclimatology*. Amongst the 53 contributions were some very significant papers, although, as Henry Frankel has pointed out,[77] only 16 favoured mobilism. A particularly important article by Ted and his research student Jim Briden compiled global climate-sensitive data and presented it on map projections on which paleolatitudes derived from paleomagnetic studies had been drawn. The only southern continent that could be represented was Australia, because, of course, no reliable paleomagnetic results had yet been obtained from Africa or South America. Nevertheless, their results gave support to Wladimir Köppen's zonal principle, and demonstrated the incompatibility of the paleoclimatic data with their present locations. Thus, they declared, 'relative movements between the land areas in question would seem to be a necessary conclusion'.

The Exception

Following the departure of Keith the fixer and his entourage, paleomagnetic research continued in Cambridge. Brian Harland, the geologist who had failed to provide Keith with the right kind of rock samples, used the magnetometer in the shed at the bottom of Mad Rise, and subsequently a similar device constructed in the Sedgwick Museum, to measure the magnetic remanence in samples collected during expeditions in 1957 to Scandinavia and Greenland. Like Alfred Wegener, Brian felt a strong attraction to the Arctic, where he spent many happy

[76] See Frankel (2012b).

[77] Frankel (2012b).

field seasons. In 1948, he initiated the Cambridge Spitsbergen Expeditions, and eventually set up the Cambridge Arctic Shelf Programme within the University in 1975. It was therefore not surprising that many of the rocks he studied in places like Spitsbergen were of the type normally associated with glaciations (i.e. tillites), since the islands, at 75°N, have been glaciated for millions of years. However, these tillites had been deposited, not during the present ice age or any of the previous ice advances known to have occurred during the Quaternary (the past 2.6 million years). They were, in fact, more than 600 million years old, having their origin in the late Precambrian. The ice age that produced them had occurred in a world very different from ours, at a time before animals, before plants, before life on land. Geologists working in the late nineteenth and early twentieth centuries found similar ancient glacial sediments on several continents, including Africa, Australia and North America. A surprising feature of these tillites, however, was their frequent occurrence sandwiched between rocks deposited under much warmer, not to say, tropical, climates. In many cases, it appeared that they had been deposited in the sea, beneath floating ice sheets, rather than being dumped by ice on land.

Having measured the remanence of rock samples of late Precambrian age from Norway and Greenland, Brian's research student Donald Bidgood plotted the results, in true Cambridge fashion, using Fisher's statistics calculated using the EDSAC 'digital computor'. The paleopoles plotted about as far from the present pole as it is possible to get—almost at the equator. If the geomagnetic field at the time (around 600 million years ago) was an axial dipole field like that of today, then a quite incredible result followed. Adjusting things so that his paleopoles coincided with the present north pole, Brian's sampling sites were simultaneously rotated almost to the equator. A literal interpretation would be that ice had extended from the poles to the tropics in the greatest glaciation ever known. Brian did not waste time before publishing articles proposing an 'Infra-Cambrian Ice Age' that occurred immediately prior to the appearance of animals on Earth. He gave a talk on the subject at Keith Runcorn's paleoclimate meeting in Newcastle in 1963, where his colleague, the paleontologist Martin Rudwick,[78] presented a complementary lecture on the possibility that

[78] Martin subsequently became interested in the history of science and has recently published a fascinating good read in *Earth's Deep History* (Rudwick, 2014).

this major glaciation—or rather the recovery from it—was somehow responsible for the so-called Cambrian explosion of life forms observed in the fossil record. This was an idea that had been advanced as early as 1864 by the great Scotsman James Croll,[79] the first person to explain the present ice age in terms of variations in the Earth's orbit.

Thus, it appeared, glaciations need not be associated with high latitudes, and Wladimir Köppen's zonal principle did not apply to the whole of geological time. If you have been keeping up with the popular scientific literature, all this will have rung a bell, for the idea of extensive glaciations in the late Precambrian has recently acquired cult status, along with the trendy title of 'snowball Earth'.

Afterthought

The proof of continental drift required the establishment of an entirely new branch of geophysics—paleomagnetism. Ironically, while the evidence gathered by British paleomagnetists was responsible for 'converting' previously diehard fixists such as Tuzo Wilson and Harry Hess to mobilism, many of the doubters were themselves paleomagnetists. People like John Graham, Allan Cox, and Bob Doell were right there on the quayside, yet managed to miss the boat in spectacular fashion. The subtle reasons for their rejection of drift could be debated at length, and whole volumes have been written on the subject, one concluding rather pretentiously that the Americans' hostile attitude resulted from 'theoretical pluralism coupled with an inductive methodology, a suspicion of highly ambitious theories and an emphasis on hard work'.[80] Regardless, we are more concerned here with the evidence provided by paleomagnetism and paleoclimatology in support of drifting continents, such that, by 1960, no reasonable and reasonably informed person could doubt it. That so many did just that says more about the human mind than about the science.

[79] See Croll (1875).
[80] Oreskes (1999).

4

Plate Tectonics by Jerks

> What made plate tectonics so immediately convincing was that it was principally designed to account for sea floor spreading, continental drift, and magnetic anomalies. With no further input, it also accounted for the distribution of earthquakes
>
> DAN MCKENZIE (2003)

The geneticist Steve Jones is one of the most successful popularisers of science today. His stock of one-liners ('my job as a genetics lecturer is to make sex boring'), though well worn, bring science as close to entertainment as anything currently broadcast on radio or television (which is, perhaps, not saying much). One of his favourite, and thus oft-repeated, witticisms concerns punctuated equilibrium in evolution—'evolution by jerks', as it is often called—that Jones inevitably, quoting the late Stephen Gould, contrasts with 'evolution by creeps'– the slow and continuous morphing of one species into another. The same concepts can be applied to plate tectonics.

Anyone living around the Pacific Ocean will be familiar with plate tectonics by jerks. That is, long periods of quiescence punctuated by very large, frequently fatal, and so far unpredictable, earth movements on a timescale of seconds. Surprisingly, careful measurements of plate motions over periods of just a few years (using geodetic methods that we'll come to in Part II) show that the plates are in fact moving almost continuously, so that what we have on all but the very shortest timescales is 'plate tectonics by creeps'. Happily, in this context, jerks and creeps are not mutually exclusive, and so, unlike the biologists, we do not need to come to blows.

*

Cold Snaps

Most of the unpleasant side-effects of great earthquakes are a direct or indirect consequence of the generation of 'surface waves'. These come in two flavours, named after the pioneer physicists who identified them

The Tectonic Plates Are Moving! Roy Livermore, Oxford University Press (2018). © Roy Livermore.
DOI: 10.1093/oso/9780198717867.001.0001

in the late nineteenth and early twentieth centuries.[1] Both involve wave-like motion of the ground surface—hence the damaging consequences for items attached to it, such as houses and nuclear power stations. Although of great interest to coastal residents around the Pacific Ocean and in Indonesia, surface waves are of less importance to seismologists studying the deep interior than the mostly harmless[2] waves that pass through the body of the Earth. Body waves also come in two flavours, 'P-waves' and 'S-waves', as you may recall from Chapter 1. The faster P-waves are simply sound waves passing through a solid or liquid, while S-waves, or 'distortional' waves as they were once called, involve side-to-side motion within solids (they cannot travel through liquids).

Ever since the great San Francisco quake of 1906, earth tremors have been associated with pre-existing fractures in the Earth's crust—known to geologists as 'faults'. In that case, the culprit was all too obvious, the break on the San Andreas fault system being visible at the surface for 330 km throughout much of southern California. The problem with faults is that they are neither perfectly flat nor perfectly smooth. So, rather than glide quietly past one another, the two sides of a fault get stuck at places where friction is high. Strain energy builds up in the rocks next to the locked zone (like pulling back the string of a bow), and released all at once when something (i.e. the fault) gives. The rocks then snap back more or less to their original shape—hence the term 'elastic' applied to both the deformation and the resulting shock waves. Usually, the rupture propagates away from a locked zone across the fault surface. For a large quake on a major fault 1000 km in length, this may take 5 minutes or more. The magnitude of an earthquake thus depends on the amount of elastic energy stored in the rocks and on the area of the fault that ruptures.

Faults can be oriented in any direction, but they are easier to picture if you think of just two ideal classes. Vertical (or near-vertical) planar faults, such as the San Andreas, where the rocks on either side slip horizontally in opposite directions parallel to the fault, are 'strike–slip' faults. Faults consisting of a plane sloping into the Earth like a ramp, in which one side slides up and the other down the dipping fault surface, are 'dip–slip' faults. Hence, Tuzo's famous transform faults are a special kind of strike–slip fault, whereas the faults observed by the FAMOUS

1 Augustus Love and John Strutt.

2 They can, however, cause significant damage close to the epicentre of a quake.

guys near the axis of the Mid-Atlantic Ridge are dip–slip. You can probably imagine that, in many cases, slip on fault planes is a bit of both.

Those unfamiliar with seismology may wonder how it is possible to work out the location of an earthquake source just from seismic body waves arriving at a recording station. The answer is that it is not. You need to combine recordings of arrivals at *many* stations distributed around the globe. Then, by a kind of glorified triangulation, you can work out where the earthquake focus must have been, and when the shock occurred. Because earthquakes occur on faults, the energy radiated from them as P- and S-waves is less like that from, say, an underground H-bomb detonation, which propagates in all directions uniformly, and more like the light from a lighthouse, with beams of energy transmitted in certain directions. By recording the arrival of seismic waves at many stations, it is possible to determine the beam pattern for a particular earthquake and, from this, establish the fault orientation and sense of motion (this is exactly what Lynn Sykes did for the transform quakes back in Chapter 2).

Unless you are uncomfortably close to the epicentre of a large earthquake, the first you will know about it will be when a P-wave arrives (assuming that you are in the habit of carrying a seismograph about with you). Once you see the kick of the pen that signals its arrival, you normally won't have long to wait until that other body wave, the S-wave, turns up, the exact delay being dependent on your distance from the earthquake focus. However, should you happen to be more than 11,450 km from the source, your wait may be very lengthy indeed. The failure of shear waves to put in an appearance at such long ranges was what led the early seismologists to conclude that somewhere, deep within the Earth, was a layer that was incapable of transmitting shear waves, signalling the presence of material with zero shear strength or, in plain English, a liquid. This was, of course, the Earth's core, or, more specifically, the outer core, now known to be composed of molten iron and nickel (plus some other stuff). Conversely, the arrival of S-waves elsewhere was strong evidence that the mantle must be solid and rigid—at least, rigid on a jerks timescale of seconds (the time it takes for an S-wave to pass).

Deep and Deeper

Herbert Hall Turner, sometime Savilian Professor of Astronomy in Oxford University, was, it seems, the Carl Sagan of his day, giving

Christmas Lectures to the Royal Institution in 1913 and 1915, and having a hand in the naming of the dwarf planet Pluto. He was also interested in the structure of the Earth, and used seismograph records of earthquakes to calculate their depths of focus, coming to the startling conclusion that some events had been triggered more than 500 km beneath the surface.[3] Party-pooper Harold Jeffreys, who later spent many years undermining Alfred Wegener's great theory, soon queered the pitch for Turner too, by pointing out that the crust in northern Europe was rising slowly after being relieved of ice several miles thick at the termination of the most recent glaciation.[4] This could only occur if temperature and pressure beneath the crust were sufficient to make mantle rock soft enough to flow under stress, which, of course, meant that it was too weak to fracture like the cold, brittle, crust above. This led Harold to conclude that 'depths of the foci of the earthquakes chiefly considered by Prof. Turner cannot exceed 35 km'. As with continental drift, Jeffreys was correct in his calculations, but wrong in his conclusions.

On the other side of the world, just before noon on 1 September 1923, a Tokyo University physics student was working happily at his desk when he heard strange sounds outside and felt something move. Within seconds, the walls began to shake and his bookshelves collapsed. Luckily, the house survived and the student, Kiyoo Wadati, rushed outside as a second tremor shook the building. Being on high ground, the effects of the earthquake were far less severe than down in the city, where fires were breaking out in the mainly timber buildings, or in the nearby port of Yokohama, which was practically destroyed by the shaking. Kiyoo had just experienced the great Kanto earthquake, the most devastating shock to hit the island of Honshu prior to the 2011 Tohoku disaster. Although of lesser magnitude[5] (7.9 versus 9.0 for the more recent shock), the 1923 quake resulted in far greater loss of life (142,800 deaths versus 15,884), partly as a result of firestorms.[6] During the next

[3] Turner (1922).

[4] Jeffreys (1928).

[5] Calculated using something known as the centroid moment tensor, which we won't bother about just now, and not the Richter scale, which was superseded decades ago.

[6] Thousands more were killed in anti-Korean violence following the quake. Overall, the death toll was about the same as that achieved by Enola Gay's 'Little Boy', twenty-two years later.

few days, Kiyoo walked through the ruins of the city, surveying the wreckage. The experience encouraged him to pursue a career in seismology in order to understand the causes of such catastrophes.

Two years later, Kiyoo joined the Central Meteorological Observatory of Japan, where he was given the rather mundane task of calculating the locations of earthquakes from seismological reports sent by local recording stations. Within two months, he experienced another tremor—a large aftershock of a quake that had taken the lives of another 40 of Tokyo's citizens. The characteristics of this shock, particularly its unusually great depth, piqued his interest in such events. Towards the end of his life, Kiyoo wrote, 'On looking back now, I realize that this was a memorable earthquake. Not only did I feel it with my body, but I also noticed its characteristic nature, and this led me to investigate deep-focus earthquakes.' At that time, Japan had the densest network of seismographs in the world, with which it hoped to develop a method for accurate earthquake prediction, an elusive holy grail still sought today. This provided Kiyoo with all the data he needed to draw some surprising conclusions.

The events recorded beneath Japan, Kiyoo found, were of two basic types. In most cases, surface waves were large and destructive right above the earthquake source, but died out rapidly with distance from the epicentre. Recordings of such events typically showed a small delay between P and S arrivals near the epicentre, the gap increasing rapidly for stations further afield. Some events, however, were quite different: surface waves were smaller and less destructive at the epicentre, and they died away much more slowly with distance. In these cases, the P–S delay was quite large, even at the epicentre, increasing only gradually with distance. This second group, roughly 20 per cent of the total, had to be much deeper than the first, supporting Turner's contention that quakes could be triggered at depths of hundreds of kilometres. Moreover, deep events tended to occur, Kiyoo found, not beneath the islands of Japan as did the shallow quakes, but to the west, beneath the Japan Sea. In fact, there seemed to be a trend, with the epicentres of the shallowest quakes lying to the east, the deepest to the west, defining a surface that, in his words, 'extends slopewise in the crust near the Japanese Islands'.

Now on a roll, Kiyoo published a series of papers between 1928 and 1935 in which, among other important observations, he remarked on the close relationship between the epicentres of deep earthquakes and the locations of active volcanoes. Showing admirable foresight, he noted that 'This tendency seems to be observable in many volcanic regions in the world. Of course, we cannot say decisively but if the theory of continental drift suggested by A. Wegener be true, we may perhaps be able to see its traces of the continental displacement in the neighbourhood of Japan.'[7] He also drew attention to 'regions where the P-wave arrived sooner than normal' in eastern Japan, which were associated with abnormally high values of gravity. This observation would turn out to play a central role in geophysical research in the next century.

Gutenberg's Bible

Ernest Rutherford, the man who discovered the proton and, for an encore, went on to split the atom, once pronounced that 'all science is either physics or stamp collecting'. The history of plate tectonics proves that even stamp collectors can make important contributions to science, good examples being Beno Gutenberg and Charles Richter (yes, that Richter), working at the California Institute of Technology, where, for many years during the mid-twentieth century, they collected, not stamps, but earthquakes.[8] That is, they kept what would these days be called a 'database' of recorded quakes, including estimates of their longitude, latitude and depth, and of their size, using the magnitude scale devised by Charles. The amount of work involved in this undertaking in the days before electronic computers, involving tables, charts, and manual calculations, is hard to imagine.

At intervals, they published their seismic stamp album, beginning in 1941 with what became known, inevitably, as the 'bible of seismicity'.[9] Even this early compilation contained significant observations prefiguring plate tectonics. For example, 'The seismic zones divide up the entire surface of the globe into blocks, the interiors of which are relatively nonseismic', and, like Kiyoo Wadati before them, they noted that, beneath Japan, 'The foci of deep and intermediate shocks appear to fall

[7] Wadati (1935).

[8] Of course, both scientists made major contributions to the subject beyond this. Nor were they the first such seismic stamp-collectors, as the distinguished seismologist Keith Bullen noted in his classic text on the subject (Bullen, 1963): it appears that 'Turner and Miss Bellamy' previously found time to catalogue seismic disturbances in the early twentieth century, noting that their geographical distribution was far from random.

[9] Gutenberg and Richter (1941). The 'bible' reference is, of course, to the famous fifteenth-century Gutenberg Bible, the first major work to be set in moveable type.

on a single smooth surface dipping at about 30 degrees under the continent'. Around 80 per cent of the energy liberated by quakes each year, they observed, was released in a narrow belt surrounding the Pacific Ocean, while most of the rest occurred in a second belt passing through the Alps and Himalayas. Outside of these zones, quakes were smaller, much less frequent, and occurred only in the uppermost layers of the Earth at depths less than about 30 km, apparently supporting Jeffreys' assertion that only these layers were sufficiently strong to allow the brittle deformation responsible for earthquakes. Within the main earthquake belts, however, seismic events in the mantle as deep as 700 km were recorded, in defiance of Harold's dogma. As the number of seismic recording stations increased and techniques improved, it became clear that a third group of epicentres could also be distinguished: small, shallow, events following the line of the mid-ocean ridges, which, in turn, were coming into sharper focus as new and more accurate bathymetric charts were generated from improved echo-sounder data collected by the expanding fleet of research ships.

It is fascinating to read Gutenberg and Richter's reports now, noting how the outlines of the plates gradually emerged as the quantity and accuracy of epicentres increased, even though the scientists concerned were unaware of the profound implications of their observations. The renowned seismologist Jack Oliver, working at the Lamont Geological Observatory[10] near New York, noted in his account[11] of developments in the 1960s that 'The successful merger of the plate tectonics model and the global seismicity patterns was on the one hand satisfying and on the other hand humbling. Our egos were alternately inflated and deflated as we found that details and features of the seismicity revealed important information on tectonics but at the same time called attention to the fact that some of those observations had been available to us earlier and ignored or overlooked.' This quotation is remarkable not only for its candour, but also for the correct use of the word 'alternately' by an American.

Once again, the US military played a major part in data acquisition. Motivated less by curiosity about the deep structure of the Earth than by a desire to detect Soviet nuclear weapons tests, the Defense Advanced Research Projects Agency (DARPA)—the same agency that was, by the

[10] Which, by now, had become the Lamont-Doherty Earth Observatory.
[11] Oliver (1996).

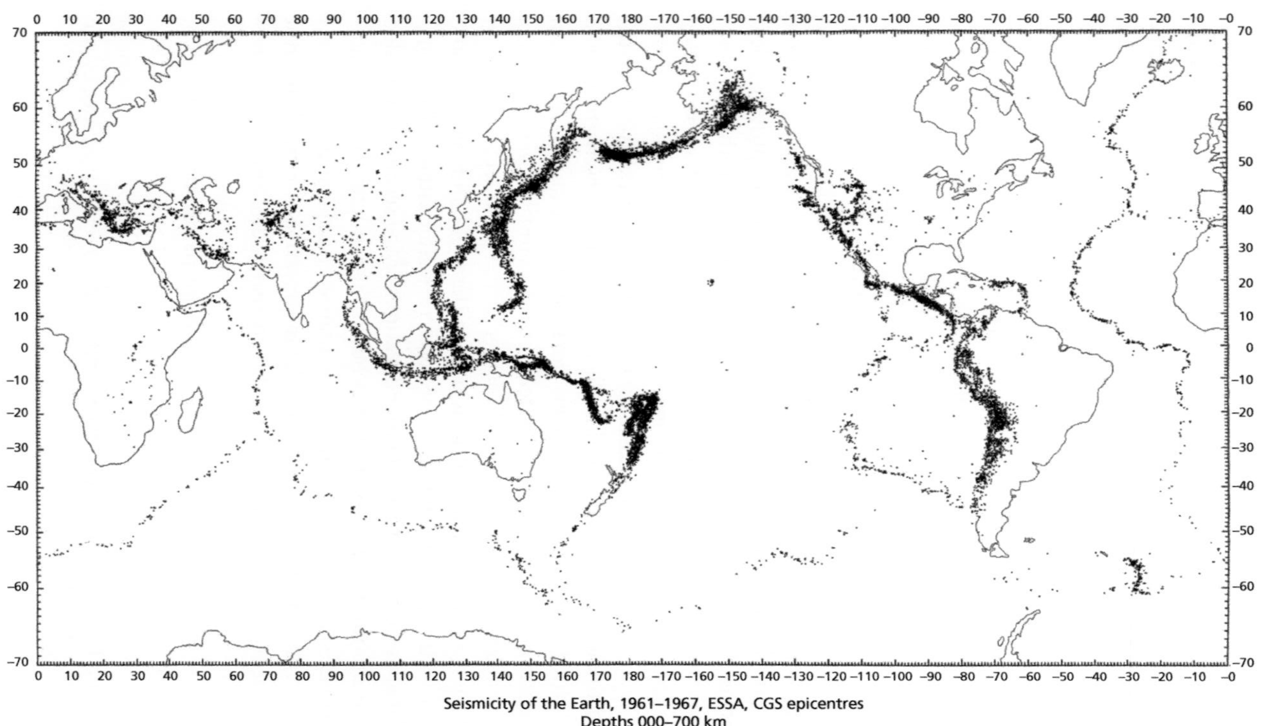

Figure 4.1 The plates defined by earthquake epicentres (Barazangi & Dorman, 1969).

early 1960s, developing satellite navigation and later invented the Internet[12]—decided to set up the World-Wide Standard Seismograph Network (WWSSN), to be installed and managed by the US Coast and Geodetic Survey. The WWSSN was exactly that—a global network of seismometers built to common standards, thus addressing some of the chief sources of error in earthquake location and measurement. By 1969, the new network had greatly improved both the quality and quantity of seismic recordings, enabling geophysicists Muawia Barazangi and James Dorman at the Lamont Geological Observatory to publish new epicentre maps that left no doubt about the geographical arrangement of the plates (Figure 4.1).

The Diving Dutchman

Now for some jargon. No apologies—here is a word you need to know. The word is 'subduction', a term introduced more than a decade before plate tectonics by the Swiss geologist André Amstutz.[13] As he defined it, the term had a much more restricted meaning than it does today, referring to the movement of rocks within mountain ranges in response to compressive forces. The existence of zones of compression within the Earth's crust had been suggested as early as 1906 by one Otto Ampferer, an Austrian geologist who proposed the term 'verschluckungs-zone' for such occurrences. Mercifully, his suggestion became lost in the geological literature.

Recognition of subduction as a phenomenon of global importance began with measurements of the gravity field. As long ago as the 1920s, one pioneering scientist developed a gravity-measuring apparatus based on an out-of-phase pair of swinging pendulums. That scientist was Felix Vening Meinesz, son of the Mayor of Amsterdam and a graduate in civil engineering from Delft University. Working for the Netherlands State Committee for Levelling and Arc Measurements,[14] he made successful measurements of gravity at various sites in Holland, following which he decided to try his luck at sea. This presented a huge

[12] Or 'ARPANET', as it was originally known.

[13] See White et al. (1970).

[14] Despite being recruited on a temporary basis, Felix remained in this post for many years, allowing Nicolaas Vlaar (1989) to observe, amusingly, that 'Vening Meinesz never did escape from the Earth's gravity field'.

problem, however, as a ship is scarcely a stable platform on which to make such delicate measurements. The biggest problem was waves, and the only way he could see to get around this was to take his device beneath them. Now it may come as something of a surprise to learn that the Dutch navy was building operational submarines as early as 1915 (no offence to the Dutch), but it was in one of these early vessels, known (like the mountain) as *K II* (the 'K' standing for 'kolonien', i.e. the colonies), that Felix set off in 1923 to try out his pendulum apparatus in the Dutch East Indies. The expedition was not, however, merely an indulgence: it had an important scientific objective, which was to use measured variations in the strength of gravity to determine the precise shape of the Earth, then only poorly known.

K II was built to a design by the Scottish firm of William Denny and powered by two German diesel engines, the delivery of which had been delayed by several years owing to the First World War. At 57 m in length, it provided a somewhat cramped home for the 6′6″ (2 m) tall Vening Meinesz and his 29 crewmates. Crucially, it was capable of diving to a depth of 40 m, deep enough to get Felix's apparatus away from the effects of the waves and achieve a precision of a few millionths of the total gravity field, not far short of modern shipborne gravimeters. Felix endured a series of later expeditions on other submarines, including one of 20,000 miles aboard *K XIII* that eclipsed even Captain Nemo's fictional exploits. During an eight-month transit westward, this vessel passed through the Panama Canal and crossed the Pacific Ocean in order to measure gravity over a range of known features, including the Mid-Atlantic Ridge and East Pacific Rise. Felix was, ever after, known by the soubriquet 'the Diving Dutchman'.

Most people are dimly aware that gravity has a value of nine point something. In fact, it's nine point something *metres per second, per second.* In other words, it is actually an acceleration. Isaac Newton's second law of motion claims that, to accelerate a given object, you need to apply a force equal to the mass of the object times the acceleration you are looking for. When the acceleration is due to gravity, which we'll call 'g', the force is known as the 'weight' of the object. The value of g varies with latitude, from about 9.78 m s^{-2} at the equator to about 9.83 m s^{-2} at the poles, owing to the rotation of the planet, and also to variations in distance from the centre of the Earth as a result of surface topography and the planet's non-spherical shape. These are things you can correct for and, like the nine in 'nine point something', are of little interest except to geodesists and astronauts. If you subtract the expected value of g at any location from the value you measure, you get the 'gravity anomaly'.[15] An anomaly close to zero tells you all is well, but departures of tens or even hundreds of milligals[16] point to something unusual. A positive anomaly of, say, 30 milligals in the middle of a continent might, for example, convince a mineral prospector that he has found a large buried ore-body, and that it is time to break out the champagne. In south-west England, the famous Cornubian granite is accompanied by negative anomalies of around −20 milligals, owing to the slightly lower density of granite compared with other crustal rocks.

By one of those wonderful strokes of fate, the Dutch East Indies, or Indonesia, as we now know it, turned out to be just about the most spectacular place on the planet in which to make gravity measurements at sea. Having corrected his readings for the values expected in that part of the world, Vening Meinesz was left with some of the largest anomalies on Earth. They were mostly negative, and often as low as −150 milligals, occasionally exceeding −200 milligals. Fortunately for Felix, the new Dutch naval vessel *Willebrord Snellius* made more than 33,000 new depth soundings in the East Indies during 1929 and 1930, allowing him to relate his giant anomalies to the equally spectacular bottom topography. The negative gravity anomalies, it soon became clear, were associated with deep ocean trenches, thousands of kilometres long and up to 7800 m deep, roughly twice the depth of the surrounding sea floor, thus providing a ready explanation: seawater is, after all, a lot less dense than crustal rocks. But simple calculations showed that even such deep trenches were insufficient to account for the huge anomalies.

Another explanation, Felix suggested, could be the same one that had been used to account for negative anomalies over mountain ranges: the low-density crust was thickened into a kind of root that descended into the mantle. This 'downbuckling',[17] as he described it, involved the upper 25 km of the crust, which might, he surmised, be forced down to a depth of 50 km or more. He explained downbuckling in terms of

15 For reasons we won't go into, this is also known as the 'free air' anomaly.

16 A milligal is, of course, one-thousandth of a Gal. A Gal is a unit of acceleration named after Galileo, equal to 1 centimetre per second, per second, or about one-thousandth of the total acceleration due to gravity on Earth. A special SI derived unit was introduced for gravity work but, being highly conservative folk, geodesists ignored it.

17 Also given the ugly name of 'tectogene'. This was a time of ugly names in geology.

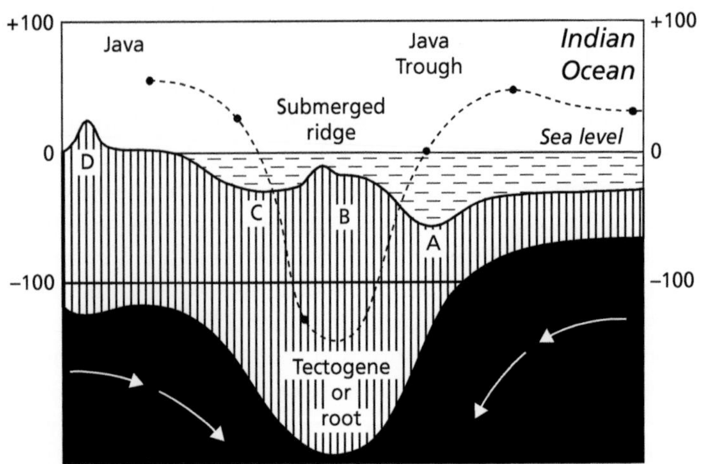

Figure 4.2 Interpretation of the Java trench by Felix Vening Meinesz. The dashed line indicates the isostatic gravity anomaly (in milligals), and white arrows show converging convection currents that result in 'down-buckling', producing a crustal root or 'tectogene'. [From Holmes (1965).]

horizontal compression, deforming the crust into waves, one of which eventually collapsed to produce a downward-pointing fold (see Figure 4.2). It was already known that the mantle was capable of flowing on longer timescales, and a theory known as 'isostasy' had been erected to explain how the crust 'floated' on the solid, yet fluid,[18] substratum. The increased thickness of low-density crust, plus the mass deficiency of the trench, explained the huge gravity anomalies, but replaced one conundrum with another: why did the low-density crust beneath the trench not simply pop up like a cork, or at least like the European crust following the ice age? Here Felix resorted to the geologist's old standby, invoking 'subcrustal convection currents' that produced the compression and also held the thickened crust down.

Having made such surprising discoveries in the East Indies, it was only logical that Felix should repeat the exercise in the *West* Indies, which he did between 1926 and 1937, using Dutch and US submarines in collaboration with American colleagues, notably Harry Hess at Princeton, who, in 1938, described Vening Meinesz's discovery of huge negative anomalies as 'probably the most important contribution to knowledge

[18] This oxymoron will be resolved in Chapter 5.

of the nature of mountain building made in this century'.[19] Yet again, the US Navy came up trumps, not only supplying submarines, such as the USS *Barracuda*, but also providing numerous new soundings via the Naval Hydrographic Office. A very similar pattern of negative anomalies and ocean trenches was revealed: a linear zone of large negative gravity anomalies stretched from Hispaniola, around the Lesser Antilles, and then west as far as northern Venezuela. Deep trenches followed the same path, leaving no doubt that the gravity lows and submarine topography were closely related.

In both the East and West Indies, the trenches contained ridges composed of deformed sedimentary rocks that often rose above sea level, forming islands such as Timor, Nias, Barbados, and Trinidad. Parallel to the long negative gravity anomalies were linear belts of gravity *highs*, corresponding to strings of volcanic islands such as Java, Sumatra, and the Lesser Antilles. These strings tended to be curved when viewed on a map and were therefore referred to as 'island arcs'. Seaward, the trenches were flanked by another strip of gravity highs, the interpretation of which would become clear only much later. Each of these phenomena, the island arcs, the deep trenches, the piles of sediments within the trenches, and the outer gravity highs, would turn out to be of fundamental significance in plate tectonics.

Hess went on to collaborate with the US Navy in the production of new bathymetric charts of the western Pacific, stimulated by global events between 1941 and 1945, during which many naval ships collected new echo-sounder data in regions far from shipping lanes. In publishing these charts, the US Navy once again provided a catalyst for scientific advances, and, in a paper written to accompany the charts, Hess established the existence of a chain of deep trenches from Japan to Palau and the Philippines, including some of the greatest depths in the oceans.

Finding Fault

The next major step was taken by another Caltech seismologist, the son of nineteenth-century Russian and Swedish migrants to the USA who, somewhere along the line, had developed an admiration for the works of a famous French writer, appropriating his name for their firstborn.

[19] At this time, of course, the century was still young, but Hess repeated this view in 1956.

He wisely dropped the leading part of his given name, 'Victor', emerging as, simply, 'Hugo Benioff'. For many years, Hugo worked for the Baldwin Piano Company, where he developed electric versions of musical instruments, including a cello, a violin, and a piano. If only he had thought of a guitar![20] But it was a different kind of vibration that brought him immortality—earthquakes. And not just any old earthquakes: Hugo's interest was in the largest and most destructive quakes on Earth: the ones that occur in the two active belts identified in the Gutenberg bible, one around the Pacific Ocean and the other in Southeast Asia.

The Tonga–Kermadec region in the western Pacific (between Samoa and New Zealand) experiences more deep earthquakes than any other volcanic arc, and more quakes deeper than 300 km than all other regions put together.[21] In 1949, Benioff contributed a paper to the *Bulletin of the Geological Society of America*, in which he made use of earthquake data from the Tonga–Kermadec trench, published that same year by his colleagues, the seismic philatelists Gutenberg and Richter. When he plotted the hypocentres[22] of these events on a cross-section of the trench zone, he found that they tended to align themselves along a steeply sloping path, about 60° from the horizontal. This, of course, was exactly what Kiyoo Wadati had found for quakes beneath Japan 25 years earlier,[23] and, it seemed, similarly sloping planes of hypocentres were to be found beneath all of the world's deep-sea trenches. At the time, Kiyoo's work had been overlooked in the West, and these zones of earthquake foci became known as 'Benioff zones'. Later, during the 1970s, the offence was recognized and corrected, and the dipping planes renamed 'Wadati–Benioff zones'.

Hugo interpreted his eponymous zones as 'great faults' (not to be confused with Jason Morgan's great faults, which were unknown at the time), unlike the folds or 'downbuckles' of Vening Meinesz's hypothesis. The scale of the proposed dip–slip faults was larger than anything

[20] Ironically, the Baldwin Piano Company was taken over in 2001 by the Gibson Guitar Corporation.

[21] Frohlich (2006).

[22] 'Epicentre' is the point vertically above the earthquake focus, 'hypocentre' is the actual location of the quake.

[23] Benioff cites only a limited number of references in his reports, not including Wadati's work. However, he does cite a Tohoku University report on the mechanisms of earthquakes in Japan, showing that he was aware of the important work done in Japan.

known previously, either on land or in the oceans, extending laterally for over 2500 km in the case of Tonga–Kermadec, and penetrating to depths of over 650 km. The sense of motion was always that of a 'reversed fault' or 'thrust'; that is, the rocks above the dipping fault slipped *up* the fault plane with respect to those below. When traced 'upslope', the Benioff zone of hypocentres invariably intersected the ocean floor at the steep inner slope of the trench, as others had noted previously, giving a strong hint as to the origin of both. Hugo pointed out that the vertical movement of the sea bed associated with large shocks on this giant fault plane would generate the destructive seismic sea waves known as 'tsunamis' that periodically inundate Southeast Asia. His explanation for the existence of these giant faults involved compression between continental and oceanic crust on either side of an initially vertical boundary or fault. As the fault evolved and rotated, depression of the oceanic 'block' created the deep oceanic trench that invariably accompanied the seismic zones. His interpretation turned out to be wrong, but the giant faults were real enough.

Going Down

Serendipitously, as Jack Oliver put it,[24] the seismologists at the Lamont Geological Observatory undertook something known as the 'Tonga–Fiji deep earthquake project'. Beginning in 1964, seismometers were installed on the islands of Tonga and Fiji to record earthquake activity, focusing on quakes deeper than 30 km. The project was not, as you might expect, designed to develop the exciting new theory of the Earth by following up the earlier work of Hugo Benioff, but was simply 'rooted in observation of the unknown'. Yet, by choosing this remote (to the USA) stretch of the western Pacific, they did more than just develop the theory—they made what was, perhaps, the most incredible discovery of all.

Jack Oliver and colleague Bryan Isacks interpreted the new data using electronic computers to fix locations with much higher precision than previously. From their results, it now became clear that the Wadati–Benioff earthquake zone was thinner than previously supposed (around 20 km, as it later transpired) and, as Kiyoo Wadati had found previously for Japan, seismic waves passing immediately beneath this zone travelled slightly faster than through the surrounding mantle.

[24] Oliver (1996).

But what really caught the eye was the size of the P and S body waves that arrived at the seismographs. These were tens to hundreds of times bigger than for body waves traversing adjacent mantle. To put it in a more scientifically correct way, the high attenuation normally expected for seismic waves passing through typical mantle was not observed in the Wadati–Benioff zone.

In a surprisingly cautious paper, considering its momentous content, Oliver and Isacks described an 'anomalous zone' about 100 km thick, with the seismic (i.e. Wadati–Benioff) zone forming just the uppermost 20 km or so, pointing out the contradiction between these proposed zones and old seismic models of the Earth consisting of concentric homogeneous shells. They included a cartoon showing the lithosphere of the Pacific plate bending downward beneath the trench and volcanic arc, and descending into the mantle (Figure 4.3), an idea that came to Jack during informal discussions with Bryan.[25] They were seriously suggesting that the plate forming the floor of the Pacific Ocean was plunging into a deep trench and disappearing into the mantle, its descent tracked for the first 650 km or so by the Wadati–Benioff zone. According to their interpretation, Benioff's giant fault zones were much more than faults: they represented the downward movement of vast sheets of rock—the 'surface' plates were foundering bodily into the Earth's interior!

You have to wonder how it was that a story that makes the sinking of the Titanic seem like a few ripples in the baby's bathwater was never communicated to the world in the way that, say, the putative detection of a Higgs boson was trumpeted a few years ago. The failure was, perhaps, symptomatic of the general inability of Earth scientists to extract such exciting advances from the boring detail and present them in a

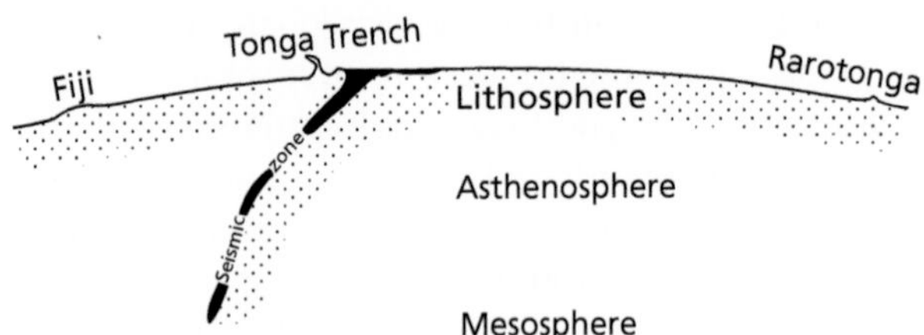

Figure 4.3 Cartoon showing lithosphere bending down into the mantle (mesosphere). [From Oliver and Isacks (1967).]

[25] Oliver (2003).

manner that might appeal to everyone else—a communications breakdown that persists today.

Less than six months after Oliver and Isacks' paper was published, Hugo Benioff succumbed to a heart attack, but those months were just long enough for him to appreciate the resolution of the problem of dipping earthquake zones. Oliver, on the other hand, was still working as recently as 2011, when, at 87, he suddenly announced in an unconcerned sort of way, 'I'm done', and expired.[26] Curiously, while the discovery itself was so grievously underplayed by the press, Oliver's death was reported in the *New York Times* under a slightly exaggerated headline that described him as the man 'who proved continental drift'.

Studies of the seismic wave beam patterns of these earthquakes, using the approach employed previously by Lynn Sykes to confirm the existence of transform faults, soon showed that Wadati–Benioff zones were not quite as simple as they seemed. Whereas shallower quakes near the trench could be explained simply by the grinding or 'shear' of one plate against another, shocks produced at depths greater than 100 km did not fit this simple-minded picture.[27] In fact, these events seemed to occur *within* the downgoing plate rather than at the interface between the two plates, and their slip vectors were not parallel to the Wadati–Benioff Zone. Their mechanisms reflected stresses within the slab rather than the sliding of one plate past the other, suggesting that plates were being stretched as they entered the subduction zone, but then encountered resistance to sinking at greater depth, resulting in compression. A jump in seismic wave speeds within the mantle at depths of around 660 km, observed worldwide, had been attributed to the transformation of mantle minerals to a more compact and dense structure under the influence of increasing confining pressure. The denser material below was very likely the cause of the additional resistance to sinking, and could also explain why no earthquake foci deeper than 700 km had been observed.

The Lamont boys had used deep earthquakes to draw the startling conclusion that Wadati–Benioff zones were not merely (merely!) giant thrust faults, but marked the descent of entire plates into what popular science writers like to call the 'roiling interior' of the planet.[28] But that

[26] Molnar (2011).

[27] Isacks, Oliver, and Sykes (1968).

[28] The term 'subduction' did not actually appear in the papers published by the Lamont group at this time. They referred to 'underthrusting', while others talked of

was not all. There were still more conclusions to be drawn, including one that would turn the nascent theory of plate tectonics on its head, so to speak. It was expressed succinctly in the one-line abstract of a *Science* article written by Bryan Isacks and Peter Molnar[29] in 1969: 'Downgoing Slabs of Lithosphere May Exert a Pull on the Portions of Plate Left at the Surface.' Could it be that a driving force for plate motions had been found in these sinking plates? Could it be that the 'immersed part of the crust tends to pull down with it the horizontal part to which it is solidly attached'? If that last quotation seems a little quaint in its syntax, then that is because it was written way back in 1923 by Harvard geologist, Reginald Daly.[30] Given the general opposition to 'mobilism' at the time, together with his complete ignorance of the plates, it seems almost miraculous that Reginald could have come so close to the concept of subduction so long before anybody else. The perspicacious Daly also recognized that, once started, sinking of oceanic crust is likely to continue, owing to its excess density, whereas foundering of the lighter, granitic, continental crust was unlikely. Hence, contrary to çontemporary belief, plates were not so much *pushed* apart at mid-ocean ridges, as *pulled* along in the wake of the sinking slab. Presumably this was not the only driving force, since some plates had no subduction boundaries yet still managed to move, albeit slowly, but it did go a long way towards an explanation of plate movements in the fixed-hotspots frame of 'absolute' motion. The subject of plate driving forces became a major topic in second-generation plate tectonics, as we will see in Part II.

The ultimate fate of subducting slabs was also to become a central question in the next generation of plate tectonics. In the early days of the theory, it was generally assumed that plates met their doom at subduction zones, being somehow recycled into the mantle whence they came. Nani Toksoz, a seismologist at the Massachusetts Institute of Technology, wrote in 1975: 'As the formerly rigid plate descends it slowly heats up, and over a period of millions of years it is absorbed into the general circulation of the earth's mantle.' Toksoz also calculated

---

'underflow'. 'Subduction' entered common usage around 1970, having been introduced (or rather re-introduced) by Esso research geologists at a 1969 Penrose Conference of the Geological Society of America at Pacific Grove in California. See White et al. (1970).

[29] Another Lamont geophysicist. By the way, the laboratory was now known as the Lamont-Doherty Geological Observatory.

[30] Daly (1923).

that, if the overall global rate of subduction has remained steady, then, since Pangea first began to fragment in the Mesozoic, an area equal to the entire surface of the Earth has been subducted into the mantle. Hence, there could be more plate material in the mantle than exists at the surface. As we shall see much later, some of these previously hidden plates are now being revealed in much the same way as brains and other human organs.

Out of the Arc

That volcanic arcs were generally associated with deep-sea trenches was remarked by Felix Vening Meinesz back in the1930s, while, in the Gutenberg bible, it was stated that 'There is regular association, with notable exceptions, of earthquakes at various depths with volcanoes, gravity anomalies, and oceanic troughs'. The association of major eruptions and great earthquakes could now be explained by the recognition of subduction as the process responsible for both phenomena.

While the plate doing the subducting is always oceanic, the overriding plate could be either oceanic or continental. In the former case, a line of volcanic islands—an island arc—is created, such as the Lesser Antilles or Tonga arcs. In the latter case, an active continental margin, like the Andean coast of South America, results. In both settings, the range of lavas erupted is much greater than that emerging at mid-ocean ridges, and, while the characteristic lava type of a mid-ocean ridge is basalt (or MORB), that of a volcanic arc is known as andesite, a rock with more silicon and less magnesium than MORB. In fact, all these rocks contain a lot of silicon, usually referred to by geochemists in terms of its oxide, silica (SiO_2). Even basalts, at the low end of the spectrum, contain around 50 per cent silica, while lavas known as rhyolites are pumped up to over 69 per cent silica. Andesites are somewhere in between.

Andesite lavas typical of volcanic arcs are derived from primary basalts by a process in which early-formed crystals of dense minerals like olivine and pyroxene sink to the bottom of a magma chamber, leaving a magma depleted in elements like magnesium and iron, and enriched in silica, potassium, and a range of misfit elements that can't be incorporated into those minerals. As the formation and removal of crystals proceeds, so the remaining magma becomes more 'silicic' (i.e. the proportion of silica increases), erupting to produce basaltic andesite,

andesite, and finally dacite or rhyolite lavas. A whole series of different compositions can thus be generated from a single parental basaltic magma. The situation is complicated further by a tendency for magmas to mix before eruption, and for bits of the surrounding 'country rock' (you just can't avoid it) to be incorporated into the melt.

In 1950, Hisashi Kuno of Tokyo University noticed that the andesites of the Japan island arc occurred in two distinct series that he named 'pigeonitic' and 'hypersthenitic', after the characteristic pyroxene minerals they contained (pigeonite and hypersthene). Each series comprised rocks with a wide range of silica concentrations, from primitive basalts, via intermediate andesites, to highly evolved dacites, but they differed in their concentrations of certain elements, notably potassium and iron. The same two series were later observed at other volcanic arcs, such as the Melanesian, Central American, and Aleutian arcs, and were renamed, even more confusingly, as the 'calc-alkaline' and 'tholeiite'[31] series. In one of the worst examples of loose terminology in a subject notorious for ugly jargon, the definitions of these terms remain imprecise and even conflicting, producing what Richard Arculus[32] at the Australian National University (ANU) in Canberra, described as a 'classificatory mess'. For the sake of sanity, we'll just refer to them as the 'medium-K' (K = potassium) and 'low-K' series, respectively.

Ten years after Kuno's observations, his colleague, Arata Sugimura, noted lateral changes in the chemistry of rocks across the Japan arc. Low-K rocks occurred in the east, close to the Pacific Ocean, while the arc rocks further west had higher concentrations of potassium and lower concentrations of silica. This variation he ascribed to variations in the depth at which melting and magma generation had occurred. With the advent of plate tectonics, William Dickinson[33] at Stanford University related these lateral variations to the increase in the depth of the Wadati–Benioff zone. At first sight, it seems paradoxical that shoving a cold plate into the mantle would lead to melting and volcanism—surely, the sinking plate is even less likely to melt than the much hotter mantle around it? Yet this was the explanation settled on initially: the 'roiling interior' would cause the slab to heat up until it became soft and started to melt. The oceanic crust, constructed at mid-ocean ridges

[31] Pronounced 'tho-lee-ite'.
[32] Arculus (2003).
[33] Dickinson (1970).

from the lowest-melting-point fraction of the mantle, naturally began to melt at lower temperatures than mantle rocks. And if heat from the surrounding mantle were not enough to promote melting, then friction along the Wadati–Benioff zone would surely do the trick.[34] The resulting buoyant magma would rise to the surface to feed volcanoes, while the heated plate would gradually merge with the surrounding mantle, so that the words 'resorbed' and 'recycled' were commonly associated with subduction during those early years. This explanation would also neatly explain why earthquakes did not occur beneath about 650 km or so—this must be the depth at which the heated slab lost its rigidity and was no longer able to snap. Like Joe Cann's onion and Gabriel Oak's dog, this explanation was far too good to survive for long.

Unfortunately, while offering a neat solution to the geophysicists, this hypothesis caused problems for the geochemists. Lavas erupting in volcanic arcs tended to be andesites, but they included a wide range of other types, all of which appeared to be derived from primitive basaltic melts by the settling out of denser minerals. The primitive magma must have been produced, like MORB, by melting of mantle rock beneath the plates. Obviously, if you melt a rock completely, you get magma with an identical chemical composition—you melt basalt and you get a basaltic magma. Total melting is, however, unlikely, even in the upper mantle, since you need exceptionally high temperatures to completely melt a cold plate (or hot mantle, for that matter). Beneath mid-ocean ridges, melting of up to 20 per cent of mantle peridotite occurs owing to the reduction in pressure, and the minerals that tend to melt first are naturally those with the lowest melting points, resulting in a magma with the composition of basalt. In turn, *partial* melting of basalt rock, such as that constituting the ocean crust, tends to produce a composition more like andesite or granite, which should thus be the most 'primitive' compositions if the sinking slab did, indeed, melt.

However, during its holiday at the Earth's surface, the descending ocean plate had assimilated a great deal of water. Some of this was incorporated into the thick sediments that accumulated during perhaps 100 million years since its creation at a mid-ocean ridge, while some was incorporated into the crystal structure of rock minerals. As it descended beneath the arc, the slab was subjected to increasing pressures and temperatures, causing much of this water to be expelled into the surrounding

[34] Oxburgh and Turcotte (1968).

mantle. Water, even in low concentrations, was known[35] to have a dramatic effect on the melting point of rock minerals, greatly increasing the amount of melting at a given temperature. Thus water released from the sinking plate would act as a flux in the wedge-shaped region of mantle peridotite rock above the slab, producing large amounts of (partial) melt. The resulting basaltic magma, being of lower density than the surrounding rock, would rise buoyantly into the crust, forming magma chambers in which the denser minerals crystallized and sank, resulting in magmas with ever higher silica content. These magmas erupted periodically in spectacular style, owing to their high water and gas content and high viscosity, constructing large 'Fujisan' stratovolcanoes. Explosive volcanoes of this type are responsible for most of the world's apocalyptic eruptions, for example, Santorini, in the Mediterranean, which exploded 3600 years ago, much to the annoyance of the Minoans, whose civilization, it seems, was ended abruptly by the event,[36] and Toba, in Indonesia, which erupted around 73,000 years ago, leading to a possible 'bottleneck' in human evolution. More recent spectacular eruptions have included those at Mount St Helens in the Cascades and Pinatubo in the Philippines.

Having explained the 'volcanic' nature of volcanic arcs, the next question was 'why arcs?' It is not clear why—perhaps he was a keen player of table-tennis, or perhaps it was just something to think about in the bath—but, in 1968, Sir Charles Frank, a physicist at Bristol University, was considering just this problem when he came up with a brainwave. He dashed off a terse letter to *Nature*, in which he imagined the Earth as a punctured ping-pong ball, and noted that, if you were to push the surface downwards at any point, you would get a curved depression with a radius of curvature of half the angle of depression. Some arcs, he pointed out, had a radius of curvature of around 20°

[35] Ringwood (1974).

[36] This idea has been challenged by archaeologists, some arguing that Minoan civilization continued to thrive on Crete and elsewhere after the eruption on Thera. However, recent work suggests that the eruption was probably the largest of the past 10,000 years, and more devastating than previously supposed. In any case, the expulsion of 100 km^3 of lava and pumice together with large volumes of toxic gases, followed swiftly by multiple tsunamis, could not have been much fun for the poor old Minoans. Some folk have even argued that, by ending Minoan dominance in the region, the eruption opened the way for the rise of Greek culture upon which our western civilization is largely based.

while subducting slabs seemed to dip at around 45°, apparently fitting his hypothesis, which still appears in textbooks today. If you're struggling with the geometry involved, don't worry: Frank's explanation was faulty. For a start, arcs exhibit a wide range of curvature, from nearly straight, like the Tonga arc, to highly curved, like the Marianas, while sinking slabs descend at angles that vary widely along a single subduction zone. Peter Vogt, a geophysicist at the US Naval Oceanographic Office in Washington, D.C., made an alternative suggestion. He pointed out that 'aseismic ridges' of thickened (and thus more buoyant) crust occurred in the oceans. Where these ridges reached deep trenches, they tended to put up more of a fight than normal oceanic crust, and thus resisted sinking. This produced cusps along subducting margins, between which the trenches formed arcs.

Despite their variable curvature, volcanic arcs do exhibit remarkable similarities. For example, they generally lie between about 125 and 250 km from the associated trench, and the 'volcanic front' (i.e. the line along which volcanism commences), nearly always lies above the point at which the Wadati–Benioff zone of intermediate-depth earthquakes reaches about 100 km depth.[37] The latter, presumably, reflects the depth at which pressure and temperature conditions are sufficient to cause reactions that release the bound water from the sinking slab, promoting melting in the wedge of mantle above.

More Surprises

It won't have escaped your attention that the construction of volcanic arcs, involving (partial) melting of mantle rocks and the rise of the resulting magma towards the surface, implies the transfer of large amounts of material from the mantle to the crust. Nor did it escape the attention of that perceptive theoretician Tuzo Wilson, who, as long ago as 1950,[38] suggested that this could provide a mechanism for the growth of continents. The significance of this realization cannot be overstated. After all, the continental crust provides the real estate on which terrestrial life, including bipedal apes like us, depends, and its origin was a matter that had been debated for many years prior to the advent of plate tectonics. Of course, material is transferred upwards on an industrial

[37] Dickinson and Hatherton (1967).

[38] Wilson (1951).

scale at mid-ocean ridges, too, producing the basaltic oceanic crust that surfaces most of the planet, but which has, geologically speaking, a limited shelf-life of less than 200 million years. Processes at volcanic arcs, however, create andesitic crust similar in composition and, crucially, density, to continental crust—crust that will not sink into the mantle and thus remains at the Earth's surface, preserving a convenient record of the planet's history since shortly after its formation.

Continental crust is very different from what you might expect on a cooling rocky planet such as the Earth or Venus. In fact, it is so different that, as late as the mid-1960s, geologists were suggesting that it might have come from outer space![39] While Dan McKenzie and his friends were solving the problems of the construction of the world in the 1960s, geochemists in the antipodes were making headway on fundamental questions concerning the origin of the continents. At first sight, the answers appeared to lie more in the rapidly diminishing realm of traditional geology than that of plate tectonics. Work at the ANU during the 1960s and 1970s by S. R. Taylor, A. J. R. White, P. Jakeš, R. J. Arculus, and M. R Perfit (they were a very formal lot) showed that this was not so.

The ANU group found the same relationship of low-K and high-K rocks within the New Guinea–New Britain arc that Kuno and Sugimura had discovered in Japan. They interpreted the low-K series as the 'earliest manifestation of arc development',[40] noting that many island arcs, such as the Tonga, Izu–Bonin, and Scotia arcs, consisted mainly of low-K rocks. They proposed that such arcs are immature, and that, with time, the proportion of medium-K series rocks increases until it becomes dominant in mature arcs like the Andean arc. This trend could also be viewed as an evolution from an 'oceanic' to a 'continental' character, suggesting that, in the context of plate tectonics, subduction zones, and volcanic arcs in particular, are the 'factories' in which continental crust is manufactured. Pointing to the gross similarity between the chemical composition of andesites and that of the continental crust, S. R. Taylor concluded, in 1967, that 'Continental areas grow mainly by the addition of andesites and associated calc-alkaline rocks in orogenic areas'. At the time, there was a long-standing argument between those geologists who believed the continents had formed early in the life of the Earth and those who thought they had grown progressively over

[39] See, for example, Donn et al. (1965).
[40] Jakeš and White (1969).

time. Studies of the chemistry of crustal rocks were now beginning to show that slow and continual growth had occurred throughout geological time by the addition of andesitic rocks at 'orogenic belts', that is, volcanic arcs associated with subduction zones.[41]

Belts and Traces

Possibly the most arcane branch of geology is 'metamorphic petrology'. By studying minerals with curious names that occur within rocks that have been altered by changes in temperature and/or pressure, geologists are able to identify the conditions under which such rocks were altered. This may not sound all that exciting, but it provides valuable clues to the processes involved in subduction.

Back in Tokyo, in 1961, geologist Akiho Miyashiro noticed that metamorphic rocks forming the crust of Honshu constituted two distinct belts running parallel both to each other and to the offshore trench.[42] One of these contained minerals formed by *high* pressures but relatively *low* temperatures. This was the Sanbagawa belt, running from Kanto, the region in which the epicentre of the great 1923 earthquake had been located, down the southeastern margin of Honshu to the island of Kyushu. Before alteration, the rocks here had been oceanic types, including gabbros and serpentinites, as found in ophiolites, but were now characterised by the presence of a rather pretty blue amphibole mineral known as glaucophane.

Just inland, a second belt, the Ryoke–Abukuma belt, contained very different rocks that had formerly been dacites and rhyolites—volcanic rocks that occur mainly on the continents. Their metamorphic minerals formed at *high* temperatures but relatively *low* pressures. Akiho pointed out that similar pairs of belts occurred elsewhere, such as in California, where the Franciscan complex and Sierra Nevada were the equivalents of the Sanbagawa and Ryoke–Abukuma belts, respectively. In every case, the high-pressure oceanic 'Sanbagawa' belt lay on the oceanward side of the high-temperature 'Ryoke–Abukuma' belt. In Japan, the two belts had similar ages and appeared to have been formed in close proximity to one another, implying a high thermal gradient between the

[41] This did not, however, preclude a phase of differentiation early in the Earth's history, during which the ancient nuclei of the continents were formed.

[42] Miyashiro (1961).

two. Miyashiro's discovery of these paired metamorphic belts thus provided the link between processes active at volcanic arcs today and those at ancient subduction zones.

In 1973, Akiho[43] put his discovery into the context of plate tectonics by correlating the high-pressure Sanbagawa belt with the low thermal gradient of the trench region in which the oceanic crust of the down-going plate disappeared beneath the forearc. Here, the cool subducting plate was subjected to pressures as high as 1 gigapascal[44] as it entered the subduction zone, creating conditions for the formation of the pretty blue amphibole mineral characteristic of this belt. Further landward, the volcanic arc was a zone of thickened crust heated by magmatism, producing the high-temperature, low-pressure, conditions of the Ryoke–Abukuma belt. It seemed that the trench and arc represented by these belts had become inactive during the Cretaceous (around 80 million years ago) as subduction migrated further offshore to the present-day Japan Trench. Recognition of paired belts in other regions of the world thus provided evidence not only of past subduction, but also of the 'polarity' of the former subduction zone. That is, where parallel Sanbagawa and Ryoke–Abukuma belts could be recognized, the subducting plate must have been moving from the former towards the latter.

Push-ups

Having spawned the idea of horizontal compression and given birth to the term 'subduction', continental mountain belts such as the Alps and Himalayas were almost forgotten in the initial excitement of plate tectonics. As a consequence of their rejection of mobile continents, geologists had previously left themselves with only vertical movements to explain possibly the most baffling observation in geology: the presence of great thicknesses of deformed rock containing fossils of extinct sea creatures, outcropping within mountain ranges miles above sea level. The poor old fixists had, almost literally, dug themselves into a hole by assuming that these rocks must have been deposited as sediments in deep (or possibly shallow) subsiding troughs they called 'geosynclines'.[45] Some

[43] Miyashiro (1973).

[44] 1 gigapascal (GPa) equals 10 kilobars, or roughly the pressure exerted by 2500 Empire State Buildings.

[45] See Chapter 2. A good review of this theory, originally introduced by James Hall in 1857, was given by Dickinson (1971).

mysterious force then crumpled and uplifted them, forming the mountains we see today. Based on rocks mapped in the Appalachians, an inelegant terminology had been developed to encompass this ad hoc theory, including such terms as orthogeosyncline, eugeosyncline, miogeosyncline, leptogeosyncline, exogeosyncline, zeugogeosyncline, and more: an etymological orgy happily rendered obsolete by plate tectonics. Often linked to the equally barmy shrinking Earth hypothesis, the theory of geosynclines was, incredibly, the staple of conventional geology until the 1960s. Since the eighteenth century, the subject of geology had been founded on the mantra 'the present is the key to the past', known to geologists, who did not flinch at long words, as 'uniformitarianism'. How embarrassing then, that the central concept in their great theory, the geosyncline itself, was so notable by its absence anywhere on the surface of the present Earth.

Back in 1948, Harry Hess had noted[46] that Alpine belts and volcanic arcs might be equivalent, the former found in continental settings, the latter oceanic. There certainly are similarities between the two. For example, the Himalayas have an arc-like shape, and giant thrust faults mark the boundary between India and Asia. But there is no evidence for a deep Wadati–Benioff zone, and, in place of a string of active volcanoes, there is a vast region of over two million square kilometres, elevated to an altitude of over 5000 m—the Tibetan Plateau. Paleomagnetic studies of the Deccan Traps had already shown that India had migrated thousands of kilometres northward, implying extensive subduction in the Tethys Ocean, which had eventually been squeezed out of existence as India barged into Eurasia.[47] Arthur Holmes, in his 1965 *magnum opus*,[48] referred matter-of-factly to the northward movement of India, noting that seismic evidence suggested that it was continuing today. Once again, his interpretation, though predating plate tectonics, was close to the mark: 'One can picture a northerly extension of Peninsular India bending down beneath the depressed floor of the Indo-Gangetic Trough and so passing under Tibet, which has thus received a double thickness of continental crust and so become the world's loftiest plateau.'

In 1970, Lamont seismologist Thomas Fitch examined the limited catalogue of large earthquakes in the Himalayas, concluding that the

[46] Hess (1948).

[47] An interesting account of this is given in Dorrik Stow's book *Vanished Ocean: How Tethys Reshaped the World* (Stow, 2010).

[48] Holmes (1965).

focal mechanisms of four large quakes that had occurred at shallow depth near the Himalayan Front indicated thrust faulting on gently dipping fault planes, supporting Holmes' notion of continental 'underthrusting'. He also calculated slip vectors for these events indicating northward or north-eastward movement of India. Beneath the Tibetan Plateau, at that time inaccessible to western geologists, earthquakes were of varied types, some indicating extension within the plateau, others horizontal movement on strike-slip faults. This pattern was later interpreted by Paul Tapponnier and Peter Molnar, at the Massachusetts Institute of Technology, as resulting from the impact of the rigid Indian plate on a deformable Asia, a bit like pushing a piece of wood into Plasticine modelling clay[49]. Accepting that continental crust is too buoyant to sink into the mantle led to the inevitable conclusion that subduction at an active margin (like the west coast of Central and South America) continues until a second continent arrives at the subduction zone, following which all hell breaks loose, geologically speaking. The situation was summed up more soberly by geologists John Dewey and John Bird in 1970:[50] 'The arrival of a continental mass, with its continental margin sediments, at a trench results in collision and an orogen,[51] which may suture continents together.' Mountain ranges were, they observed, the result of collision between two continental margins, one of which was either a volcanic active margin, like the Andean margin of South America, or else bordered by an island arc—in other words a subduction zone. The other was like the margins of continents around the Atlantic, where both continent and adjacent ocean floor form part of the same plate, and are thus free of large earthquakes—in other words, a passive margin. Sediments deposited on the continental shelf and slope of the latter would be caught up in the collision, deformed, altered by heat and pressure, and elevated to form the mountain range.

The great thicknesses of marine sediments now deformed in mountain belts were not, as previously believed, deposited in sagging troughs or downwarps within continents or oceans, but simply represented sediments deposited on the continental shelf and slope of a passive margin, such as you would find off the Atlantic coasts of North America, Europe, or Africa. The mysterious crumpling force was simply the vice

[49] Tapponnier and Molnar (1977).
[50] Dewey and Bird (1970a).
[51] An orogen is a mountain range.

of a closing ocean that deformed and altered these rocks through excess heat and pressure, finally elevating them far above sea level.

Studies of Indian plate motion from sea-floor data collected in the Indian Ocean[52] showed a sudden drop in its formerly very rapid—about 150 mm per year—northward rate of movement around 55 million years ago, decelerating to 100 mm per year until 38 million years ago, with continued migration at a much lower rate of around 50 mm per year to the present day.[53] The collision, mighty as it was, had not stopped India in its tracks. The subcontinent had continued northward for another 2000 km, pushing up the Himalayas and driving beneath Asia in a kind of continental subduction, just as Holmes had noted. Any ocean trench had, of course, been erased by the collision of continents, destroying all the intervening ocean crust. Well, perhaps not all, for splinters of the oceanic plate had been preserved as ophiolites dotted along the suture zone. Further west, in the collision zone between Africa and Europe, some ocean crust remained *in situ* in the Mediterranean, while other splinters were festooned around the Middle East as ophiolites. Collisions such as this corresponded with stage six of Tuzo Wilson's ocean cycle, of which the Himalayas were the type example.

Afterthought

Finally, forty years after Felix Vening Meinesz first descended beneath the waves with his pendulum gravimeter, the penny had dropped. The Indonesian, or Sunda, arc (which, by the 1960s, was no longer a Dutch possession) could now be seen for what it was. The huge gravity anomalies and deep trenches marked the locations where the Indo-Australian plate arched its back, whale-like, and dived beneath South East Asia, its passage marked by a stream of earthquakes. Ridges of deformed rocks, emerging as the islands of Flores and Nias, were stacks of sediment scraped from the floor of the Indian Ocean as it entered the trench. To the north, the line of active and extinct volcanoes, including Sinebung, Merapi, Krakatoa, Tambora, and Toba, marked the zone of mantle

[52] Molnar and Tapponnier (1975).

[53] Later work suggested that the initial drop in rate occurred suddenly around 50 million years ago, after which motion was erratic at rates of less than 100 mm per year until 44 million years ago, when it dropped below 50 mm per year (Patrait and Achache, 1984).

melting induced by the release of water from the doomed plate, a hundred kilometres beneath. Christmas Island was part of the outer rise or hinge that emerged above sea level.

What was going on beneath deep-sea trenches was not 'downbuckling' of the crust, or even thrusting on a giant crustal fault, but a truly unbelievable phenomenon operating on a global scale, a phenomenon at which even the renowned geologist and pioneer drifter Sam Carey, who found little difficulty believing that the Earth had expanded like a balloon since the Jurassic, drew the line: *subduction*—regarded by Carey as 'a myth'.[54] A sane man might accept that the ground beneath his feet could, through some freak of nature, give way, or that entire houses occasionally collapse into sink holes, but who, in his right mind, would believe it possible that the floor of the deep ocean could simply founder and plunge into the planet's interior?

Yet, so far from being a myth, subduction is the process, above all others, that characterizes global tectonics on Earth, such that one of today's leading geophysicists[55] has suggested that 'subduction tectonics' would be a better name than 'plate tectonics', and you can see his point. Not only does the still poorly understood process of subduction drive the plates, generate most of world's large earthquakes and volcanoes, cool the mantle, and create the continental crust, we now know that it also plays a major role in the Earth 'system' (for example, in regulating the carbon cycle) that could not have been imagined by Amstutz and his contemporaries. Subduction thus represents recycling on a planetary scale, while, at the same time, providing the world's largest landfill.

[54] Unfortunately for Sam, establishment of this unlikely phenomenon as a major, if not *the* major, process on the Earth, rendered planetary expansion unnecessary, and much of his later scientific output redundant.

[55] Stern (2004).

5

Plate Tectonics by Creeps

> Our main need is for making the two-dimensional picture of surface observations three-dimensional by combining it with such knowledge as we have about the internal structure of the earth.
>
> Walter Elsasser (1968)

By now, you will have noticed a paradox. Earthquakes are caused primarily by the movement of plates. Plates move over the mantle, and the mantle is solid rock. So why aren't there earthquakes *beneath* the plates? If all plates are moving (and they are), there should be earthquakes just about everywhere, at a depth corresponding to the base of the plates. The answer is that while plate boundaries are the natural home of jerks, the mantle beneath the plates is the preserve of creeps. Except where cold slabs penetrate to depths of 650 km or more (producing the deep earthquakes beloved of Kiyoo Wadati and Hugo Benioff), the mantle moves only very gradually by *creep*, a phenomenon that occurs in many solids, including metals, when subjected to stress. Creep can be very inconvenient, particularly where a metal happens to form part of, say, a turbine blade in an aero engine. On the other hand, creep is the process that transformed the Earth into a living world.

*

Disturbing Matter

One of the most remarkable non-observations ever made was reported in the mid-eighteenth century by Pierre Bouguer,[1] who, by good fortune and nepotism, had assumed the position of professor of hydrography at Le Havre University in 1730. Between 1735 and 1745, Bouguer led (along with Ch. M. de La Condamine and L. Godin) an ill-tempered

[1] For non-Francophiles, pronounced 'boo-gay'. His famous book was entitled *La Figure de la Terre* (1749).

The Tectonic Plates Are Moving! Roy Livermore, Oxford University Press (2018). © Roy Livermore.
DOI: 10.1093/oso/9780198717867.001.0001

Franco-Spanish expedition to the Andes of Peru (now Ecuador) to measure the length of a meridian degree of latitude. The idea was to compare the lengths of arcs in different latitudes to distinguish between various suggestions for the Earth's shape—that is, to determine whether it was strictly spherical, oblate (flattened slightly at the poles, as Isaac Newton thought), or prolate (flattened slightly at the equator, as most French scientists believed). While there, Pierre calculated the gravitational attraction of Chimborazo, a dormant Andean stratovolcano, the summit of which, it emerged later, has the distinction of being the furthest point on the Earth's surface from the geocentre, by virtue of the equatorial bulge upon which it sits.

In his attempt to test Newton's ideas about gravity, Pierre carefully observed the deflection of a plumb-line both to the north and to the south of the volcanic cone. If it seems odd that a mere mountain could have a noticeable effect on gravity in the presence of a planet the size of the Earth, you have to remember that he was 3000 times closer to Chimborazo's centre of mass than he was to that of the Earth, and the effect declines rapidly—in proportion to the square of distance. For some strange reason, the deflection he measured was very much less than his calculations predicted, leading him to wonder whether there was not some deficiency of mass beneath the volcano. He later generalized his finding by stating that the gravitational attraction of the entire Andes 'is much smaller than that expected from the mass of matter represented in those mountains'.[2] The continuing squabbles between Bouguer and his colleagues make an entertaining story that we will, reluctantly, have to forego in the interests of brevity.[3]

Now, if ten years of your life seems like an excessive sacrifice for the dubious reward of knowing the length of an arc in some remote corner, then have sympathy for the Surveyor General of India a century later. George Everest[4] spent 25 years surveying arcs in India, from Cape Comorin in the south to the Himalayas in the north, a distance of over 2400 km, completing the 'Great Trigonometrical Survey' begun by William Lambton in 1806. During this prolonged and tedious survey, latitudes were fixed by two different methods. First, measurement by meticulous triangulation gave the 'geodetic latitude', based on distances

[2] See Daly (1940), p.36.

[3] If you are interested, see Ferreiro (2011).

[4] Pronounced '*eve-rest*'. George had spent several years in South Africa prior to working in India, honing his geodetic skills.

and angles determined using traditional methods of ground surveying. Second, direct measurement of the zenith using the stars and a plumb-line gave the 'astronomic latitude'. In just about every case, the two disagreed, the astronomic latitude being consistently lower (further south) than the geodetic. Moreover, the difference, known as the 'deflection of the vertical', increased as the survey migrated northward. The difference in geodetic latitude between stations bounding the northernmost arc of the survey between Kaliana and Kalianpur was 5° 23′42″, whereas the difference in astronomic latitude was 5° 23′37″, and five seconds of arc was a big difference that needed explaining. Everest incorrectly attributed it to errors in the triangulation and computation of the geodetic latitude, assuming the direct astronomical measurements must be correct.

The problem was soon addressed by the Venerable John Henry Pratt, M.A., Archdeacon of Calcutta.[5] In 1855, the Archdeacon wrote a contribution to the *Philosophical Transactions of the Royal Society*,[6] in which he challenged Everest's interpretation, suggesting that the discrepancy resulted from 'the attraction of the superficial matter which lies in such abundance on the north of the Indian Arc', or, in plain English,[7] the Himalayas. This 'disturbing cause', he noted, would tend to 'draw the lead of the plumb line at the northern extremity of the arc more to the north than at the southern extremity, which is more removed from the attracting mass'. The effect of this would be to lessen the difference in computed astronomic latitudes between the two stations, as observed. He presented detailed calculations, based on the known and estimated topography of northern India, demonstrating that the attraction of the Himalayas was, indeed, an important factor—more important, in fact, than he had originally suspected. If his sums were correct, then the expected difference in latitudes at the two stations should have been, not five seconds of arc, but fifteen. Why, therefore, was the effect only a third of that predicted using Isaac Newton's laws?

[5] I don't suppose there are many archdeacons conducting research in physics these days (indeed, many of us are uncertain what an archdeacon—or, for that matter, a deacon—actually does), but, in the mid-nineteenth century, it was not unusual for authors of scientific papers to sport religious or military titles. A single issue of the *Quarterly Journal of the Geological Society* for 1847 included no less than four Reverends, including the great Adam Sedgwick, as well as two Captains and a Lieutenant Spratt.

[6] The paper had been read before the Society in December 1854.

[7] Or, strictly, Sanskrit.

Another great Essex man,[8] George Biddell Airy, had the answer. Despite an impoverished start to life, George had, thanks to a supporting uncle and prodigious mathematical talent, navigated a course through the British class system to the highest academic offices,[9] including Plumian Professor of Astronomy at Cambridge, President of the Royal Society, and Astronomer Royal.[10] Not immune from the poor judgement that besmirched the careers of subsequent Plumians, George famously described Charles Babbage's plans for a 'computing engine' (to assist in calculating the positions of stars) as 'useless', and dismissed its proponent by saying 'I think it likely that he lives in a sort of dream as to its utility'. Regrettably, George's advice to the Government resulted in the scrapping of Babbage's groundbreaking Difference Engine No. 1 in 1842, although, to be fair, it had already cost British taxpayers as much as two new battleships.[11]

To return to the burning question of why it was that mountains failed to deliver on their promise when it came to gravitational attraction, Airy, on learning of Pratt's explanation, published a four-page article in the same issue of the *Philosophical Transactions* as the Archdeacon's original 1855 paper. Demonstrating a remarkable lack of tact, George began by saying 'it appears to me, not only is there nothing surprising in Archdeacon Pratt's conclusion, but that it ought to have been anticipated; and that, instead of expecting a positive effect of attraction of a large mountain mass upon a station at a considerable distance from it, we ought to be prepared to expect no effect whatever, or in some cases even a small negative effect'. He went on to elaborate his theory, explaining how such 'disturbing masses' as mountain ranges are 'compensated' by the presence of crustal 'roots' extending into the 'substratum'—a

[8] Being educated at Byatt Walker's School, Colchester, and Colchester Grammar School.

[9] Ariel and Berger, in their account of Airy's academic achievements in *Plotting the Globe* (Ariel and Berger, 2006), quote the 'affurism' of Josh Billings: 'poverty is the stepmother of genius'.

[10] Airy's alloyed brilliance was recognized by Geoffrey Davies, a distinguished geophysicist at ANU, who wrote, 'What I had never appreciated until I read Daly's extensive quotation from Airy's short paper was how penetrating and far-reaching was Airy's thinking' (Davies, 1999).

[11] A trial portion of the engine exhibited by the London Science Museum is now regarded as 'a celebrated icon in the pre-history of the computer'. Difference Engine No. 2 was constructed in 2002 and shown to work correctly.

theory that later became known as 'Airy isostasy'.[12] The essence of his theory was that the Earth's hard crust floats on a fluid substratum much as a berg floats on seawater: thicker ice rides higher, but also penetrates deeper into the supporting fluid. And, just as freshwater ice is less dense than seawater, so the continental crust is less dense than the fluid layer beneath, a layer that Airy referred to, not being a geologist, as 'lava'. He also added the important caveat that 'This fluidity may be very imperfect; it may be mere viscidity', opening the way to our present understanding of solid-state creep in the mantle. Applying his theory to a hypothetical 'table-land' (i.e. plateau), he went on, 'the disturbance [to the gravity field] depends on two actions; the positive attraction produced by the elevated table-land; and the diminution of attraction, or negative attraction, produced by the substitution of a certain volume of light crust [...] for heavy lava'. The result of one counteracting the other would be that the anomaly observed in gravitational attraction would be very small.

A few years later, Archdeacon Pratt responded to Airy's insults by declaring the Astronomer Royal's theory 'untenable' on three counts.[13] Firstly, he accused Airy of assuming an unrealistically thin crust in his calculations, claiming that the true thickness was 'about 800 or 1000 miles at least'. Of course, the continental crust, even at its thickest, is no more than 60 miles thick, so John was wrong on this count. Secondly, he claimed that the cool crust should be heavier than the substratum, also incorrectly, since the crust is not just solidified mantle but chemically differentiated from it (and significantly less dense). Thirdly, he noted that Airy's explanation involved crustal thinning where a 'hollow' (e.g. an ocean basin) existed at the surface, claiming that such thickness variations could not be produced by cooling. As twentieth-century geophysicists later showed, the oceanic crust is, in fact, very much thinner than the continents and has, indeed, formed by cooling.[14] The Archdeacon went on to propose an alternative mechanism by which the attraction of surface topography might be compensated. In his version, the crust was regarded as a series of narrow vertical

[12] The term 'isostasy' was introduced in 1889 by USGS geologist Clarence Dutton. Airy's model was first quantified by Weikko Aleksanteri Heiskanen, and so it is referred to by geodesists as the Airy–Heiskanen model of isostasy.

[13] Pratt (1859).

[14] Although not, perhaps, in the way imagined by nineteenth-century scientists.

prisms, extending from the ground surface to some unspecified 'depth of compensation'. The density of rocks in each column would be different, such that the mass in every column above the depth of compensation was equal. While this made the task of calculating the isostatic anomaly[15] much simpler, you did not have to be a geologist to realize just how unrealistic the idea was. It also appears that John believed that compensation had been achieved early in the life of the Earth and that the crust had remained static ever since. All in all, it seems that the Archdeacon was aptly named.

Weighty Arguments

While all this unpleasantness was going on, an Aberdeenshire farmer was busy investigating rocks in the valley of the Ythan river, close to his home in Ellon. In spite of day-jobs as a manager of a large estate[16] and a university lecturer on agriculture,[17] Thomas Jamieson found time to make extensive studies of the glacial geology of Scotland, including ancient rocks known to geologists as 'gneisses' that had been scored by the movement of glaciers. These rocks formed the basement for an overlying sequence of much younger strata, including marine clay, coarse gravel, and peat beds, demonstrating a wide variety of conditions during and after the last ice age. Jamieson pointed to the existence of old pebble beaches and fossil marine shells found halfway up the sides of hills at elevations of over 1000 feet in Britain, and over 600 feet in Scandinavia, as evidence that sea level had apparently fluctuated dramatically throughout this period. On the face of it, this was counter-intuitive, since the formation of large ice sheets could only be achieved by abstracting water wholesale from the oceans, resulting in a general *fall* in sea level. However, while global sea levels undoubtedly fell during the period of maximum ice advance, the record seemed to show that regions covered by thick ice followed a distinct pattern of subsidence and then uplift, while unglaciated areas remained largely unaffected.

In 1865, Jamieson published an article[18] in which he argued that the land surface of Scotland had indeed been depressed at the time of the

[15] That is, the difference between the observed attraction of gravity and that expected if the topography was fully compensated.

[16] This was the Ellon Castle estate.

[17] A job for which, incidentally, he was given a personal reference by Charles Darwin.

[18] Jamieson (1865).

most recent glaciation, confessing that 'It has occurred to me that the enormous weight of ice thrown upon the land may have had something to do with this depression. Agassiz[19] considers the ice to have been a mile thick in some parts of America; and everything points to a great thickness in Scandinavia and North Britain. We don't know what is the state of the matter on which the solid crust of the earth reposes. If it is in a state of fusion, a depression might take place from a cause of this kind, and then the melting of the ice would account for the rising of the land, which seems to have followed upon the decrease of the glaciers.' In this, Thomas was invoking a mechanism similar to that offered by George Airy a few years earlier, but operating on a much shorter timescale of millennia. In fact, considering the youth of the geological features involved, it was likely that the land surface was still rising.

Jamieson reckoned that an ice sheet 3000 feet thick would weigh around 2000 million tons per square mile,[20] a formidable load even for the Earth's crust. Some elastic behaviour—that is, downward bending—of the crust would be expected, but, beyond this, it would yield by means of fluid motions in the layer beneath, which he supposed might be composed of 'melted matter'. Echoing Airy and Pratt, he suggested that 'The surface is in a state of delicate equilibrium, and [...] any considerable transference of pressure will cause a re-adjustment of levels'. Also, like Airy and Pratt, he demonstrated remarkable insight into the mechanism behind isostatic compensation, noting that 'Bodies that seem absolutely rigid to pressure applied for a short space of time yield perceptibly to a force which is long continued'. Thus, he argued, the mantle or 'substratum' need not be either totally rigid nor molten: it could be solid, yet subject to flowage over thousands or millions of years. In a later article in the *Geological Magazine*, published in 1882,[21] Jamieson gathered evidence from around the world in support of these ideas, summing up his theory with admirable brevity: 'The land is first heavily loaded with ice, then it goes down and the ice clears off, after which the land comes up again.'

In that same year, 1882, the rector[22] of Harlton, a village near Cambridge, published a letter,[23] also in the *Geological Magazine*, in which he quoted

[19] That is, Louis Agassiz, the famous Swiss geologist.

[20] 7×10^8 tonnes km^{-2} in today's money.

[21] Jamieson (1882).

[22] A rector was a parish priest in the Church of England.

[23] Fisher (1882).

Jamieson's 1865 paper and calculated that, in order to depress the crust and raise shorelines to 700 feet (the present elevation of shell beds in Scandinavia), a thickness of 2310 feet of ice would have been required. The rector, the Reverend Osmond Fisher, a leading authority on the geology of East Anglia and author of a remarkable book on the physics of the Earth's crust,[24] also argued in 1881 that the 'crust is thin and floats in equilibrium upon a slightly denser substratum of molten rock, the elevations on the surface of the crust being due to compression, and being supported through flotation by corresponding protuberances (which I call "roots of the mountains") projecting downwards into the denser liquid—a mode of support long ago suggested by Sir G. B. Airy'.[25] Siding with the field geologists, he observed, scornfully, that 'With respect to the yielding of the crust, I think we cannot but lament, that mathematical physicists seem to ignore the phenomena upon which our science founds its conclusions, and, instead of seeking for an admissible hypothesis, the outcome of which, when submitted to calculation, might agree with the facts of geology, they assume one which is suited to the exigencies of some powerful method of analysis, and having obtained their result, on the strength of it bid bewildered geologists to disbelieve the evidence of their senses'.

This attack was directed chiefly at W. Thomson, LL.D., F.R.S., Professor of Natural Philosophy in the University of Glasgow, later 'Sir William Thomson' and, still later, 'Baron Kelvin, of Largs in the County of Ayr'.[26] Assuming that the Earth had cooled from a completely molten state early in its history, Thomson calculated the time it would take for the newly solidified planet to cool to its present temperature by the conduction of heat through its surface. This, he thought, would give an estimate of the age of the Earth. His calculation used assumptions about the unknown physical properties of material inside the Earth, in particular, its thermal conductivity, that is, the rate at which heat would flow through the rocks towards the surface.[27]

[24] *Physics of the Earth*, published in 1881, with a second edition in 1889. Osmond also published articles on rose-growing, and gave sermons that were, on occasion, highly antagonistic towards Rome.

[25] Fisher (1882).

[26] Or Lord Kelvin, as he is now known, and after whom the unit of absolute temperature is named in the Système International d'Unités.

[27] It is difficult for us now to appreciate the Victorian scientist's limited knowledge of such fundamental quantities as the Earth's age. For example, it was seriously considered

To any reasonable person, the numbers William came up with were big—vastly, hugely, mind-bogglingly big. The maximum age of the Earth, he pronounced confidently, was 100 million years (revised to 20–80 million years), give or take.[28] This was roughly 15,000 times longer than the theological estimates, and put paid to any ideas that Man had been around soon after the first Act of Creation. On the other hand, it was a grievous blow to geologists and natural historians. As William and his allies were keen to point out, it required much more time than this to turn jellyfish into Man,[29] and hence his estimate counted as an argument against biological evolution and in favour of supernatural beliefs in a creator. During his 1871 Presidential Address to the British Association for the Advancement of Science, Thomson opined, somewhat ungrammatically, 'I feel profoundly convinced that the argument of design has been greatly too much lost sight of in recent zoological speculations. [. . .] But overpoweringly strong proofs of intelligent and benevolent design lie all round us, and if ever perplexities, whether metaphysical or scientific, turn us away from them for a time, they come back upon us with irresistible force, showing to us through nature the influence of a free will, and teaching us that all living beings depend on one ever-acting Creator and Ruler'. Geologists, noting the extreme slowness of the processes responsible for the erosion of mountains and the accumulation of sedimentary rocks, also demanded billions of years rather than the millions granted by William. The biologists and the geologists, it seemed, needed to cash cheques on the Bank of Time[30] for much more than its reserves permitted.

---

by some that the Earth might always have existed, and there might actually be 'no vestige of a beginning, no prospect of an end', as suggested in the eighteenth century. On the other hand, James Ussher, a bishop in the Church of Ireland, famously carried out meticulous calculations based on the Christian bible to show that everything began around teatime on Saturday, 22 October, 4004 BC. The Bishop was not alone in his delusion, similar estimates, give or take a few years, having been made by Johannes Kepler and Isaac Newton, amongst others.

[28] His estimates tended to shrink with time, but were always within the range 20–400 million years.

[29] This misconception regarding evolution was voiced by ex-Prime Minister, the Marquis of Salisbury, who pointed out in his inaugural address as President of the BAAS in Oxford in 1894, 'The biologists do well to ask for an immeasurable expanse of time, if the occasional meetings of advantageously varied couples from age to age are to provide the pedigree of modifications which unite us to our ancestor the jelly-fish'.

[30] See, for example, Jamieson (1882) and Chamberlin (1899).

Unfortunately for Kelvin (formerly Thomson), and even more so for science, his calculations were flawed. Embarrassingly, he was challenged in his beliefs by his former assistant, John Perry.[31] Like Thomson, Perry originated from what is nowadays Northern Ireland, and regarded himself, at least in 1890, as an 'affectionate pupil' of the great man. Five years later, however, John was publicly contradicting his former mentor and publishing articles critical of Thomson's widely accepted conclusions in that influential organ *Nature*. It seems that John tried buttonholing William at the 'after-dinner table' in Trinity College, and later wrote to him on the subject, all to no avail. Eventually, he decided to go public. Three articles in 1895 presented, not just Perry's objections to Thomson's methods, but also private correspondence with Kelvin and his ally Peter Tait that poor old William must have found severely discomfiting.

Rather than beat around the bush, Perry came right out into the open with his twelve-bore and gave the old bird both barrels. He pointed out that Thomson's calculation of 20–80 million years for the age of the planet, while much more reasonable than the theological estimates, was nonetheless based on a series of weak assumptions. In his calculations, Thomson had assumed that the entire planet is not only solid, but of a composition similar to that of surface rocks, making no allowance for density changes (thereby ignoring an increase of more than 150 per cent), nor, crucially, of variations in thermal conductivity. His estimates were based on the time it would take a body the size of the Earth to cool by conduction alone. Conduction, as any budding physicist will tell you, is a slow and inefficient way to transfer heat, involving as it does the knocking together of vibrating molecules and other small particles. On the other hand, a fluid is able, by definition, to flow, taking its properties with it, as can easily be observed by stirring your cup of tea. Left to its own devices with a sufficiently high temperature gradient, a fluid will spontaneously begin to stir itself, dissipating heat in a phenomenon we know as convection. Now if, as Perry proposed, the interior of the Earth was at least partly fluid, as required by Airy's theory of isostasy, then its effective thermal conductivity would be a lot higher than if it were entirely rigid and able to transport heat only by conduction. This would result in much more heat being supplied to the surface, making the Earth appear appreciably younger than if it really had cooled by

[31] Several fascinating articles have described how. See Richter (1986), Burchfield (1998), and England et al. (2007).

conduction alone. Assuming an increase in thermal conductivity with depth therefore led to a calculated age far greater than Thomson's estimates. Perry argued that 'if there is even only ten times the [thermal] conductivity inside, it would practically mean that Lord Kelvin's age of the earth must be multiplied by 56'.[32] As Perry emphasized, in order to invalidate Thomson's claims, one merely needed to show that greater ages than Thomson's estimates were *possible* without flouting the laws of physics.[33] It therefore appeared that the biologists' and geologists' time cheques might be honoured after all.

Curiously, despite appearing in such a prominent journal, Perry's objections went largely unheeded.[34] Acceptance of the notion of fluidity in the mantle and the existence of thermal convection would have provided a means of overcoming some of the objections of that other stick-in-the-mud, Harold Jeffreys—specifically, the lack of a driving mechanism for continental drift. As it was, the fixists' views continued to take precedence for another half-century. Yet, although Thomson and his miserly dates managed to avoid Perry's buckshot for a time, old William's footwork was far too tardy to save him from the broadside that followed.

Yet another great Essex figure, Robert John Strutt, eldest son of John Strutt (the man who discovered the Rayleigh surface waves that proved so useful in Chapter 4) and inheritor of the title 'fourth Baron Rayleigh', was instrumental in finally downing Thomson. In 1904, Strutt junior published a classic monograph[35] on what he referred to as 'Becquerel rays'—i.e. radioactivity. Radium, a highly unstable metal of the alkaline earth group that includes calcium and magnesium, was known to produce radioactive 'emanations', this having been discovered, to their cost, by Pierre and Marie Curie in 1898. Present in minor amounts in the igneous rocks of the Earth's crust, radium decayed to helium, giving out heat. Simple calculations showed that this far exceeded that needed to account for the observed flow of heat through the surface of the

[32] It is interesting to note that, taking Kelvin's estimate as 80 million years, this would give a calculated age of 4.48 billion years for the Earth—not so far from the 4.67 billion years established by later radiometric methods.

[33] The confrontation of Perry and Thomson is described in Shipley (2001).

[34] The wider significance of this failure was recognized in an article by geophysicists Philip England, Peter Molnar, and Frank Richter: 'John Perry's Neglected Critique of Kelvin's Age for the Earth: A Missed Opportunity in Geodynamics' (England et al., 2007).

[35] Strutt (1904).

planet, leading young Strutt to conclude that 'It seems highly probable, therefore, that the internal heat of the earth is due to radium'.

Assuming that all the helium in mineral samples was derived from the decay of radium, Robert calculated the time taken to produce this helium, given the known rate of decay, and hence made estimates of the mineral age. These were subject to large errors, however, resulting from the escape of volatile helium gas during the lifetime of the mineral, and also to his failure to take into account helium produced by radioactive decay of other elements, specifically, uranium and thorium. It was soon realised that lead was a better decay product to use, since it tended to stay where it was within the mineral. In 1907, the Yale chemist Bertram Boltwood used the decay of uranium to lead, rather than radium to helium, to date a suite of rocks, finding ages as great as 2.2 billion years. One of Robert Strutt's students, Arthur Holmes, also became interested in the subject, and followed Boltwood by using the U/Pb ratio to determine mineral ages exceeding 1 billion years for Precambrian rocks. He also published a book entitled *The Age of the Earth* in 1913.[36] In the same year, it was discovered that several different isotopes of uranium, thorium, and lead were involved in radioactive decay, raising doubts about the technique and the dates based on it, such that many continued to cling to Thomson's discredited estimates. Nevertheless, despite the continuing controversy, it became possible, by incorporating corrections for these new complications, to calculate much more accurate estimates of the Earth's minimum age. A committee (including Arthur Holmes as a member), set up in 1926 by the National Research Council in the USA, was instrumental in achieving universal acceptance of radiometric dating and an age of the Earth of a few billion years. Later, Arthur redetermined the Earth's age using all the new advances, concluding that it must be about 3.35 billion years. By 1956, measurements on meteorites had established that the Earth was somewhat older even than this, with an age of 4.55 billion years.[37]

For a long time, it was widely believed—and in some quarters still is—that William Thomson's neglect of radioactive heating (which, to be fair to Thomson, only became known in 1896[38]) was his biggest error. In fact,

[36] Holmes (1913). Nine years earlier, Robert Strutt had, as we have seen, published his own book on radioactivity, the phenomenon that Holmes used to determine that the Earth was at least 1.6 billion years old.

[37] Patterson (1956).

[38] Henry Becquerel (1896); Ernest Rutherford (1902).

even had he taken into account the additional heat produced by decay of unstable isotopes, his estimate would still have fallen far short of the billions of years required by geologists. The real culprit, as Perry showed, was the fluidity of the 'substratum', that is, the mantle, and the effect of this on the rate at which heat could be dissipated. It seems strange to us now that theoretical arguments, based on broad assumptions and a crudely simplified model, should win the day over observational evidence, but that was clearly the situation in Victorian England. The trap was exposed by the great American geologist Thomas Chamberlin in an eloquent review of Thomson's address to the British Association in 1899: 'There is, perhaps, no beguilement more insidious and dangerous than an elaborate and elegant mathematical process built upon unfortified premises'. Sixty-five years later, Teddy Bullard echoed this sentiment in putting down the latter-day continental drift naysayer, Gordon MacDonald, with typical eloquence: 'Many precedents suggest the un-wisdom of being too sure of conclusions based upon supposed properties of imperfectly understood materials in inaccessible regions of the earth'. You can't help wondering whether Kelvin's rarefied status also played a part. A comment in one of Perry's letters to Peter Tait suggests this: 'I have been so accustomed to look up to you and Lord Kelvin that I think I must be more or less of an idiot to doubt when you and he were so "cocksure".'[39] The error of supporting an idea simply because of the standing of its proponent was highlighted by Mark Twain (née Samuel Langhorne Clemens) in his sublime essay 'Was the World Made for Man?',[40] in which he suggested, with no shortage of irony, that 'As Lord Kelvin is the highest authority in science now living, I think we must yield to him and accept his view'. In addition, it is more than likely that the arguments were coloured by politics and religion almost as much as science.

Nevertheless, Thomson's flawed approach eventually found relevance in explaining the relationship of heat flow to crustal age beneath the oceans. In 1986, Frank Richter (no, not that Richter) noted,[41] in this

[39] Perry (1895). This same hubris is probably what led Thomson to make such prophetic statements as 'heavier-than-air flying machines are impossible', 'radio has no future', 'X-rays will prove to be a hoax', and 'There is nothing new to be discovered in physics now'. Although unsupported by published sources, these pronouncements are believed to have been made in one or more addresses between 1895 and 1900. At any rate, they are just the kind of big-headed claims the old sod would have made.

[40] Twain (1903).

[41] Richter (1986).

context, that it appears 'To a very good first approximation the appropriate model is the one used by Kelvin, except that now time is equated with age of the sea-floor'. That is, heat flowing through the deep sea floor depends on the time elapsed since the formation of the ocean crust at a mid-ocean ridge, rather than the age of the entire Earth. So, while Thomson's age estimate was invalid, his method remained applicable on a more local scale to the thermal evolution of plates. As Frank observed, 'The key to understanding the near surface thermal structure of oceanic regions is plate tectonics'.

More Depressing Thoughts

Originally established by Thomas Jefferson in 1807 as the Survey of the Coast, and described as the USA's first civilian scientific agency, the US Coast and Geodetic Survey (now the National Geodetic Survey) made important contributions to the science of geodesy during the early twentieth century by setting up a network of stations throughout the USA to collect precise data on the size and shape of the country and, indeed, of the planet, but also bearing on the isostatic compensation (or otherwise) of the North American continent.

At the back of the Survey's 1894 report was an appendix. In this, 'Mr G. R. Putnam', described only as 'Assistant', presented measurements of gravitational acceleration made by him using a pendulum apparatus at twenty-six stations from the Atlantic coast to the Colorado Plateau, the first large-scale gravity survey ever carried out in the USA. The strength of the gravity field is related directly to the amount of matter beneath the measuring site. If perfect isostatic equilibrium prevailed (all loads fully compensated), then the attraction would be just what you would expect on a rotating, homogeneous, planet of the same mass as the Earth. Uncompensated loads on the crust, such as an ice sheet or mountain range, would result in an excess of matter, and thus a slightly greater attraction than normal—a positive gravity anomaly. Conversely, without compensation by higher-density material below, a deep valley would produce a negative gravity anomaly.

The Assistant, George Rockwell Putnam, noted that his results showed that, while the interior plains of the USA were close to isostatic equilibrium, there existed 'a considerable defect of gravity beneath the western mountains and plateaus having a striking relation in amount to the average continental elevations'. He concluded:

'The results of this series would therefore seem to lead to the conclusion that general continental elevations are compensated by a deficiency of density in the matter below sea level, possibly in much the same way that the portion of an iceberg standing above sea level is compensated by the difference in density of ice and water below the surface, but that local topographical irregularities, whether elevations or depressions, are not compensated for, such irregularities being maintained by the partial rigidity of the earth's crust.' In other words, while continents and ocean basins are compensated on a gross scale, smaller topographic features are supported by the strength of the crust and thus are not fully compensated.

Also contained in that appendix was an article entitled 'A Geologic Report on Some Coast and Geodetic Survey Gravity Stations', by G. K. Gilbert of the United States Geological Survey. Grove Karl Gilbert was a Senior Geologist with the USGS, who had, a few years earlier, published an article entitled 'The Strength of the Earth's Crust', in which he asserted, like George, that mountain ranges are supported by the rigidity of the crust, while continents and oceans exist by virtue of isostatic equilibrium. In a study of former Lake Bonneville (occupying the area of the present Great Salt Lake) in 1890, Gilbert had calculated that the uplift indicated by ancient shorelines now located way up on the sides of hills was much less than expected to compensate for the draining of a 300 m-deep lake, following the melting of the ice sheets, which he attributed to the strength of the crust. Gilbert was thus the ideal man to provide geological interpretations of the new gravity measurements, and had been invited[42] to visit the sites at which gravity measurements had been made by Putnam, in order to obtain suitable samples for density measurements and so provide a geological context for the results. In his report, he went further than George, asserting that the entire Rocky Mountains and Appalachian ranges, as well as the Wasatch plateau, 'appear to be of a nature of added loads', and were thus out of isostatic equilibrium. It seemed that the isostatic compensation of North America was far from perfect.

[42] The invitation was issued by Putnam's boss, Dr T. C. Mendenhall, Superintendent of the Coast and Geodetic Survey. Thomas Corwin Mendenhall, a far-sighted ex-military man did much to influence the adoption of metric units in the USA, notably by means of the 'Mendenhall Order' of 1893, although Americans never did manage to spell the word 'metre' correctly.

In the years following, a large database was built up by the Coast and Geodetic Survey containing data of two kinds: measurements of the *strength* of gravity using a pendulum apparatus and measurements of the *direction* of gravity based on plumb-line observations of the deflection of the vertical. The results were interpreted by the Survey's Inspector of Geodetic Work, John Fillmore Hayford, in a series of 'splendid'[43] reports. In the first two, published in 1909 and 1910, he used 733 plumb-line observations of the vertical at fifty-three stations to show that the entire continental USA was, in gross terms, in isostatic equilibrium with the adjacent oceans. Being a geodesist rather than a geologist, Hayford adopted Archdeacon Pratt's rather unrealistic form of isostasy in an attempt to discover how such equilibrium is achieved. He followed Pratt in assuming, for ease of computation, the crust to be composed of vertical columns, imagining that, at some fixed depth, pressure was equal everywhere, and that compensation for surface elevation was achieved through changes in average density of the columns. For example, a high mountain peak would be underlain by rocks of lower average density than a deep valley, while a continent would be underlain by lower-density crust than an ocean basin. Hayford then proceeded to use the extensive new data gathered by his organisation to discover what this fixed depth of compensation might be, deciding that it must be between 100 and 140 km, most probably around 122 km.[44]

I'm sure you've spotted the weaknesses in this line of enquiry. If it turned out that compensation was incomplete, as Putnam and Gilbert had suggested, or there was no single depth of compensation, then Hayford's results, which were based on a local (Pratt) model of isostasy, would be largely meaningless. Like Kelvin's age of the Earth, Hayford's 122 km depth of compensation carried an air of authority, but was fatally undermined by the questionable assumptions upon which it was based.[45] And it was not long before his assumptions *were* questioned. In 1911, Harmon Lewis at the University of Wisconsin published a critique of Hayford's work. Demonstrating, if anything, even less tact than George Airy, Lewis referred carelessly to 'the possibility that Hayford

[43] So described by MacMillan (1917), although, in truth, rather tedious, being burdened with extensive calculations.

[44] Hayford (1910).

[45] Interestingly, it has been pointed out to me that this is quite close to present-day estimates of the thermal thickness of oceanic lithosphere.

had made an error in his geodetic work', even subtitling one section of his article, somewhat insensitively, 'Criticism of Hayford's Work'. He then proceeded to demolish Hayford's ideas, pointing out that 'The theory of isostasy cannot account for the general uplift of sediments without folding'. Lewis' argument was that the presence of more-or-less undisturbed sediments at great elevations in mountain regions was contrary to the theory of isostasy, which predicted that the addition of sediments should *depress* the crust rather than leading to uplift.

His criticisms drew a barely civil response from the wounded Hayford, who declared, ominously, 'The critic [i.e. Lewis] may not reasonably object to having his article treated in the manner in which he has treated the publications criticized'. Referring to himself in the third person, Hayford then, seemingly, answered Lewis' criticisms point by point, declaring that the assumption of 'perfect compensation' did not affect his estimates of the depth at which compensation was complete. Curiously, he claimed that the flow or 'undertow' responsible for isostatic adjustment occurred above the depth of compensation, and it was this that gave rise to the folding and faulting of rocks in mountain ranges. Somehow, material was, he claimed, transferred horizontally between the vertical columns in his model. At the same time, he refuted Lewis' much more reasonable suggestion that any such flow takes place at greater depth.

So far, only deflections of the vertical measured with a plumb-line had been used by Hayford. Now, direct measurements of the gravity field, obtained using a pendulum apparatus similar to that used by George Putnam, were employed by Hayford and his colleague and successor William Bowie[46] to test the isostatic theory further. Hayford and Bowie noted that, since plumb-line and pendulum measurements are entirely independent, the latter would provide a test of the isostatic theory erected on the basis of the former. By 1912, there were 124 gravity stations in operation,[47] the results from which Hayford used to determine the degree of isostatic compensation across North America. Despite shortcuts and simplifying assumptions, this meant an awful lot of calculations, so it was a good thing that another part of John's job

[46] By 1909, Hayford had moved on to become Director of the Department of Engineering at Northwestern University, Illinois, and William Bowie had taken his place as Inspector of Geodetic Work, and later Chief of the Division of Geodesy.

[47] Plus another ten in Alaska. See Bowie (1912).

title was 'Chief, Computing Division'. Unfortunately, in the absence of what we would today understand as computers (something for which George Airy must receive a share of the blame), the work had to be performed manually by batteries of human beings, such as 'Miss Sarah Beall', who was singled out for special praise in the 1912 report.[48] To calculate the theoretical attraction at each station, the effects of the surrounding topography out to a distance of over 4100 km were included, ensuring that the full width of the continent, from coast to coast, was taken into account for each site. Hayford and Bowie eventually came to the conclusion that, contrary to the earlier findings of Putnam and Gilbert, the isostatic anomaly across much of the continent was close to zero. That is, compensation was nearly perfect.

As Alfred Bull (who we're coming to) noted later,[49] Hayford and Bowie based their work on four assumptions: '(1) that isostasy was complete; (2) that the compensation was under the topographical feature concerned; (3) that the compensation was uniformly distributed vertically; and (4) that it extends to a uniform depth below sea level'. Unfortunately, all four were later shown to be incorrect by rivals jostling one other to demolish Hayford's theories. His methods nevertheless became the norm in the USA, and the 'Hayford anomaly' was adopted as a standard method for reducing gravity data. For years, he and Bowie continued to argue for 'perfect isostasy', that is, complete compensation of all topographic loads by a matching local mass deficiency, regardless of scale. This, as their critics were not slow to point out, implied zero strength for the crust, rather like jelly.

As in the case of poor old William Thomson, it was a former minion who finally put the boot in. George Putnam, the former lowly 'Assistant', became a proponent of 'regional isostasy' during the 1920s, developing his idea that compensation involved flexure of the crust, extending laterally beyond the bounds of the crustal load. Amongst other criticisms, George noted that Hayford's method of 'correcting' for local topography based on Pratt's model involved assumed crustal columns just 4 metres in diameter, but 113.7 *kilometres* in depth, assuming these to be fully compensated, something which could not possibly be the case in reality.

[48] See Hayford and Bowie (1912). An earlier human computer, or 'computor', Radhanath Sikdar, working for the Great Trigonometrical Survey of India, had been responsible for calculating the height of peak XV in the Himalaya, showing it to be the highest on Earth. The peak was subsequently named after George Everest.

[49] Bull (1927).

Since assuming the role of Superintendent in 1909, Hayford had channelled much of the Coastal and Geodetic Survey's resources into the construction of tables for the calculation of topographic corrections using his model, so Putnam's fundamental doubts about their usefulness were scarcely welcomed. As Tony Watts, Professor of geophysics at Oxford University, has noted,[50] we do not know what relations between the two were like while Putnam worked at the Survey, but we do know that in 1910, just a year after Hayford assumed the role of Superintendent, George was put in charge of lighthouses.

The problem was, as Putnam later pointed out,[51] that 'Isostasy is in large part a geological problem'. Geodesists, being primarily mathematicians and therefore highly impractical folk, saw nothing wrong in idealized models such as Hayford's that were geared towards ease of computation rather than reality. While they were delighted with their conclusions regarding isostatic equilibrium, they almost completely neglected field observations of real rocks and were satisfied with explanations involving vertical prisms of 'crust' riding on subterranean 'lava'. Particularly unrealistic was the notion that every little bump and dip in the landscape could be matched by a corresponding change in crustal density beneath, since this would imply, as already noted, a jelly-like crust with no strength whatever. Clearly, the geologists needed to rescue the theory of isostasy from the geodesists, and turn it into something capable of explaining the behaviour of the Earth on geological time scales. This they duly did.

Solid Evidence

It was time to take stock. And there was no better man to do this than the great American geologist, Joseph Barrell. A descendent of seventeenth-century immigrants from Suffolk, Joseph was rolled out in December 1869. The family originally earned their living, as you might guess, as coopers, but, by the late eighteenth century, Joseph's great-grandfather (and namesake) had not only accumulated significant wealth, but also represented the city of Boston in the State Legislature and hosted George Washington during his visits to the city. He also found time to father twenty children within three marriages. In spite of his surname, Joseph

[50] Watts (2001).
[51] Putnam (1930).

junior was what would nowadays be regarded as 'wiry': lean, but surprisingly strong for his size. In his late forties, he 'prided himself on the longevity of his ancestors, and from a careful study of insurance tables held the unshakable belief that some thirty years of active life were ahead of him'.[52] Not long after this, Joseph died of meningitis.[53]

In 1914 and 1915, Barrell, then an Associate Professor at Yale University, published an extended article in the *Journal of Geology*, entitled 'The Strength of the Earth's Crust'. In eleven instalments totalling more than 240 pages, Joseph codified the subject of isostasy, reconciling it with geologists' theories concerning the Earth's internal properties. Referring back to the earlier results of the Coast and Geodetic Survey, he remarked that the conclusions of Putnam and Gilbert 'have been thought to be superseded and controverted, however, by much more elaborate and complete geodetic studies, first by Hayford, and later by Hayford and Bowie, which went to show that the crust was very much weaker and in much more perfect static equilibrium'. Reinterpreting the large mass of data acquired by the Survey, Barrell proceeded to expose the weaknesses in John Hayford's work from a geological perspective, noting that surface features such as plateaus or basins larger than about 25 km across were, in Hayford's model, compensated by variations in density extending uniformly to a depth of 122 km. However, since such topographic features commonly result from differences in the resistance of surface rocks to erosion, this could not be true. So, while it seemed that Hayford's work, being 'so much fuller', had superseded Putnam and Gilbert's earlier efforts, this was actually not the case. Indeed, one might almost conclude that Hayford had led the science of isostasy and, by its wider influence, geology, down a blind alley.

Joseph recognized that, despite appearances, the interior of the Earth, from the surface to the core, must be solid. Cavalier extrapolations of near-surface gradients by earlier theoreticians led to the conclusion that, at a depth of 1000 km, the temperature must be around 30,000°F (at which temperature rocks and metals are gases). It had therefore been suggested[54] that the interior of the Earth was filled with 'rocks' in the gaseous state. Pointing out the foolishness of such notions,

[52] Schuchert (1925).

[53] Nevertheless, in his 49 years, Joseph's achievements were sufficient to elevate him to the front rank of American geologists. His interests were wide-ranging, and he was a pioneer in studying past climates.

[54] Arrhenius (1900).

Joseph wrote: 'direct and positive evidence from several independent sources has forced on geologists the belief that the earth is not only solid throughout, but, as a whole, is more rigid than steel'.

Possibly the most important instalment in Joseph's great part-work, insofar as the interior of the Earth is concerned, was Part VI. Here, Joseph noted that, through time, mountains are eroded and the detritus carried to the sea by rivers, where it is dumped as sediment. The crust beneath mountains therefore becomes relieved of some of its weight while the adjacent ocean crust is loaded with an increasing thickness of sediment. To maintain isostatic or 'flotational' equilibrium, mass had to be transferred in the opposite direction, that is, from somewhere beneath the oceans to somewhere beneath the continents, implying 'some lateral counter-movement in the earth below'. This could not be within the crust, as Hayford had suggested, because the crust was too strong—evidenced by its ability to be faulted and deformed in brittle fashion. In fact, it was not only the crust that was too strong: a thick upper layer consisting of crust together with underlying mantle was cool enough to exhibit such strength. But beneath this cool, rigid, layer, which he referred to as the 'lithosphere', must be a hotter, weaker, layer of mantle that he christened the 'asthenosphere'. And, although not actually fluid, the asthenosphere must be capable of flowing horizontally as a result of the increased pressure caused by the deposition of sediments. The asthenosphere, he concluded, must be *below* the level of compensation and of great thickness.

In a later paper on sea-level changes consequent on the growth and melting of ice sheets[55], Joseph referred to his previous series of articles, arguing that 'If the hypothesis set forth there is valid—that a thick and strong lithosphere rests upon a thick zone of comparative weakness, an asthenosphere; then the weight of a continental ice sheet should tend to depress the crust into this weak zone. The crust up to a certain limit would yield as an elastic plate, the asthenosphere would reach its elastic limit at a far earlier stage and from that point yield by flowage, a flowage, it is thought, which is akin to the recrystallization which explains glacial flow.' Joseph was on the right track. His definition of the lithosphere had gone halfway to defining the plates (he even used that very word)—they are, after all, simply fragments of lithosphere. And, by introducing a weak layer or asthenosphere, he had (unwittingly) provided

[55] Barrell (1915).

the means by which those lithospheric plates could move over the mantle beneath.

Fortunately for us, Joseph Barrell wrote a non-technical summary[56] of his famous part-work in connection with a series of lectures he gave at Columbia University in the spring of 1916.[57] Here, Joseph referred to the asthenosphere as the 'shell of weakness revealed by isostasy', which must have 'very fundamental relations to the architecture of the earth, the mode of expression of its internal forces, and to the origin of magmas'. He explained the apparent contradiction that the weak asthenosphere is capable of flow yet, at the same time, transmitting earthquake waves: this was a result of the confusion between rigidity and strength. Rigidity increases with pressure (and therefore depth), and, since seismic velocities depend on rigidity, earthquake waves must travel faster with depth. On the other hand, under a permanent load and high temperatures, solid rock will reach its elastic limit and then 'flow by recrystallization without the moduli of elasticity being changed'. Although non-intuitive, the recognition that solid rocks could flow, given enough time, was fundamental to progress in understanding mantle convection.

For physicists, the idea that the solid mantle could be treated as a very viscous fluid was a real boon. It meant that the theory of fluid dynamics developed over the preceding century could be applied to the Earth's interior and conclusions drawn regarding its response to stresses, such as those produced by surface loads like mountain ranges. If you put enough stress on a rigid solid, it will deform. Likewise with a fluid, except that you won't need to exert much effort to deform it and it won't return to its original shape. Also, in a rigid solid, the amount of deformation is proportional to the amount of pressure you exert. That is, if you double the pressure, you get twice the deformation. In a fluid, on the other hand, doubling the pressure will result in doubling, not of the *amount* of deformation, but the *rate* of deformation. That is, the deformation will be twice as fast. In theory.

Getting Warm

The next advance arrived in 1921, with a remarkable article by Alfred Joseph Bull, an amateur geologist who served for two years as President

56 Barrell (1919a).

57 This summary being published shortly after his death in 1919.

of the Geologists' Association and gave his professional address as the 'School of Photo-engraving, Fleet Street, London'. Alfred's ideas were summed up succinctly in the first paragraph of his paper, 'It is here suggested that the folding of mountain ranges may be produced by the frictional drag of moving portions of the asthenosphere, and that these movements may be convective and result from its unequal heating by radio-active elements'.[58] In other words, the convecting asthenosphere exerted a drag force on the crust above, resulting in horizontal motions that led to the creation of mountains. Alfred's drag force thus provided a mechanism by which surface motions could be coupled to mantle convection cells beneath. He went further, suggesting that the fragmentation of the supercontinent of Gondwana had been accomplished by the divergence of convection currents in the substratum driven by the heat from radioactive decay. Like Barrell and others before him, he also noted that, while transmission of shear waves (that is, S-waves) required that the mantle be rigid on very short timescales, the theory of isostasy demanded the ability to flow under the application of limited, yet long-continued, stresses. He also concluded that isostatic compensation was incomplete, owing to the strength of the crust.[59]

Alfred quickly dismissed the contemporary notion of Earth contraction as a cause of mountain building, on the grounds of the obvious facts that mountain ranges form locally over limited periods of time, rather than continuously all over the globe as you might expect if the planet were shrinking, and that much of the surface was actually in tension rather than compression. Warming to his subject, Alfred continued, 'Should a relatively cooler portion of the asthenosphere be situated between two hotter portions, all being at about the same level, then, when movement commenced, there would be two opposing currents dragging the under surface of the crust, which would be crumpled along a line perpendicular to the currents. There would thus arise a line of folding above the region where the asthenosphere was sinking, and the crust outside the moving regions would be in tension. Now the folded mass, being lighter than the asthenosphere, would float in it and project above the general level of the crust. It would thus be a mountain chain with light roots.' Bull's 'drag' hypothesis was developed in a

58 Bull (1921).

59 Despite the previous recognition of the 'lithosphere' as the rigid outer layer, the term 'crust' has continued to be applied in this context.

series of later papers, and soon picked up by Arthur Holmes, the grandfather of plate tectonics, as the answer to his problems concerning the mechanism for continental drift.

Free convection was first studied in detail at the turn of the twentieth century by Henri Bénard at the Collège de France. By 1899, Henri was conducting experiments on this intriguing phenomenon using equally intriguing substances, such as spermaceti, heated from below—experiments that he continued until his death in 1939. During this period, he developed optical techniques to observe patterns of honeycomb-like cells in convecting fluids, now known as Bénard cells. He noted that the fluid rose in the centre of each polygonal cell and descended along the boundaries between cells. A theoretical basis for Bénard's observations was soon provided by Robert Strutt's dad, John. Inheriting the title 'third Baron Rayleigh', John made the most of his aleatory good fortune, becoming one of England's (and certainly Essex's) greatest scientists during a distinguished career in which he held the posts of Chancellor of the University of Cambridge, President of the Royal Society, and (slightly less illustriously) President of the Society for Psychical Research. Most of his work was carried out in laboratories built in the stables at the family estate in Essex, although, during the agricultural depression of the early 1880s, he spent a few years at the Cavendish Laboratories in Cambridge (where Crick and Watson later made their breakthrough). As we saw previously, Strutt senior made important contributions to the study of waves, in particular the understanding of seismic waves propagated along the surface of the Earth, now known as Rayleigh waves. He also wrote a famous book on the theory of sound, explained how seabirds soar 'without doing a stroke of work', and found the answer to that long-standing conundrum, 'why is the sky blue?' It was therefore no surprise when John was awarded a Nobel Prize in 1904, though he received the prize, not for any of these great insights, but instead for the rather mundane-sounding achievement of isolating the inert gas argon from the atmosphere, which perhaps says something about the Nobel committee. John Strutt is described on the Nobel website as 'one of the very few members of higher nobility who won fame as an outstanding scientist', which perhaps says something about the nobility.

In a groundbreaking paper published in 1916, just three years prior to his demise, John investigated the conditions necessary for convection in a fluid layer heated from below. In this case, expansion of heated fluid at the bottom lowers its density, leading to a reduction in the pull of gravity relative to the surrounding, cooler, fluid, and thus creating an upward 'buoyancy force'. When this force becomes strong enough to overcome the fluid's resistance to flow, that is, its viscosity, the hotter material will rise to the surface, where it cools and eventually sinks once more, a phenomenon that has become known as 'Rayleigh–Bénard convection'. Interestingly, this was not actually what Henri Bénard had observed. Since he had used a very thin layer of fluid with a free upper surface in his experiments, the convection patterns he observed were dominated, not by buoyancy, but by surface tension. Nevertheless, Strutt's theory seemed to represent the situation within the mantle quite nicely, with heat supplied from the cooling core and lost through an upper boundary with the lithosphere.

John concluded that there exists a critical vertical temperature gradient below which convective motion is prevented by the fluid's viscosity. The temperature gradient, in turn, depends on the physical properties of the fluid and its thickness. Obviously, the less viscous a fluid, the more likely it is to convect. If it conducts heat well (i.e. it has a high thermal diffusivity), heat will be removed more rapidly, delaying or preventing the onset of convection. A high thermal gradient produced by heating from below will encourage convection. The transition from 'heat loss by conduction' to 'heat loss by convection' was, Strutt found, determined by the ratio of buoyancy forces to resistive forces resulting from these effects, the ratio being expressed later (by others) as a dimensionless number (the 'Rayleigh number'). Convection begins when the Rayleigh number exceeds a critical value that, according to John Strutt's theory, was close to 657.5. In fact, depending on the assumptions made, the critical Rayleigh number for convection in the mantle is somewhat higher: 1000–2000. As you might expect, the greater the amount by which the Rayleigh number exceeds the critical Rayleigh number, the more vigorous the convection. John also found that the convection pattern was dependent on the small initial perturbation that started the whole thing off, so that convection could develop in several different modes and although the system tended towards a stable regular polygonal pattern, it never quite reached it. Within a few years, Arthur Holmes applied John Strutt's new theory to the question of drifting continents, just as he had applied Robert Strutt's work on radioactivity to the problem of dating rocks, concluding that the movement of continents was related to thermal convection in the underlying mantle. Of course, since almost no-one believed in continental drift at the time, almost no-one took this seriously.

Holmes Solves the Mystery

A Geordie from a modest home in Gateshead, Holmes struggled to make ends meet as a young man. His straitened circumstances were not much helped by a meagre (even for those days) annual salary of £200, earned as a demonstrator in London's Imperial College of Science between 1912 and 1920. In order to remain solvent, he later took a disastrous job with an oil company in Burma,[60] following which he became, like many oil industry geologists before and since, unemployed. After this, it was with 'great relief' that he was glad to accept a post in Durham University's new geology department in 1924, becoming professor of geology the following year. Here he remained until 1938, developing the fundaments of continental drift and mantle convection theory.

In one of the greatest papers in the history of science, entitled 'Radioactivity and Earth Movements', published in 1931,[61] Arthur Holmes set out his theory of continental drift driven by mantle convection. Presenting simple back-of-an-envelope calculations, he argued that even very modest contributions of heat energy from the radioactive decay of the unstable elements uranium, thorium, and potassium in mantle rocks would have dramatic effects on the amount of heat flowing out at the surface. Using a very conservative value of heat production, he concluded that, to allow the heat accumulated over a period of 200 million years to escape through the floors of the oceans, 'it would be necessary for one third of the whole of the ocean floors [...] to be engulfed and heated up to 1000°C., and replaced by magma which cooled down to form new ocean floors'. He suggested an explanation for how this might be accomplished: 'A process competent to bring about this result on the scale indicated would be some form of continental drift involving the sinking of old ocean floors in front of the advancing continents and the formation of new ocean floors behind them'. He had arrived at the very essence of the theory that would later be known as plate tectonics. Indeed, Tuzo Wilson, in his breakthrough *Nature* paper of 1965[62] (see Chapter 2), expressed exactly the same idea: 'Large

[60] While in Burma, he lost his young son to dysentery, and returned penniless.
[61] Although presented to the Glasgow Geological Society in January 1928.
[62] Wilson (1965).

areas of crust must be swallowed up in front of an advancing continent and re-created in its wake', although, like Harry Hess and Bob Dietz, he failed to cite Holmes' work in his article.[63]

Arthur had come to the conclusion that 'Admitting that the continents have drifted, there seems no escape from the deduction that slow but overwhelmingly powerful currents must have been generated in the underworld'. He imagined an ordered mantle circulation, with rising currents beneath the equator and sinking currents beneath the poles. Such an arrangement, he thought, might explain the northward movement of Africa and India, resulting in the closing of an ocean, the Tethys,[64] that formerly separated the southern continents from Eurasia. In slightly less rigorous fashion than Strutt senior, he linked convection to the rate at which temperature increases with depth, noting that 'In the earth, the critical gradient appears to be about 3°C per km. So long as this is exceeded, convection must go on, provided that the material has no strength, and that the viscosity is not too high.' Applying Alfred Bull's hypothesis, he went on: 'As a result of the powerful drag thus imposed on the lower part of the crust the latter will flow with the currents but with a lower velocity', and explained that 'the ultimate effect on continental blocks originally over the equator will be to drag them apart, leaving a depressed geosynclinal[65] or oceanic belt along the equator'. A figure showed the gap between separating continental blocks being filled with 'new ocean' (Figure 5.1). He conjectured: 'Where the ascending currents turn over, the opposing shears and the resulting flowage in the crust would produce a stretched region or a disruptive basin which would subside between the main blocks. If the latter could be carried apart on the backs of the currents, the intervening geosynclines would develop into a new oceanic region. The formation of a new ocean floor would involve the discharge of a great deal of excess heat.' Rough estimates, based on currents flowing at 50 km per million years associated with an excess temperature of 10°C, gave a minimum age for the Atlantic Ocean of 100 million years. He also concluded that the new ocean crust would be 'more nearly basaltic' than the substratum.

[63] Given that this was written whilst Tuzo was on sabbatical in Cambridge, it is even more surprising that Holmes' priority was not brought to his attention.

[64] Named by the geologist Eduard Suess in 1893.

[65] This unfortunate term was explained in Chapter 2.

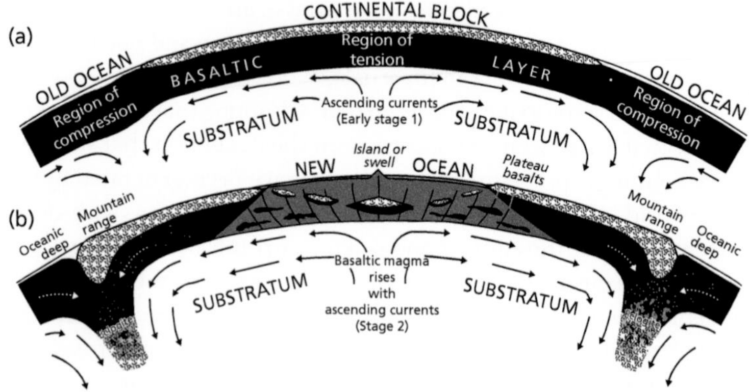

Figure 5.1 Holmes' mechanism for 'engineering' continental drift. Note in (b) the creation of 'new ocean' crust from basaltic magma at the mid-ocean ridge. Note also the initiation of subduction and creation of mountain ranges at continental margins, associated with downward convection currents. [Figure 262 from Holmes' *Principles of Physical Geology* of 1944.]

He went further, considering what would happen where two currents met and turned downward. In these circumstances, he predicted, compression and thickening of the continental crust would take place, along with the large-scale alteration of ocean crust at high temperatures and pressures into eclogite, a rock with the same chemical composition as basalt, but different mineralogy (chiefly garnet and pyroxene). This was a highly significant step, for eclogite, at around 3400 kg m^{-3}, is substantially denser than basalt (2850 kg m^{-3}), and, crucially, denser than the rocks of the upper mantle (density 3300 kg m^{-3}), with the result that the altered crust would naturally sink into the mantle. 'Such foundering', he continued, 'would effectively speed up the downward current' and provide an explanation for the existence of ocean deeps, such as those in the East Indies visited by Felix Vening Meinesz. Here, forty years before it was even given a name, was the prototype of the concept of subduction.

In this single paper, Arthur had explained both the creation and destruction of ocean crust, and also established a mechanism for 'engineering' continental drift, as he put it, thereby overcoming the objections of that old curmudgeon Harold Jeffreys and his ilk. As he noted in his conclusion, his theory also explained, amongst other things, the distribution of earthquakes and volcanoes, mountain-building, and the formation of ocean deeps. To their eternal shame, most geologists and geophysicists failed to appreciate the immense significance of Holmes' theories, condemning themselves to a further quarter of a century of fruitless fixism. Not that they were able to erect convincing arguments against convection in the mantle. Even Jeffreys had to admit, grudgingly, that 'So far as I can see there is nothing inherently impossible in it', before pouring cold water on the idea: 'but the association of conditions that would be required to make it work would be rather in the nature of a fluke'.

Somewhat later, in his great textbook of geology, published in 1944, Holmes recognized the role of Joseph Barrell's 'lithosphere' in the drift of the continents and oceans: 'More probably it is the crust and an appropriate thickness of the upper mantle that together constitute the layer that glides over the interior, the zone of gliding being within the low-velocity channel.' The low-velocity channel was thus, he concluded, a zone of 'flowage', corresponding to Barrell's 'asthenosphere'. He summed up his ideas thus: 'during large-scale convective circulation the basaltic layer becomes a kind of endless travelling belt on the top of which a continent can be carried along, until it comes to rest (relative to the belt) when its advancing front reaches the place where the belt turns downwards and disappears into the earth'. Arthur was now prefiguring many of the ideas that evolved during the 1960s.

Some of Holmes' revolutionary ideas on ocean-floor creation were revived in the 1950s by Harry Hess, the Princeton geophysicist we met in Chapter 1, and subsequently rebranded as 'sea-floor spreading'.[66] In a much-quoted article written on the eve of plate tectonics,[67] Hess wrote that mid-ocean ridges 'are interpreted as representing the rising limbs of mantle-convection cells'. He also followed Holmes in believing that 'continents ride passively on the convecting mantle'. Somehow, these two ideas, brilliant as they were when advanced by Holmes in the 1930s, have, like tuberculosis, persisted, despite half a century of scientific progress. In fact, such notions were already superseded by the late 1960s, when it was realized that volcanism at mid-ocean ridges results, not from rising convection currents, but from the decompression of shallow mantle as plates are pulled apart (the reduction in pressure is accompanied by a reduction in melting point) and, moreover, that the plates are able to drive themselves.

[66] The name was actually applied by Hess' rival Bob Dietz in 1961.
[67] Hess (1961).

His Last Bow

In his *opus magnum* of 1965, Holmes himself spotted a fly in the ointment. He noted: 'One obvious complication arises. If Africa moved away from the Mid-Atlantic Ridge along the route indicated by the Walvis Ridge, it cannot at the same time have moved away from the Mid-Indian-Ocean Ridge.' What he was getting at was this: if mid-ocean ridges are the surface expression of rising convection columns, they would be more or less fixed with respect to each other, making it impossible for Africa to be carried simultaneously eastward away from the Mid-Atlantic Ridge, and westward away from the Central Indian Ridge. This is an objection we came across in Chapter 2, an objection that was terminal for the notion of rising mantle convection currents beneath mid-ocean ridges. Nevertheless, if subduction zones mark the downward movement of material from the surface, it stands to reason that mantle must ascend somewhere to balance: somewhere—but where? Forced to abandon the cherished notion that mid-ocean ridges marked the position of rising currents, researchers had only two explanations left: either the mantle rose gradually everywhere other than at subduction zones, or there existed localized columns of upwelling.

Recall that, during the early 1970s, Jason Morgan had taken the bold, even rash, step of proposing that hotspots were manifestations of tall 'plumes' of hot rock, rising from the bottom of the mantle, nearly 3000 km beneath the Earth's surface. Twenty such 'pipes to the deep mantle', as he described them, acted as conduits through which hot mantle rock ascended to the surface at an assumed rate of 2 m per year, bringing around 500 km^3 of material to the surface annually. Such a flow, occurring beneath mid-ocean ridges, might, he believed, actually provide the driving force for plate tectonics. A more likely rate of upward flow, it turns out, is closer to 20 mm per year, so that the net force exerted by this means would be far too small to power the plates. Nevertheless, such pipes, if they existed, could form an important return flow from the deep mantle. During the 1970s, simple laboratory experiments using viscous silicone oil, together with early computer simulations, suggested that rising and falling currents in a convecting fluid could become focused into narrow columns according to the values of viscosity and other properties chosen, and whether heat was supplied from within (i.e. by radioactive decay of the isotopes ^{40}K, ^{238}U, ^{235}U, and ^{232}Th). These studies began to give an insight into what is a complex, not to say chaotic, phenomenon. Major progress, however, had to await the advances in computer technology and physics theory that occurred during the second generation of plate tectonics after 1980.

Pitch Fever

In 1927, a year before Holmes presented his audacious theory to the wide-eyed members of the Glasgow Geological Society, Thomas Parnell, at the University of Queensland in Brisbane, set up a remarkable experiment. He filled a glass funnel with pitch (bitumen) and then waited. The pitch, to all intents a solid that you could smash with a hammer, would, he thought, flow, given enough time, and eventually drip out of the bottom of the funnel, the time between drops being a measure of the viscosity of the pitch. The first drop was recorded in 1938, but, being busy, Professor Parnell missed it, and also the one that followed. Thomas died in 1948, a disappointed man. However, in 1961, his successor, John Mainstone, discovered the experiment still 'running' in a cupboard and began to take a lively interest in it. Having failed to witness the instant of a drop actually falling in 1962, 1970, 1979 and 1988, John set up a webcam to record the climactic event in 2000, only to be frustrated at the crucial moment by a power cut. The 'pitch drop' experiment has now become the longest-running laboratory experiment on Earth (verified by the *Guinness Book of Records*), earning an Ig Nobel Prize for Parnell (posthumously) and Mainstone in 2005. By a cruel twist of fate, Mainstone died in 2013, just months before the ninth drop was recorded on 24 April 2014, making headlines in newspapers around the world.[68]

The results of this thrilling experiment allowed Queensland University physicists R. Edgeworth and B. J. Dalton[69] to calculate the viscosity of the pitch as 2.3×10^8 Pascal-seconds (Pa s),[70] although the actual value varied widely with temperature, as you would expect (honey, for example, becomes runnier when you heat it up). To put this into some sort of context, water (at room temperature) has a viscosity of 0.001 Pa s, motor oil (at room temperature) somewhere in the range 0.065–0.319 Pa s, and (bearing in mind the likely readership for this book) ketchup, 50–100 Pa s.

[68] This could be a slight exaggeration, but see, for example, http://www.dailymail.co.uk/sciencetech/article-2609460/.

[69] See Edgeworth et al. (1984).

[70] The 'Pascal-second' is the modern SI unit of viscosity, equivalent to kg m^{-1} s^{-1}.

Beyond its role as a PR exercise, the pitch-drop experiment demonstrates two very important principles. Firstly, human life is short, making it difficult to appreciate many natural phenomena. Secondly, an apparently brittle solid can flow if given enough time, such that it has a viscosity. These principles are fundamental to an understanding of the mechanism driving plate tectonics.

Mantle rocks, however, are not like bitumen. Whereas the latter is a complex organic semi-solid, igneous rocks are aggregates of interlocking mineral crystals that a physicist might refer to as 'pseudoviscous polycrystalline solids'. In the mantle, the crystals are likely to be silicate and oxide compounds forming minerals such as olivine, pyroxene, and various less familiar species predicted from high-pressure laboratory studies. Mantle rocks are thus definitely solid and crystalline. However, when heated to near their melting temperatures, they become weakened and able to 'flow' by a number of mechanisms classified under the heading of 'creep'. Mineral physicists, partly galvanized by the rise of plate tectonics itself, developed theories of deformation and flow by creep during the 1960s and 1970s. When the applied stresses are low, crystalline rocks deform by the diffusion of 'vacancies' (a 'vacancy' is an empty location in a crystal lattice that would normally be occupied by an atom) between grain boundaries. The application of stress creates new vacancies and causes atoms to jump across them in a direction controlled by the stress, resulting in a painfully slow[71] migration of matter known as 'diffusion creep'. The speed of deformation, or 'strain rate',[72] becomes much less for rocks in which the 'grains' are larger, but the deformation may itself cause a reduction in grain size, thereby resulting in an increase in strain rate in a kind of positive feedback.

Diffusion creep is very convenient for theoreticians, since the rate of deformation is proportional to the stress applied (i.e. a linear relationship), according to a theory that originated back in the seventeenth century with one Isaac Newton—thus any fluid behaving in this way is known to physicists as a 'newtonian fluid'. Unfortunately, Nature, as

[71] The slow rate of migration is the reason that seismic waves, travelling on a timescale of minutes, are not affected by creep.

[72] Strain rate is measured as 'proportionate change in dimensions per second', but, since 'proportionate change' is dimensionless, the units are 's^{-1}'. This is somewhat cumbersome for something deforming as slowly as the mantle, resulting in typical strain rates of $10^{-12}\ s^{-1}$. Nevertheless, this is fast enough to permit mantle rocks to carry appreciable amounts of heat along with them.

always, refuses to be constrained by the laws of Newton or anyone else, and it was discovered[73] that other, 'non-newtonian', forms of creep occurred under higher stresses. It was found that 'dislocation creep', whereby planar defects in the lattice migrate through the crystal, tended to dominate at stresses greater than 1 bar and, since this process involved a non-linear relationship between stress and strain rate, it was difficult to incorporate into theoretical models. Nevertheless, since dislocation creep was effective over a wider range of temperatures, it was evident that flow could occur even in the lower mantle where diffusion creep was inhibited.

Convection Turned Upside-Down

Clarence Dutton, the USGS geologist who, in 1882, coined the term 'isostasy', had, a few years earlier, visited Hawaii and observed lava lakes at Kilauea. He noticed that 'At brief intervals the surface darkens over by the formation of a black solid crust, with streaks of fire round the edges. Suddenly a network of cracks shoots through the entire crust, and the fragments turn down edgewise and sink'. A slightly more poetic version was given by that highly respected geologist Mark Twain,[74] who also witnessed the spectacle: 'Every now and then masses of the dark crust broke away and floated slowly down these streams like rafts down a river. Occasionally the molten lava flowing under the superincumbent crust broke through—split a dazzling streak, from five hundred to a thousand feet long, like a sudden flash of lightning, and then acre after acre of the cold lava parted into fragments, turned up edgewise like cakes of ice when a great river breaks up, plunged downward and were swallowed in the crimson cauldron.' A century afterwards, another USGS geologist, Wendell A. Duffield, published further observations of the Kilauea lava crust that he identified explicitly with plate tectonics,[75] although, as he admitted, the resemblance was somewhat superficial.

Just a few months later, however, Douglas Oldenburg and James Brune, at the University of California at San Diego, reported results of fascinating laboratory experiments designed to replicate sea-floor

[73] Gordon (1965); Weertman (1970).
[74] Twain (1872), p.189.
[75] Duffield (1972).

spreading using 'plates' made of paraffin wax.[76] By applying tensional stress to the wax as it was cooled by a fan, they were able to generate the characteristic pattern of ridge segments and transform offsets observed in the oceans. They found that this pattern could be established and maintained purely as a passive response to the tensional stresses placed on the wax layer. In other words, given the right physical conditions, you only had to pull the wax plates apart to generate the ridge-transform pattern that characterized mid-ocean ridges like the East Pacific Rise. Spreading was thus a result of the plates being pulled by stresses from afar, rather than being pushed apart and carried along by convection currents rising from the deep mantle. The hot wax 'mantle' rose beneath the spreading centre simply because of the release of pressure caused by separation of the 'plates'. Applying their results to the Earth implied that there was nothing special about the mantle beneath spreading centres, and rising convection currents were not required. Mid-ocean ridges were thus freed to migrate wherever they liked over the surface of the Earth.

Walter Elsasser, who had previously made such an important contribution to the understanding of the geomagnetic dynamo (as we saw in Chapter 3), also made a major contribution to the development of mantle convection theory. In 1967, Walter gave a talk in Newcastle at one of Keith Runcorn's many scientific meetings,[77] focusing on convection in the upper mantle. Here, he declared that 'The key to all models of mantle convection is the distribution of temperature with depth'. By this time, the incisive work of people like Dan McKenzie and Frank Richter had already shown that the idea of plates being carried along by mantle convection currents was untenable. Walter now concluded that 'the convective circulation in the upper mantle is largely controlled by irregularities in its free upper surface, in a manner inverse to the way in which thermal convection of a fluid is often controlled by the corrugations of its lower boundary'. The following year, he presented a second paper, entitled 'The Mechanics of Continental Drift', during the Annual General Meeting of the American Philosophical Society in Philadelphia.[78] At that time, he still believed that mid-ocean ridges were, as promulgated

[76] Oldenburg and Brune (1972).

[77] This one was called, somewhat long-windedly, 'The Application of Modern Physics to the Earth and Planetary Interiors' and covered a wide range of topics from geomagnetism to postulated changes in the universal gravitational constant.

[78] The meeting was entitled 'Gondwanaland Revisited: New Evidence for Continental Drift', and Walter's paper was published later that year in the *Proceedings* of the Society (Elsasser, 1968). It included the sentence quoted as the epigraph to this chapter.

by Harry Hess, the loci of rising convection currents, assuming that heated mantle rose in great curtains beneath the mid-ocean ridges and then diverged, carrying the oceanic lithosphere horizontally by exerting drag on the bottom of plates. He also accepted the idea that convection occurred only in the shallower layers of the mantle above 700 km depth, which he believed was well substantiated by observations.

Three years later, Walter published a major article in the *Journal of Geophysical Research* entitled 'Sea-Floor Spreading as Thermal Convection',[79] in which his ideas had matured considerably. This paper, perhaps more than any other, summarizes the understanding reached during first generation plate tectonics. In it, Walter set out a highly modified version of plate tectonics: a version that has survived, with minor modifications, into the twenty-first century. The main points of the revised theory were:

1. The lithosphere reacts to stresses by breaking up into plates that readjust themselves according to the prevailing forces.
2. Mechanical stresses are transmitted horizontally over great distances by the rigid lithosphere (plates).
3. A low-viscosity asthenosphere decouples lithosphere from the middle mantle.
4. The driving force behind mantle circulation is thermal convection.
5. Convection is driven, not by heating from below, but by cooling from above.
6. The primary driving forces for convection are exerted by slabs of lithosphere sinking along the Wadati–Benioff zones.[80]
7. Mid-ocean ridges and transform offsets reflect tension as the plates are pulled apart, rather than being pushed apart by rising mantle.
8. Tension produces symmetrical spreading unrelated to relative motion between the lithosphere and asthenosphere.
9. Material gained by the lithosphere at mid-ocean ridges is balanced by material lost by subduction at trenches.
10. The Atlantic and Indian oceans have opened since the Mesozoic (i.e. during the past 250 million years), and the Pacific has shrunk by a similar amount.

[79] Elsasser (1971).
[80] Referred to by Walter as 'Gutenberg–Benioff faults'.

11. Subducting slabs retreat as they sink and the hinge of subduction migrates seaward.

The most fundamental changes were items 5 and 6. Walter now asserted that, unlike more familiar examples of convection in which heating occurs from below, mantle circulation is 'a sort of upside-down convection in which motion is engendered by a cold top layer sinking rather than by a hot bottom layer rising'. He explained that 'most of the convection-generating forces are exerted by parts of the oceanic lithosphere that have been cooled at the surface, [...] this relatively cold material then sinks down and thereby serves as the primary driving agency of the circulation'. In this refurbished theory, the low-viscosity asthenosphere no longer drives the plates by exerting drag, but now serves to decouple the lithosphere from the mantle beneath, allowing the plates to move with minimal resistance (item 3). And, far from being mere passengers on a convecting mantle, the plates form a fundamental component of the convection system itself. As David Bercovici, a geophysicist and prime mover in second-generation plate tectonics, later put it,[81] 'In the end, that "active" plates cool, thicken, subside and eventually sink as subducting slabs that cool the mantle is tantamount to saying they are convective currents, and that, in effect, plate tectonics is the surface expression of mantle convection'.

In 1972, Dan McKenzie noted that, while the kinematic theory of the plates appeared to be largely complete, a full dynamic theory involving the mantle was still a long way off. In 1974, he published a paper[82] describing groundbreaking research carried out by himself and colleagues in Cambridge, in which a much-simplified two-dimensional model was used to generate a set of partial differential equations describing convection confined to the upper mantle. Being non-linear, these equations were not amenable to analytical solution, and so numerical solutions were sought using an IBM 360 mainframe computer. In view of the complexity of the problem, Dan and colleagues were forced to make 'many drastic and unrealistic assumptions' in order to achieve a result with the limited computing power available, so that the relevance of their results to plate tectonics was 'therefore in some doubt'. Nevertheless, they had introduced an approach that would eventually

[81] Bercovici (2003).

[82] Co-authored by Jean Roberts and mathematician Nigel Weiss. See McKenzie et al. (1974).

unite the observation and theory of complex natural systems like convection.

A very significant step on the road to explaining plate tectonics as the surface manifestation of mantle convection was the introduction of the concept of a boundary layer in 1967, by Don Turcotte and Ron Oxburgh at Oxford.[83] During convection caused by heating from below, a cold, unstable, layer would be formed at the surface, while a hot, likewise unstable, boundary layer formed at the bottom. The upper, cold, boundary layer represented the plates, of course, while the lower, hot, boundary layer would become a focal point in twenty-first-century plate tectonics (see Chapter 10). In a review of the subject in 1976, Dan McKenzie and Frank Richter[84] noted that earlier theoretical work, such as that by John Strutt, had used highly simplified systems, assuming a convecting fluid of uniform density,[85] heated from below. Under these conditions, convection would have vertical symmetry—you could turn the convection pattern upside-down and it would look the same. In reality, the density of mantle rocks is affected by increasing pressure, while their viscosity is strongly dependent on temperature, particularly at low temperatures. If these effects were taken into account, then material near the bottom of the mantle would have reduced viscosity and the lower boundary layer would be thinner, whereas viscosity near the top of the mantle would be higher and the boundary layer would be thicker. Heat would be supplied to the lower boundary layer and lost from the upper boundary layer by conduction, leading to instability (i.e. the top boundary layer sinks while the bottom boundary layer rises, forming a convection cell).

Another crucial omission from Strutt senior's convective theory was the heat produced internally by the phenomenon described by Strutt junior—radioactivity. As Dan and Frank pointed out, this implied a circulation more complex than simple Rayleigh–Bénard convection. The horizontal dimensions of convective cells in the bottom-heated case was predicted to be about the same as the depth of the convecting fluid. Plates, however, have widths of thousands of kilometres (the Pacific plate is over 10,000 km wide), whereas the depth of the mantle is less than 3000 km. It was found that such large aspect ratios, with cells that

[83] Turcotte and Oxburgh (1967).

[84] McKenzie and Richter (1976).

[85] Except where this involved gravity. This is known as the Boussinesq approximation.

are much wider than they are deep, could be generated if most of the heat necessary to drive convection is generated *within* the convecting layer itself (i.e. by radioactive decay, as in the Earth). At that time, it was widely believed that convection was confined to the upper mantle (above 670 km) or else separate convection systems existed in the upper and lower mantle. In 1977, however, Geoffrey Davies, a geophysicist then working in New York, also became interested in computer simulation of mantle convection, presenting evidence that the chemistries of the upper and lower mantle were similar and that viscosity was actually fairly uniform throughout the mantle, leading to the conclusion that convection probably involved the entire mantle, as envisaged by Arthur Holmes. The base of the convecting mantle was therefore the core–mantle boundary. As a result, in 1979, Walter Elsasser[86] added the following items to his shopping list:

12. The Rayleigh number of the mantle is large: the mantle therefore convects throughout its entire depth.
13. A thermal boundary layer exists at the bottom of the mantle.

Thermal boundary layers, as Walter explained, were of fundamental importance. For example, if there was more heat flowing out of the molten iron core than could be dissipated by conduction through the mantle above, then a layer of anomalously hot mantle would develop just above the core boundary. Further heating would create buoyancy in this layer, so that eventually it would detach from the boundary and rise vertically as a 'plume' towards the surface like the hot wax in a lava lamp. Similarly, a cool surface boundary layer would result in a sinking column of viscous fluid. Because the top boundary layer has to conduct not only the heat input through its base, but also the heat generated within the convecting mantle, it will have a greater temperature contrast than the lower boundary layer, leading to larger, more numerous, 'downwellings' as the cooled material sinks. These cool downwellings dominate the convection pattern, while the relatively small amount of heat input through the base leads to narrow, rising, plumes. Rising and sinking columns would transmit a torque to the surrounding mantle, providing the driving force for flow throughout the mantle. This, of course, was the opposite of the situation assumed previously, in which convecting mantle drove the plates (Figure 5.2).

[86] Elsasser (1979).

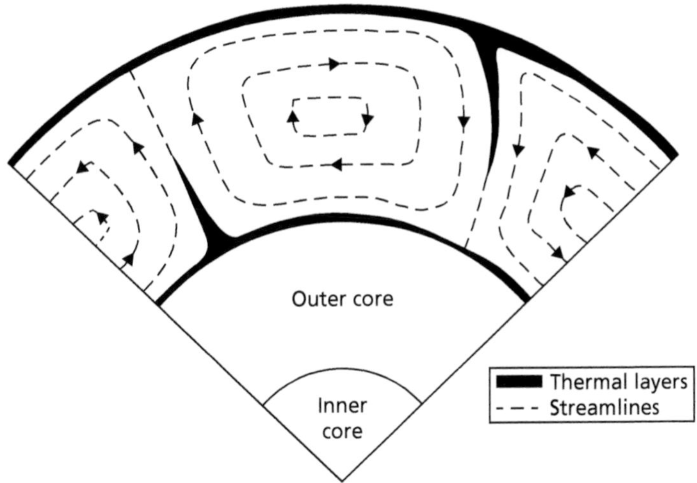

Figure 5.2 Elsasser's 'grossly oversimplified' mantle convection model. The heavy black regions are boundary layers. [From Elsasser (1979).]

By 1980, the obstacles placed in its path by Harold Jeffreys and the fixists had been overcome, and a basic theory of mantle convection by creep established. It was clear that plate tectonics and mantle convection were intimately associated, but, contrary to earlier ideas, it now appeared that convection was driven by plate motion, rather than the reverse. And, far from being mere passengers, the plates formed an integral part of the convection system. Somehow, these surprising advances failed to be transmitted to the wider public, so that, even today, you can find school and even university textbooks showing plates driven by mantle convection rather than the other way around. Likewise, despite being recognized during the nineteenth century, the rigid layer, or lithosphere, continues to be confused with the chemically distinct crust. By definition, the plates are the cool, rigid, outer layer of the Earth composed of the uppermost mantle and crust (more mantle than crust, in fact): talk of 'crustal plates' is therefore highly misleading—but not uncommon.

A related misconception that has persisted is that mid-ocean ridges are the sites of rising limbs of convection currents. Within a few years of Tuzo's landmark article on transforms, it became clear to geophysicists that the divergence of plates at spreading centres must result in passive upwelling of mantle rocks from shallow depths, with concomitant partial melting in consequence of the reduction in pressure. Magma erupted at mid-ocean ridges is therefore derived from whatever patch of shallow mantle the ridge happens to be passing over, rather than a rising current from the deep mantle.[87] The upward component of mantle circulation had to be somewhere else, and that somewhere else was very likely a set of plumes rising from the lower mantle, numbering between a dozen and 122.[88]

Afterthought

Within a single generation, the new theory of the plates had been assembled and applied to the mass of new geological and geophysical data acquired in the oceans. Initially limited to kinematic processes at the surface of the Earth, the concept was soon expanded to encompass whole-mantle convection, of which the plates formed the surface component. At last, Earth Science had its great unifying theory. Observations that had perplexed geologists for decades, for example the locations of earthquakes, volcanoes, and mountain ranges, together with the surprisingly young age of ocean crust, now fell into place like the pieces of a jigsaw. The inspired insights of Arthur Holmes, a man many years ahead of his time, had provided the foundations upon which the mighty edifice of plate tectonics was erected.

In 1968, A. A. Meyerhoff, despite his misplaced scepticism concerning drift, published a short article in the *Journal of Geophysical Research*,[89] bringing Holmes' theories of sea-floor spreading, continental drift, and mantle convection to the attention of a wide audience of Earth Scientists, inadvertently precipitating an unseemly fracas. In reply to Meyerhoff's article, Harry Hess and Bob Dietz, both of whom had claimed the sea-floor spreading hypothesis as their own, published self-serving and inaccurate notes underplaying Holmes' contributions and claiming, incorrectly, that he had only suggested *stretching* of continental crust rather than spreading of new oceanic crust. Hess also made the astounding assertion that 'Holmes' work and sea-floor spreading have very little in common'.[90] He conceded that Dietz had coined the

[87] Nevertheless, it should be borne in mind that mid-ocean ridges are still the primary mechanism of heat loss from the Earth's interior.

[88] Burke and Wilson (1976).

[89] Meyerhoff (1968).

[90] Hess clearly had unconventional views on the pioneers of continental drift and, in 1968, rejected a review paper by Tuzo Wilson, partly on the grounds that he had been too generous to Alfred Wegener.

name 'sea-floor spreading' for what he described as 'my concept', but, privately, he resented the fact that Dietz had beaten him into print on the subject, and accused him of plagiarism: 'In November 1960 I explained to Dietz in rather great detail the ideas in my paper that appeared in the Buddington Volume.[91] I had just completed the manuscript. He rushed off and had these ideas published in *Nature* early in 1961. Even the phraseology which I used in describing the ideas is the same. There was no hint of any acknowledgment to me. Being as generous as I can I might suppose he forgot the source of "his" ideas. But I believe this is a bit too generous.'[92]

Bob flatly denied that he had any such meeting with Harry in 1960, and claimed that he had been a mobilist since 1953, despite never bothering to read Holmes' seminal publications (neither he nor Hess made a single reference to the great man's work in their original articles of the early 1960s). Dietz's account of his contributions at that time was described by historian Henry Frankel thus: 'It is as if he were recounting what he would have said had he been a mobilist while working on the origin of trenches and seamounts in the early 1950s.' Bob the confabulator later claimed that it was his paper[93] that converted Tuzo Wilson to the concept of mobilism in 1961. On the question of priority, William Menard, the only colleague acknowledged by both Harry and Bob in their papers on sea-floor spreading, wrote later, 'If there was any priority, it may have belonged to Arthur Holmes for the ideas in his textbook of 1944.'[94]

In fact, Holmes' model of sea-floor spreading was much closer to reality than the one promulgated by Harry. Arthur believed the oceanic crust was produced by mantle melting and volcanism, thus having a basaltic composition, whereas Harry proposed that 'The oceanic crust is serpentinized peridotite, hydrated by release of water from the mantle over the rising limb of a current'. That is, he suggested that spreading involved mantle rocks ascending from great depth and releasing water that had been bound into their minerals, which then reacted with near-surface peridotite, turning it into the soft green rock known

[91] Hess (1962). This is the article referred to by Hess as 'geopoetry', a term filched from the book *The Pulse of the Earth*, by the Dutch geologist Jan Umbgrove (Umbgrove, 1947).

[92] Allwardt (1990).

[93] Dietz (1994).

[94] The quote is from Menard's aptly named book *The Ocean of Truth* (Menard, 1986).

as serpentinite. This was, according to historian Alan Allwardt,[95] 'the one unique feature of Hess's hypothesis', and it was wrong. As we saw in Chapter 1, dredging, drilling and seismic studies soon showed that the oceanic crust is basaltic, pretty much as Holmes predicted, being formed by melting and volcanism at mid-ocean ridges. The idea that the ocean floor is carried along by convection currents, an idea derived from Alfred Bull's 'drag' hypothesis and adopted by both Harry and Bob, also turned out to be incorrect, and was consigned to the waste bin of Earth Science.

By 1968, plate tectonics finally achieved general acceptance, prompting Walter Elsasser to write, 'We are at present in what is perhaps the most remarkable age of discovery since the days of Columbus.'[96] Yet, just months later, Harry Hess, nearing the end of his life, declared that the good times were over for plate tectonics, leaving only a mopping-up exercise. This was his biggest mistake of all.

[95] Allwardt (1990).
[96] Elsasser (1968).

PART II

SECOND GENERATION

The acceptance of continental drift has transformed the earth sciences from a group of rather unimaginative studies based on pedestrian interpretations of natural phenomena into a unified science that holds the promise of great intellectual and practical advances.

Tuzo Wilson (1976)

6

Scum of the Earth

> Give the geologists ten or twenty years, they'll bugger it all up again.
>
> TEDDY BULLARD[1]

The geologists' ship was now under full sail. Ahead lay an exciting expedition to reinterpret the vast store of geological treasure gathered on the continents since the days of James Hutton and William Smith. However, having spent centuries constructing their cathedrals (to mix metaphors), these men were not going to abandon them so easily. Well into the 1970s, the poor old geologists were still refusing to mend their ways, despite what they regarded as a 'reign of terror' by geophysicists. 'Plate tectonics is fine', they admitted grudgingly, 'but it does not work in my area'. One of the most progressive, John Dewey, later recalled[2] that, on being shown a long marine magnetic anomaly profile in 1965, and having its implications spelled out to him by geophysicists, he was only mildly impressed and remarked 'Interesting, but keep it in the oceans and don't let it onto the continents.' The reaction of the geophysicists, who 'muttered darkly about the ignorance and narrow-mindedness of geologists', was, he recollected, 'slightly scathing'.

*

Land Ahoy!

Thanks to the buoyancy of continental crust, the Earth sports extensive areas of dry land on which a wide variety of organisms, including wise apes, can evolve. The greater elevation of continents relative to the ocean basins is a consequence of isostasy acting on the lighter, thicker, continental crust, which, despite representing only 41 per cent of the planet's surface, provides 96 per cent of its land area today (even though

[1] Quoted by Vine (2010).
[2] Dewey (2003).

The Tectonic Plates Are Moving! Roy Livermore, Oxford University Press (2018).
DOI: 10.1093/oso/9780198717867.001.0001

nearly a third of the continental surface is covered by sea). Continental plates consist of 100–200 km thickness of mantle rock (peridotite) upon which rests 20–80 km of continental crust,[3] the two separated by a sharp boundary easily seen on seismic records as the 'Moho'.[4] The precise composition of continental crust, however, remains the subject of active research.

When talking about continental crust (or any rocks, for that matter), it is important to make a distinction between chemical composition and mineralogy. The former refers simply to the proportions of the chemical elements from which a rock is constituted, whereas the latter is the collection of minerals actually making up the rock, and is a function not just of chemistry, but of temperature and pressure, too. For this reason, mineralogy can provide a great deal of information about the conditions deep within the Earth, as we shall see. In fact, rocks are classified primarily by the minerals they contain, identified in hand specimens or by making thin sections and looking under a microscope. More precise classification is achieved by grinding a rock into powder and then using X-ray fluorescence or mass spectrometry to give the proportions of different elements present. From the results, calculations are made to establish the minerals that ought to be present (which sounds a bit weird to non-geochemists).

Geochemists have all kinds of clever techniques for using the chemical composition and mineralogy of a rock to establish its history. In the case of crystalline continental rocks, the concentrations of silicon and magnesium are particularly important. The former is usually presented as 'silica' (SiO_2) concentration, while the latter, also presented as an oxide (MgO), is often converted to a ratio with iron, known as the 'magnesium number'.[5] As magma cools and crystallizes within a crustal magma chamber, minerals rich in magnesium and iron (i.e. 'mafic' minerals) tend to form first and, owing to their higher density, sink to the floor of the chamber, leaving the remaining melt with reduced MgO, but higher SiO_2. Minerals crystallizing from this more 'evolved'

[3] The global average thickness of the continental crust is close to 37 km.

[4] The crust–mantle boundary has long been equated with the Moho, but this is now being questioned in some quarters. It may also, at least in certain regions, be gradational rather than sharp.

[5] For pedants, the formula is $100 \times Mg/(Mg + Fe^{2+})$, where Mg and Fe^{2+} are atomic proportions calculated by dividing the percentage by weight (wt%) of oxides by their molecular mass.

magma thus include much higher percentages of feldspar (silicates containing aluminium plus calcium, sodium, and/or potassium) and silica (quartz), resulting in rocks such as andesite and rhyolite (if they erupt) and diorite and granite (if they don't). These are given another portmanteau that is equally useful in describing rocks of the continental crust: 'felsic'. It is the lower density of these minerals that, crucially, makes crystalline rocks of the continents lighter than mafic oceanic rocks, and thus resistant to subduction. The difference between the thin (6–7 km) oceanic crust, composed of basaltic rocks with a density of 3300 kg m^{-3}, and thick (20–80 km) continental crust, with an average density of 2900 kg m^{-3}, results in the greater buoyancy of the continents, which thus have an average elevation of 800 m above sea level, compared with the average depth of oceanic crust of 3700 m below sea level.[6] If the continents were to disappear overnight, the entire Earth, with the exception of volcanic islands like Hawaii and Iceland, would be submerged to a mean depth of over 2000 m, making this truly the blue planet.

Boreholes might seem the obvious means of sampling the continental crust, but current drilling technology cannot cope with temperatures above 300°C, encountered at depths of just 10 km or so. Because of this, our current knowledge is based on seismic studies, fragments of rock coughed up in volcanic eruptions (xenoliths), and rare outcrops thought to be representative of deeper rock types. This evidence shows that, whereas the oceanic crust is remarkably uniform worldwide, continental crust is highly variable, both horizontally and vertically. We can see for ourselves that the *upper* crust varies from place to place—this is, after all, the rationale behind geological mapping. Generally speaking, it consists of a variety of sedimentary rocks deposited by water or wind, igneous rocks crystallized from magmas, and metamorphic rocks formed by burial and heating of the other two. Overall, it has a composition similar to a granitic rock known as granodiorite. On the other hand, a typical chunk of the *lower* crust might be classified as a 'granulite': a purely crystalline metamorphic rock consisting of the minerals quartz, feldspar, and pyroxene, possibly with dull red garnets. A middle-crustal rock might consist of amphibole and feldspar minerals, and thus would be classified as an 'amphibolite'. Many of these minerals are rich in magnesium and iron, and therefore both amphibolite and

[6] Eakins and Sharman (2012).

granulite are classed, like basalt, as 'mafic'. Mantle peridotites, containing large proportions of olivine, are even richer in these two elements, and are therefore classed as 'ultramafic'. Taken as a whole, however, the bulk composition of the continental crust is remarkably similar in composition and physical properties to andesite, the characteristic rock type of volcanic arcs. When this composition is added to that of the depleted subcrustal mantle part of the lithosphere, the result is a composition very similar to that of chondrites, that is, to the original composition of the Earth. Hence, the present continental crust was most likely derived from the mantle by partial melting.

The most ancient rocks are found in regions of cold, strong, crust, known as 'cratons',[7] some of which are exposed at the surface while others are buried beneath a layer of younger sedimentary rocks. The former are referred to by the unhelpful term 'shields', while the latter are known, equally unhelpfully, as 'platforms'. Many cratons are of Archean age (i.e. older than 2.5 billion years), and form the nuclei of present-day continents, frequently surrounded by less rigid rocks formed during later periods (Figure 6.1). The rocks found in cratons reflect their long and painful history, and commonly consist of dull grey sodium-rich granites known as tonalites,[8] together with bands of more colourful rocks referred to as 'greenstone belts'.[9] Apart from their interest as recorders of the earliest chapters in the Earth's history, these rocks also host important mineral deposits,[10] along with most of the world's diamonds. The main point, however, is that, although they contain rocks deformed and altered by ancient tectonic events, cratons have resisted further deformation for the past billion years or more. During this time, much of their former topography has been removed by erosion, so that they tend to form landscapes with low relief. It also appears that each craton carries its own bit of mantle along with it, forming a sort of root or 'keel' that extends to 200 km depth or more. As with plates, nobody knows exactly how many cratons there are, but

[7] Pronounced 'kray-tons', the term is derived from the Greek word *kratos*, meaning 'strength'. 'Kraton' was introduced by Hans Stille, following Leopold Kober's use of 'kratogen' in the 1920s, and later anglicized by Marshall Kay to 'craton'.

[8] To be precise, three related rock types are characteristic of these terranes: tonalite, trondhjemite, and granulite.

[9] The green colour is a result of various greenish amphibole minerals, including chlorite, actinolite, and epidote.

[10] Underlined by the fact that greenstone belts are frequently referred to as 'gold belts'.

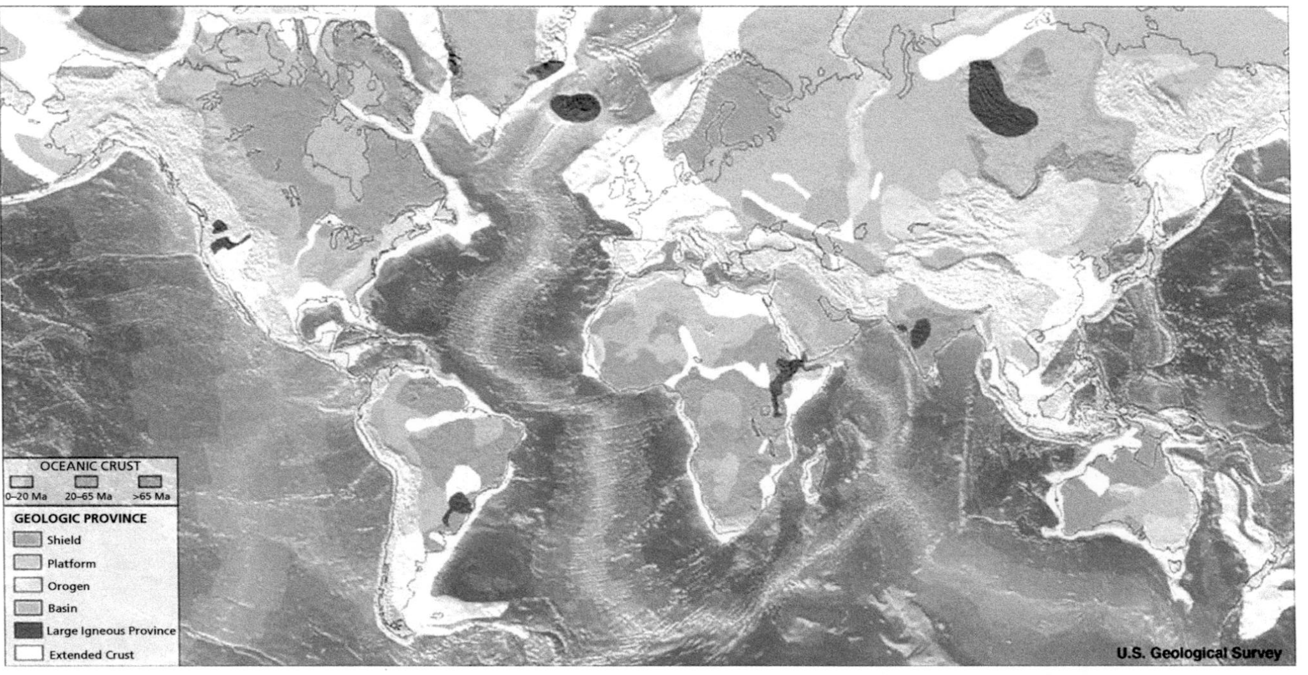

Figure 6.1 Geological provinces of the world. [Courtesy of the USGS.]

35 seems a good number. Notable examples include the Canadian Shield, the Kaapvaal craton of southern Africa, and the Yilgarn and Pilbara cratons of western Australia.

The second major component of continents consists of the orogenic belts,[11] forming up to half the total surface area in some cases, such as North America. These are mosaics of younger rocks that have been welded to the cratonic cores mainly during the closure of former oceans. The Alpine–Himalayan mountain range represents such a belt in the making, a consequence of the closure of the Tethys Ocean. The origin of orogenic belts has been a preoccupation of geologists for more than a century, but only with the arrival of plate tectonics have they begun to make sense of the role of these belts in the evolution of continental crust.

Suspect Terminology

Not all US geologists during the 1960s were luddites. In 1966, Warren Hamilton and Bradley Myers, at the United States Geological Survey (USGS), attempted a reinterpretation of the geology of the western USA in terms of moving plates.[12] They fully accepted major plate motion on the San Andreas fault system, even while many of their colleagues did not, concluding that this was only one aspect of deformation in the western USA. Major faults divided the region into what they referred to as 'terranes',[13] each having a different history of movement. For example, they noted 'The volcanic terrane of northwestern Oregon and southwestern Washington forms new volcanic crust in a region which was oceanic before Cenozoic time'. Warren and Bradley went on to suggest that 'fragments of the continental plate are drifting independently over the mantle and oceanic crust', a phenomenon they attributed (incorrectly) to the rotation of the Earth, concluding (also incorrectly) that 'continental pieces are not being carried passively upon moving mantle material'. Despite these errors, the recognition of discrete

[11] An 'orogenic belt', or 'mobile belt', is a linear or arcuate zone, on a regional scale, that has undergone compression (Allaby, 2013). In everyday language, this is a mountain belt.

[12] Hamilton and Myers (1966).

[13] A 'terrane' is defined as a fault-bounded area or region with a distinctive stratigraphy, structure, and geological history.

geological entities as 'terranes', each with its own history, would later become a fashionable approach by geologists worldwide.

Now, it is just possible that you may not have come across fusulinids. In any case, it is a certainty that you will never have seen these microscopic plankton in the flesh, since the last one gave up the ghost 250 million years ago, during the most thorough clear-out of species ever experienced on Earth (at least, until the arrival of *Homo sapiens*). Nevertheless, their calcite shells, produced in a similar way to living foraminifera,[14] are common in marine rocks of Permian and Carboniferous age (359–252 million years old), sometimes constituting entire beds of limestone. One such limestone has been well known from the Cache Creek formation of British Columbia for many years. In 1950, a group of geologists sampled an outcrop of this formation near Fort St. James, a few hundred kilometres north of Vancouver, and were surprised to find that the fusulinid species preserved within them bore little resemblance to species identified elsewhere within North America. Instead, they belonged to a genus called *Akiyoshiella*, known (at the time) only from south-west Japan. Rocks of similar age and lithology, also containing Asian fauna, extended in a belt as far north as the Yukon and as far south as northern Washington State, presenting a major conundrum for any interpretation based on fixed continents. The anomaly did not escape the attention of that great synthesizer, Tuzo Wilson, who, fresh from his success with his proto-Atlantic hypothesis, naturally applied the same reasoning to the Pacific coast. While recognizing the potentially foolhardy nature of the exercise, Tuzo suggested that a 'western land' (eastern Asia) had collided with the former passive margin of western North America between 400 and 100 million years ago, leaving behind slivers of crust now incorporated into the Cordillera. The idea was little more than speculation, however, and occupied only a few paragraphs of his article.[15]

A little later, in 1971, James Monger and Charles Ross, of the Canadian Geological Survey, published an article pointing out another strange anomaly concerning the fusulinid microfossils. While the Cache Creek belt correlated with a 'Tethyan' assemblage found in Japan and elsewhere in Asia, the fossil assemblage in the adjacent eastern belt

[14] Foraminifera are nanoplankton that produce tiny shells of calcium carbonate. We will come across them again in Chapter 8.

[15] Wilson (1968).

matched those of similar age in the rest of North America. Changes in fossil assemblages are not, of course, unusual: indeed, this is the very basis of biostratigraphy (the dating of rock sequences using fossils). This assumes, however, that different fossil assemblages are of different ages and can thus be used to date the rocks in which they are found. In this case, it seemed that the two very different assemblages were of much the same age. Even more surprising, a third band of rocks to the west of the Cache Creek belt contained the same North American assemblage as the one in the east. The origin of this fossil sandwich was a mystery. Recognizing the uselessness of geosynclinal theory[16] in explaining this geology, James and Charles suggested two explanations: an unconvincing fixist interpretation involving different environments of deposition, and a mobilist interpretation in which the faunas 'may have been transported bodily for considerable distances and brought into contact with one another'. They proposed that the western and eastern belts had once formed a continuous offshore volcanic arc that had collided with the Pacific margin of North America, following which a sliver of ocean crust containing the Asian fusulinids was also plastered on. Later, strike–slip faulting (i.e. like the San Andreas) had shifted part of the accreted volcanic arc along the coast, enclosing the exotic ocean crust.

In papers still littered with the confused geosynclinals terminology,[17] Clark Burchfiel at Rice University and Greg Davis at the University of Southern California applied plate tectonics principles to the Cordillera further south in the western USA. They established that, inboard of these displaced belts (they did not use the term 'terranes', preferring the ugly word 'allochthons'), close to the present-day Rocky Mountains, lay an ancient passive margin formed by rifting during the Precambrian, at least 850 million years ago. A volcanic arc later formed offshore, separated from North America by a marginal basin until the Triassic, when the basin closed and the arc was sutured to the North American continent. Subduction then continued to the west of the accreted arc, forming an Andean-type volcanic margin. Clark and Greg summarized their main points thus: 'Microcontinental fragments or island arcs may have been swept locally into the continental margin, but no major continent–continent collisions occurred.' At about the same time, James Monger and colleagues at the Geological Survey of Canada applied

[16] See Chapter 2.
[17] For example, Burchfiel and Davis (1972).

a similar analysis to the complex geology of the Canadian Cordillera. They identified 'five clearly defined physiographic and geological belts', from the Rocky Mountains assemblage of thick sediments in the east, representing the shelf and slope of the ancient rifted margin, to the young volcanic rocks in the west, representing the most recent phases of subduction at an active margin. They suggested that most of the Cordillera west of the Rockies was 'allochthonous' (i.e. it had come from somewhere else), but rejected the idea that it contained a fragment of Asia.

Just south of San Francisco, ideally situated next door to Stanford University and Palo Alto, lies Menlo Park, the location of the major west coast headquarters of the USGS. Menlo sits right on one of Hamilton and Myer's 'terranes' (as they called them)—the Franciscan terrane. This is a mangled mass of thick sediments, volcanic lavas, and other rocks, including serpentinites, deposited in the trench at which the Farallon plate[18] was subducted until a major plate boundary reorganization 28 million years ago. Here, in the early 1970s, a group of terranists[19] known as the 'Menlo Mafia' was formed, the Godfather of which was the palaeontologist Davy Jones. It was the Menlo Mafia that really got the 'terranology' bandwagon on the road. And bandwagon it was, with geologists worldwide anxious to leap aboard regardless of its destination. Conferences and symposia were held with increasing frequency through the 1970s and 1980s to review progress in reinterpreting the geology of most of the world's continents. Many continental regions, it appeared, were composed entirely of terranes, including West Antarctica and nearly all of Britain.

In 1980, Peter Coney, together with Davy Jones and James Monger, wrote a bestselling article in *Nature*, drawing together previous work on what they now called 'suspect terranes'. Most of the North American cordillera was, they claimed, a vast collage of geological provinces 'swept from far reaches of the Pacific Ocean'. They were 'suspect' because their original settings relative to North America were uncertain. By this time, more such terranes had been 'discovered' in the American cordillera than there were states of the Union.[20] They constituted pretty well all of Alaska, the 300 km-wide western seaboard of

[18] The ancestor of the Juan de Fuca, Cocos, and Nazca plates.

[19] Le Grand (2002).

[20] Coney et al. (1980).

Canada and the USA, and even bits of Mexico. Prior to 350 million years ago, as Burchfiel and Davis had suggested, the Pacific margin of North America was passive, just like the Atlantic margin today, accumulating thick sediments from rivers and submarine sediment flows. This margin formed originally in the Precambrian (750 million years ago) as North America rifted from some other continental mass, giving birth to the ancestor of the Pacific Ocean. Later, a line of intra-oceanic volcanic arcs developed at a subduction zone separated from the continent by a series of marginal basins, much as in the present-day western Pacific (which we will come to). These marginal basins were themselves subducted, and the volcanic arcs, together with fragments of oceanic crust, were accreted to the North American continental margin, so forming the Cordillera. Other terranes, they noted, had 'oceanic affinities' and therefore may have originated far out in the Pacific Ocean.

1990 was a bad year for the terranists. John Dewey and his colleague, the equally modest Turkish geologist Celâl Şengör, produced a very lengthy dismantling of the terrane concept in a paper entitled 'Terranology: Vice or Virtue?',[21] presented at a Royal Society Discussion Meeting in London. They highlighted the complexity of collages such as the American Cordillera, and the various modes by which plate tectonics creates terranes. For example, long slivers of continental crust isolated in the oceans are to be found today, notably in the Arctic, where the Lomonosov Ridge consists of a submarine ribbon of continental crust rifted from Eurasia (currently the subject of a territorial dispute between Canada, Denmark, and Russia), and in the south-west Pacific, where the Lord Howe Rise, a similar, submerged[22] continental strip was rifted away from eastern Australia during the opening of the Tasman Sea. A similar feature appears to have split off from northern Gondwana and crossed the Tethys Ocean from south to north in the Jurassic, eventually colliding with southern Eurasia—this was dubbed the 'Cimmerian continent' by Celâl, and its migration compared to the sweep of a windscreen wiper. John and Celâl pointed out that the idea of fault-bounded blocks of deformed rocks was not new—similar definitions had long been used by geologists for a range of other terms, while the word 'terrane' had previously been applied, somewhat vaguely, to a

[21] Şengör and Dewey (1990).

[22] These strips were thinned during rifting, and thus, by isostasy, ride lower on the mantle.

variety of different geographical and geological entities. Terranism, as it had come to be called, thus consisted of the application of a jargon word as a blanket term for continental and oceanic blocks, slivers, or fragments that were better described as just that. At the same Discussion Meeting, Warren Hamilton took an even more hostile view of terranists,[23] comparing their 'geographic terminology' with plate tectonic interpretations, and concluding that 'members of the geographic-terminology group do not present the critical evidence needed for evaluating plate interactions because they generally do not comprehend how plates interact', leading to 'much foolish speculation'. By 2002, even James Monger was no longer using the term 'suspect terranes' when reviewing the evolution of the Canadian cordillera, but was now referring to 'morphogeological belts', while others continued to speak (seriously!) of 'allochthonous tectonostratigraphic terranes' (really, geologists are their own worst enemies).

Some geologists believed that terranes provided a new way of looking at global tectonics, a few even suggested that the idea superseded plate tectonics itself. In truth, the terranists were merely describing plate tectonics on a more regional level that allowed the details of local geology to be fitted into the grand scheme of plate motions. Perhaps more importantly, terranism restored a degree of lost pride by giving priority to local rock outcrops in preference to global understanding. Ultimately, the concept of terranes may turn out to be about as helpful as Hall's geosynclines, yet it does highlight two important aspects of continental tectonics: first, continents grow through time by the docking of slivers of crust formed elsewhere, and, second, a proportion of these additions constitute a net increase in the volume of continental crust, having formed by mantle melting beneath volcanic arcs or oceanic plateaus.

Subduction Factories

Mount Suribachi is a cinder cone situated near the southern shore of a volcanic island in the Bonin group, 1200 km south of Japan. The island is Iwo Jima, and the peak of Mount Suribachi was the spot on which US servicemen famously raised the American flag in February 1945, following a bloody battle to establish a base from which to launch attacks on

[23] Hamilton (1990).

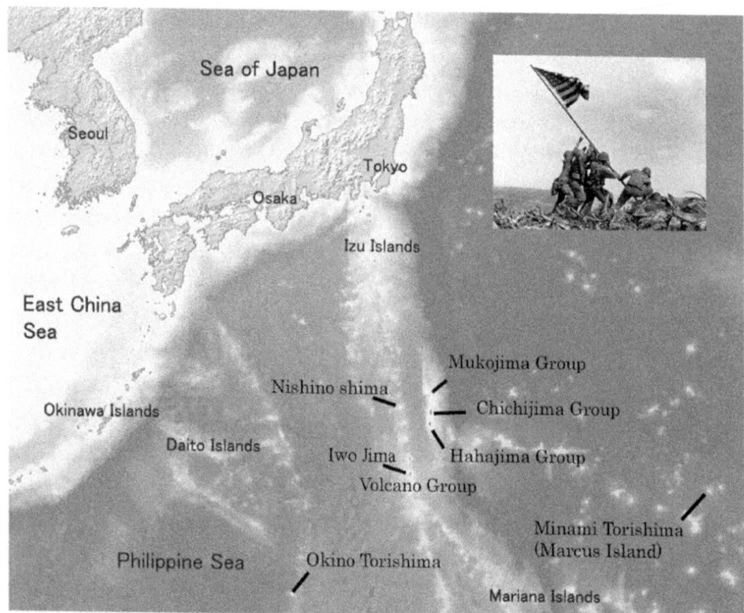

Figure 6.2 Map showing the location of Izu–Bonin islands, including Iwo Jima, the island on which the US flag was raised on 23 February 1945 (inset). [Public domain image, https://commons.wikimedia.org/w/index.php?curid=743950.]

the Japanese mainland. Iwo Jima is part of a narrow line of (mostly submarine) volcanoes stretching more than 2500 km from the Izu peninsula near Tokyo to Guam in the Mariana Islands (Figure 6.2). The group as a whole forms part of what is referred to (with apologies to Big Blue[24]) as the IBM (that is, the Izu–Bonin–Marianas) volcanic arc, a classic example of an oceanic island arc formed far from the nearest continent (in this case, Eurasia), at a subduction zone where one oceanic plate sinks beneath another.[25] The name 'Bonin', in fact, derives from the

[24] 'Big Blue' is a sobriquet applied to the International Business Machines Corporation (IBM). It is possible that younger readers may be unaware that IBM is not just a large business corporation, but also the 'inventor' of the PC, originally known as the 'IBM PC'. Still younger readers may be unaware that PC stands for 'personal computer', the ancestor to their tablets and phones.

[25] Island arcs are frequently referred to nowadays as 'intra-oceanic arcs' to emphasize this fact.

Japanese for 'uninhabited', although two islands today support modest populations. These are Chichi-jima and Haha-jima ('father' and 'mother' islands), located some 120 km east of the main line of volcanoes, in what is known as the 'fore-arc' region, close to the deep Izu–Bonin trench. Even though the islands make rock sampling much more convenient for geologists, the submarine arc has nevertheless been subjected to intensive study, beginning with the numerous depth soundings made by US warships during the Second World War and continued more recently using the Japanese *Shinkai 6500* deep-diving submersible, along with dredging and IODP[26] drilling. The results show that the volcanic arc is built on normal oceanic crust formed by seafloor spreading, and also that subduction-related lavas sampled on the fore-arc islands are of a type of basalt formed by partial melting of mantle peridotite depleted by previous melting. These basalts were named after the arc and are thus known as 'boninites'. Their occurrence at a volcanic arc is interpreted by geologists as marking the earliest stages of subduction, and thus the IBM arc attracts researchers interested in the unsolved problem of how subduction is initiated. Close similarities between basalts dredged from the fore-arc and typical ophiolites[27] have contributed to the current belief that most ophiolites represent mafic crust formed by spreading in a fore-arc setting rather than at mid-ocean ridges.

Subduction of the Pacific plate at the nascent IBM trench began between 52 and 51 million years ago, and was very likely related to the abrupt change in Pacific plate motion that produced the sharp elbow in the Hawaiian–Emperor Ridge. Following this change, the more westerly direction of Pacific plate motion resulted in convergence in the western Pacific, involving the initiation of subduction and the sinking of a slab of old Pacific plate. Current interpretations hold that, as subduction began and the Pacific plate started to sink, mantle peridotite rose from the asthenosphere to fill the space above and was thus subject to decompression and partial melting, generating basaltic magma just as beneath a mid-ocean ridge.[28] As it continued to sink, the slab released

[26] That is, the International Ocean Discovery Program that superseded the International Ocean Drilling Program, that superseded the Ocean Drilling Program, that superseded the Deep Sea Drilling Project we met in Chapter 1.

[27] See Chapter 1.

[28] Arculus et al. (2015).

watery fluids into the adjacent mantle, lowering the melting point and generating a second pulse of basaltic magma from the already-depleted peridotite, producing boninite lavas. Eventually, subduction and arc magmatism settled down into the familiar pattern, establishing a line of stratovolcanoes erupting andesitic and basaltic lavas. Dating of rock samples suggests that the switch to boninitic volcanism took around 2 million years, while the sinking of the slab to the critical 100–150 km depth range and the onset of 'normal' arc volcanism took 7–8 million years.

After 20 million years of subduction, the young volcanic arc unaccountably split right down the middle. In fact, island arcs are prone to this since (a) they are put under extensional stress by subduction and (b) the volcanic line represents the weakest part of the arc crust. A narrow ocean basin, known as a 'back-arc' basin, then formed along this line by sea-floor spreading. The western half of the bifurcated arc was thus carried away from the subduction zone and its supply of magma from the mantle cut off. This remnant arc now forms the Kyushu–Palau Ridge in the West Philippine Sea (Figure 6.3). The eastern half remained above the zone of mantle melting and developed into the modern active IBM arc. Spreading eventually ceased around 15 million years ago, leaving the Shikoku and Parece Vela basins as we know them today. More recently, around 7 million years ago, the Mariana arc also split, resulting in the creation of the Mariana Trough separating the West Mariana Ridge remnant arc from the active Mariana arc, the type example of an island arc. Unlike the Shikoku and Parece Vela basins, the Mariana Trough experiences active sea-floor spreading today. Within the past 2 million years, the Izu–Bonin arc has also begun to split, forming the Sumisu Rift.

The IBM subduction zone is by no means uniform. As we travel south from the Izu Peninsula, the subducting Pacific plate beneath the arc increases in age from around 140 million years in the north to more than 160 million years in the south (amongst the oldest oceanic crust in existence, other than in ophiolites), accompanied by a threefold increase in subduction rate from around 20 mm per year to around 60 mm per year, while the angle of descent of the slab steepens until, beneath the Marianas, it forms an almost vertical sheet sinking deep into the mantle. This is believed to be a cause of the opening of the Mariana Trough, with the near-vertical slab acting as a 'sea anchor' while the overriding Philippine plate migrates westward as it converges

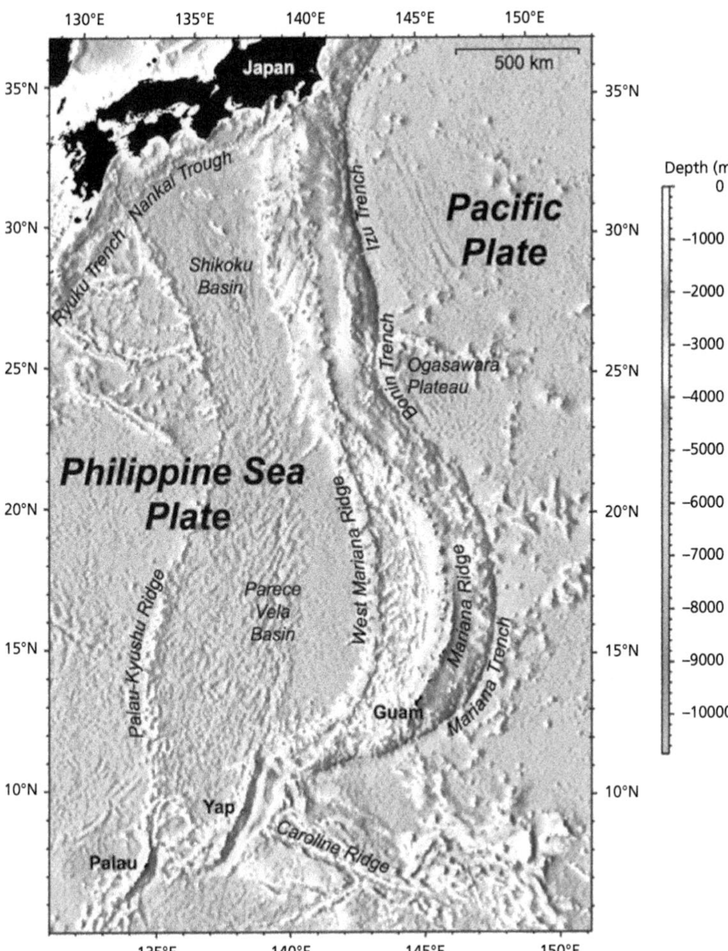

Figure 6.3 Location of the Izu–Bonin–Marianas volcanic arc and the associated back-arc basins. [Map courtesy of SOEST, Hawaii.]

with Eurasia. As is well known, the trench deepens southward, culminating in the Challenger Deep at almost 11,000 m, mentioned in Chapter 1. The subducting sea floor also changes its character, becoming spotted with numerous seamounts (extinct submarine volcanoes) east of the Marianas. The fore-arc, that is, the region between the volcanic island arc and the associated trench, is generally around 200 km in width, but narrows in the south towards Guam. A well-defined outer rise marks the region of downward bending of the Pacific plate before its descent into the IBM trench.

Geophysical work shows that the arc crust is 20 km thick, and suggests that it is composed of a basaltic and boninitic upper crust above a felsic middle crust, and an intermediate lower crust, becoming mafic in the deepest layers.[29] Other island arcs, such as the Tonga–Kermadec and South Sandwich arcs, exhibit many of the same features, including remnant arcs, back-arc basins, and boninite lavas. The crust of such arcs is widely regarded as 'juvenile' continental crust that may be incorporated into future continents after further processing in what have become known as 'subduction factories'.

Such factories, like those of the Industrial Revolution, are easily recognized by their chimneys—that is, the stratovolcanoes that form in great arcs both on land and in the oceans. A typical subduction factory is imagined to operate on two 'floors'. On the lower floor (which might be regarded as the 'foundry'), mantle peridotite is partially melted and the resulting basaltic magma transferred upwards into the crust. On the upper floor, the molten rock is stored in giant settling tanks (magma chambers), where it undergoes fractional crystallization, involving the sinking of dense crystals of olivine and pyroxene, leaving the remaining magma depleted in a range of elements that have gone to form those minerals. Rocks from the walls of the magma chamber may be assimilated into the hot melt along with further injections of fresh magma from elsewhere. As a result of all this, the magma becomes depleted in magnesium and enriched in silica (plus a load of trace elements that won't fit into the settling minerals). Unlike melt lenses beneath mid-ocean ridges, in which silica content is usually below about 55 per cent, magma within volcanic arcs may have a silica content as high as 70 per cent, resulting in the eruption of lavas that cool to form minerals such as quartz and alkali feldspar in rocks such as dacite and rhyolite, although andesite, with a silica content of around 60 per cent, is typical. Most of the magma freezes without erupting, to form giant 'plutons' that constitute the major part of the arc and thus are the principal product of the subduction factory. Above these plutons are a few kilometres of andesite lavas, and below is a layer of dense cumulate minerals.

As with manufacturing generally, subduction factories are not all the same. Volcanic arcs vary depending on three principal factors:

[29] Ichikawa et al. (2016).

(i) the rate of plate convergence, (ii) the distance from sources of terrestrial sediments, and (iii) the age of the subducting crust. Convergence rates range from 20 mm per year (e.g. in the Lesser Antilles) to 240 mm per year (at the Tonga arc, making it the fastest relative plate motion on Earth today). Dredging of the inner wall (i.e. the arc-side) of the IBM trench has produced typical arc rocks, such as andesites, boninites, and gabbros, as well as serpentinites derived from the mantle beneath the arc. While it may not surprise you that dredging a volcanic arc produces arc rocks, it turns out that this is quite unusual. Here, the incoming Pacific plate carries a sediment load just a few hundred metres thick, consisting of fine-grained carbonate or siliceous deposits, the remains of dead plankton. Most of this sediment is swallowed by the subduction zone, leaving the trench clear of sediments. By contrast, at many island arcs outside the western Pacific, the incoming plate carries a much heavier load, including large quantities of coarser sediments supplied by nearby rivers, only part of which can be digested by the subduction zone. The upper layers are bulldozed off the subducting plate by the overriding arc, forming a wedge of sediments known as an 'accretionary prism'. This is frequently imagined to operate something like a very large and very slow snow-plough, scraping sediments from the subducting plate and stacking them at the toe of the fore-arc wedge. Such accretionary prisms tend to form where the incoming plate carries 1 km or more of sediments, and the convergence rate is high. Some sediment disappears beneath the fore-arc, only to be plastered or 'underplated' onto the underside of the prism, while the remainder continues its journey down what is known as the 'subduction channel', that is, the boundary between the downgoing plate and the mantle above, where it may influence mantle melting and the generation of arc magmas.[30]

A fine example lies to the east of the Lesser Antilles arc, where the South American plate is subducting beneath the Caribbean (Figure 6.4). In the northern part of this subduction zone, the Puerto Rico Trench descends to over 6000 m, while, further south, the trench seems to disappear. Deep seismic reflection profiling reveals that the trench does, in fact, continue southward, but that it has been filled completely by sediments delivered by South American rivers, particularly the Orinoco, much of which arrived in submarine avalanches known as turbidity

[30] Remarkably, the concept of sediment subduction was proposed in 1962, four years *before* the mechanism of subduction itself.

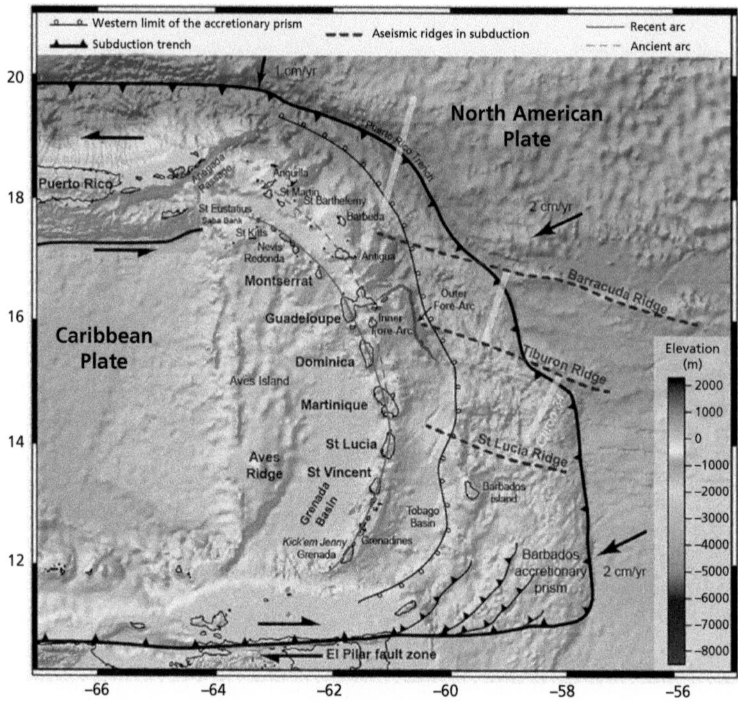

Figure 6.4 The Lesser Antilles Arc. The deep Puerto Rico Trench disappears south of Antigua, where the trench is filled with accretionary sediments. [From Gailler et al. (2017).]

flows. Some of this material has been scraped off during subduction to form the Barbados Ridge, a gigantic wedge of sediments, 300 km wide and 20 km thick, upon which sits the coral-fringed holiday isle. Similar accretionary arcs include those adjacent to Java and Sumatra, the Aleutians, those in the Gulf of Alaska, and the Nankai Trough off the coast of Japan.

Where sediments are thin and convergence rates are high, as in the IBM arc, another process becomes important in the subduction factory. As we will discover in Chapter 7, the surface of oceanic crust is by no means smooth, being covered with linear 'abyssal hills' produced by faulting and volcanism, upon which lies a thin drape of oceanic sediment. The bending of a plate prior to its descent creates more faults, increasing the roughness of the sea floor as it subducts, while, in many places, such as the area to the east of the Marianas, it is also pockmarked with lines of seamounts. When this sea floor passes beneath the toe of a subduction zone, it acts as a rasp, eroding crustal material from beneath the fore-arc and carrying it down the subduction channel. Large seamounts will remove chunks of material as they enter the subduction zone, producing re-entrants in the fore-arc, easily seen on bathymetry (Figure 6.5). Something similar is believed to have occurred as the seamount carrying Haha-jima arrived at the IBM fore-arc.[31] In addition, fluids squeezed out of subducting sediments invade the overriding arc plate, causing it to fracture and collapse into the trench, from where it is conveyed into the subduction channel. Younger, warmer, plates, subducting at shallower angles, will naturally have a greater effect in undermining the fore-arc. Removal by these processes (known as 'subduction

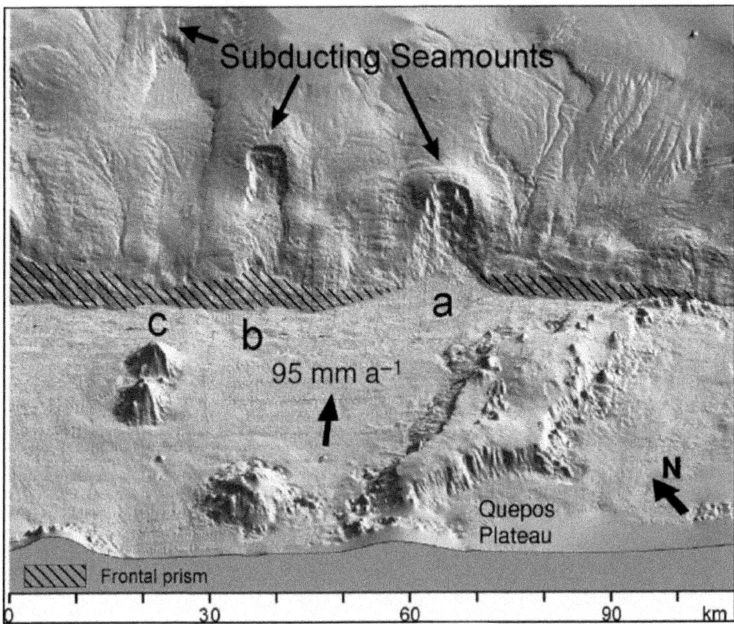

Figure 6.5 Erosion of fore-arc sediments by subducting seamounts in Central Costa Rica. [From Von Huene et al. (2004).]

[31] Subducting seamounts influence the occurrence of earthquakes by suppressing giant earthquakes. This is likely because the size of a quake reflects the surface area of fault that is ruptured, which is reduced by an irregular interface. The result is many small events rather than a few big ones.

erosion') may result in the loss of the mafic crust underlying the arc, and is an important contributor to the factory's balance sheet.

Dropping Off

If you have been paying attention, you may be wondering how it is that simply separating the lighter from the heavier components of a basaltic (mafic) magma can produce an *overall* andesitic (intermediate between mafic and felsic) crust, since the *average* composition of the new crust must remain the same as the original magma. Geochemists and geophysicists have also spotted that catch, and have suggested a number of ingenious mechanisms by which the crust of volcanic arcs may be 'processed' to produce continental crust. The most obvious and direct method is simply to lose the heavier mafic components. If it turns out that the cumulate layer of heavy minerals at the base of the crust is less buoyant than the underlying mantle from which it was derived, then a ready mechanism for getting rid of it is at hand. The bottom of the new arc crust could simply detach itself and sink into the mantle, a process dubbed 'delamination'. Such a phenomenon is made all the more likely since the density of the peridotite beneath the cumulate layer is significantly reduced by the partial melting that produced the arc magmas in the first place. Moreover, calculations show that the cumulate layer is likely to remain heavier than the mantle through which it sinks, possibly allowing it to descend all the way to the core–mantle boundary, according to some recent suggestions.

These ideas could be tested by examining the composition of the lower arc crust, but, as with continental crust, access is tricky. Luckily, a few ancient arcs have been preserved reasonably intact on land, providing geochemists with direct access to the rocks of the lower crust. The Kohistan arc, now lying on its side within the Himalayas of northwest Pakistan, is a good example, providing a more-or-less complete crustal section. Laboratory measurements of seismic wave speeds through Kohistan rock samples[32] match those observed in the IBM arc very well, reinforcing the interpretation that it formed as an intraoceanic arc within the Tethys Ocean during the Cretaceous (134 to 90 million years ago), eventually colliding with the Eurasian margin around 90 million years ago, and becoming an Andean-type continental volcanic

[32] Chroston and Simmons (1989).

arc. As the arc crust grew and thickened, so the pressure and temperature of the lower crust led to changes in its minerals, transforming the basalts and andesites to metamorphic rocks of medium to high grade. Finally, between 50 and 28 million years ago, the arc was caught up in the continental crusher as India converged on southern Tibet, and the arc rocks became folded and faulted as they were incorporated into the Eurasian plate.

Good exposures coupled with intrepid field mapping[33] have allowed the composition of the Kohistan arc to be unravelled. Overall, the andesitic composition of the arc is very similar to that of average continental crust.[34] At the top of the sequence, there are 4–6 km of lavas and associated volcanic rocks with an average silica content of 58 per cent, ranging in composition from boninite to andesite. Beneath this are 26 km of granitic rocks forming the Kohistan batholith, a large body of frozen, unerupted, magma[35] that ranges from mafic to felsic in composition, with an average silica content of 65 per cent. The lowermost Kohistan arc is composed of mafic and ultramafic rocks, including peridotites, dunites and pyroxenites. Density estimates[36] reveal that these rocks, containing dense pyroxene and garnet minerals, would indeed be less buoyant than the mantle beneath, leading to instability. One could speculate that, had the arc's evolution not been so rudely interrupted by continental collision, this dense layer might have become 'delaminated' and sunk down into the mantle, to be replaced by mantle peridotite, thereby creating a sharp Moho, as observed beneath continents. Such 'foundering' is believed to be episodic, occurring every few million years beneath typical arcs. However, even after allowing for delamination, a problem remains. The trace element composition of the lower continental crust, as determined from xenoliths and exposed crustal rocks, does not match that of Kohistan and other exposed arcs, being much more enriched in incompatible elements. Some other process is required to supply these elements and convert volcanic arcs into continental crust.

Peter Kelemen (Lamont-Doherty Earth Observatory) and Mark Behn (Woods Hole Oceanographic Institution) believe this could be a

[33] Made easier, but not necessarily safer, by the opening of the Karakorum Highway.

[34] Jagoutz and Kelemen (2015).

[35] Most magma within arcs never reaches the surface, but freezes within the crust to form 'plutons'.

[36] Jagoutz and Behn (2013).

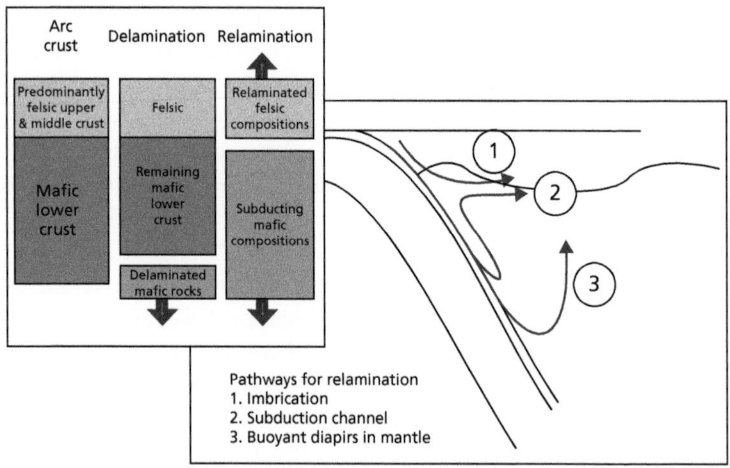

Figure 6.6 Pathways for relamination of buoyant subducted material. [From Kelemen and Behn (2016).]

'density sorting process' involving the 'relamination' of felsic material during subduction and collision.[37] They propose an addition to the subduction factory in the form of a 'crustal refinery' that operates by melting of subducted sediments and other crustal material to produce 'a mafic residue that returns to the mantle and a felsic fraction that is relaminated to the base of the crust in the upper plate'. Relamination may occur during sediment subduction, as in the southern Lesser Antilles, or following subduction erosion, as in the IBM arc. While the lower crusts of the IBM and Kohistan arcs have trace element concentrations much lower than continental crust, Peter and Mark found that, by adding the buoyant fractions to arc lower crust they produced a good match to typical lower continental crust. 'Thus', Jagoutz and Kelemen[38] conclude, 'the arc crustal volume is produced by magmatic addition, minus the material removed by delamination and subduction erosion, plus the material added by relamination' (Figure 6.6).

Thus, it seems likely that the IBM and other island arcs in the western Pacific really do represent the factories at which future continental crust is being created today, and may one day find themselves forming a 'cordillera' along the eastern margin of Eurasia, before being incorporated

[37] Kelemen and Behn (2016).
[38] Jagoutz and Kelemen (2015).

into the growing supercontinent by collision with North America as the Pacific closes.

Thick LIPs

Every so often, the Earth spews out vast quantities of lava on a scale large enough to bury a medium-sized continent beneath great thicknesses of basalt. Luckily for us, things seem quiet today, the most recent episode having formed the Columbia River Basalts of Washington and Idaho around 17 million years ago. This is, indeed, fortunate, since the global impact of such giant outpourings is frequently catastrophic, wiping out sizeable chunks of animal and plant life on land and in the oceans. Good examples are the famous outbursts that ended the Paleozoic and Mesozoic eras,[39] producing the Siberian Traps and Deccan Traps, respectively. The exceptional nature of these events can be imagined from the fact that millions of cubic kilometres of flood basalts[40] are typically erupted at rates higher than that of all the world's mid-ocean ridges combined.

Large igneous provinces, or 'LIPs', as such formations are known, occur in the oceans, too, in the shape of extensive oceanic plateaux. Like those on land, submarine LIPs are composed mainly of basalt, resulting from the partial melting of anomalously hot mantle. They tend to be even larger than the flood basalt provinces on land, and the largest of all, with an area of around 2 million square kilometres (comparable to Greenland),[41] is a submarine plateau in the western Pacific Ocean, close to the Solomon Islands (Figure 6.7). This is the Ontong Java Plateau, formed in the middle Cretaceous, between 125 and 120 million years ago. Parts of the original plateau have since rifted away to form the Manihiki Plateau to the east and the Hikurangi Plateau in the south. The surface area of the original (combined) LIP has been estimated[42] at 5 million square kilometres, representing 1 per cent of the Earth's surface. Drilling and seismic studies reveal a crust more than

[39] The Paleozoic was terminated 252 million years ago, the Mesozoic came to an end 66 million years ago.

[40] The Siberian Traps, for example, originally consisted of around 5 million cubic kilometres of mainly basaltic rocks.

[41] A realistic estimate of the volume of rock contained within the plateau was given by Chandler et al. (2012) as around 100 million cubic kilometres.

[42] Chandler et al. (2012).

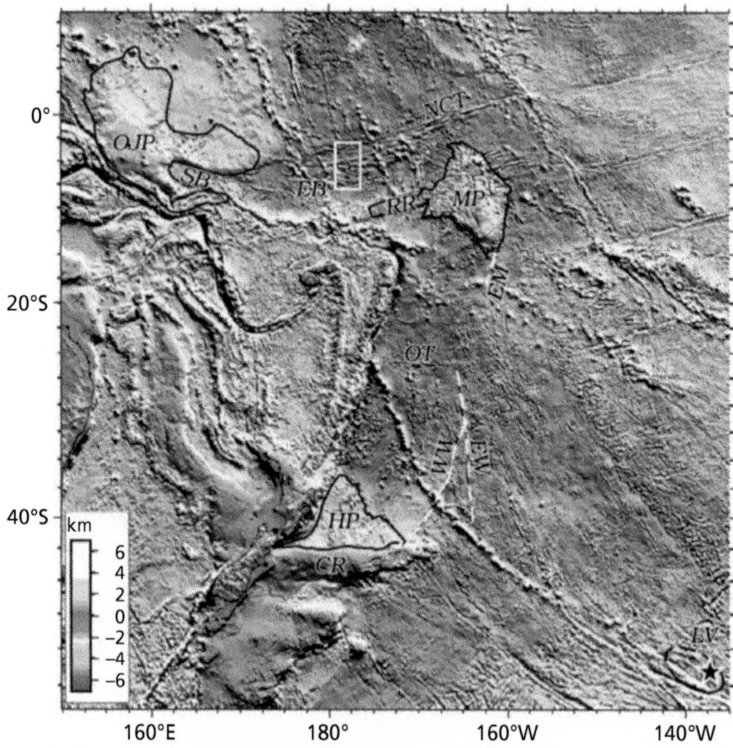

Figure 6.7 The Ontong Java (OJP), Manihiki (MP), and Hikurangi (HP) plateaux in the western Pacific Ocean. [After Chandler et al. (2012).]

30 km thick, composed of basalts with a trace element signature[43] suggesting an origin involving a rising plume of hot mantle rock. Recent work links the formation of the composite Ontong Java–Manihiki–Hikurangi LIP to the arrival of the large (roughly 1000 km in diameter) head of a mushroom-shaped plume from the lower mantle, around 125 million years ago.[44] On impacting the base of the plate, the plume spread out, forming a circular region, 200°C hotter than the surrounding mantle rock. The elevated temperature, combined with the reduction in pressure,

[43] Geochemists have never, I think, received the credit they deserve for developing clever methods for tracing the sources of magmas sampled at the surface. Unfortunately, owing to limited space, they will not receive it here, either.

[44] As usual, an origin involving an asteroid impact has also been suggested, and can be dismissed.

led to extensive melting and the rise and eruption of huge volumes of basaltic magma, forming a giant plateau of thickened mafic crust. Following the massive initial eruptions and the construction of the plateau, the tail of the plume continued to rise and melt as the Pacific plate passed over it, creating a trail of volcanic islands and seamounts—the Louisville Ridge—as the plume waned.[45]

From the location of the Ontong Java Plateau, hard against the Solomon Trench, and of the Hikurangi Plateau adjacent to the Hikurangi Trench, it is evident that these plateaux are reluctant to subduct, and may thus collide with subduction zones, with spectacular consequences. For example, a recent model[46] suggests that the arrival of the Hikurangi Plateau at the northern margin of Gondwana, around 110 million years ago, jammed the subduction zone, leading to the rifting and separation of New Zealand. Hence, if that giant plume had not created the Ontong Java–Manihiki–Hikurangi Plateau, and if the plateau had not subsequently been dissected into three separate plateaux, and if the Hikurangi Plateau had not collided with the Gondwana subduction zone, then New Zealanders would today be living in Antarctica. Nevertheless, there is evidence that a significant fraction of the plateau *has* been subducted, in the course of which chunks of the associated Kermadec volcanic arc have been torn off by tectonic erosion, as explained earlier. Likewise, the arrival of the Ontong Java Plateau at the Melanesian trench caused further mayhem. Dating of hotspot tracks shows that the Australian plate's northward motion slowed 26 million years ago, as it veered to the north-west for three million years before northward migration was resumed. This has been attributed to the Ontong Java Plateau choking the subduction zone and forcing a reorganization of plate boundaries[47] in which the volcanic arc was transferred from the Australian plate to the Pacific plate as the former began to subduct eastwards beneath the latter. Contemporaneous perturbations of Pacific plate motion suggest that the collision had even wider, possibly global, consequences.

It is easy to imagine the situation in, say, 150 million years time as Eurasia, Australia, and the Americas collide, sandwiching these plateaux and

[45] The present location of the Louisville hotspot is west of the East Pacific Rise, near 52.4°S, 137.2°W.

[46] Davy (2014).

[47] Knesel et al. (2008).

their thickened crust within a new supercontinent. However, at present, it is not known what proportion of oceanic plateaux remains at the surface and how much may subduct. Gerald Schubert (at the University of California at Los Angeles) and Dave Sandwell (at the University of Texas) estimated that, if plateaux defy subduction and survive intact at the surface, they could contribute almost 5 per cent of the volume of continental crust globally.[48] And, even if a large slice of such a plateau is eventually subducted, it is still likely that the remainder makes a substantial contribution to the long-term growth of continental crust. Paul Mann (also at the University of Texas) and Asahiko Taira (at the University of Tokyo) investigated this question on the basis of seismic reflection profiles across the Solomon arc and Ontong Java Plateau, concluding that, while the uppermost part of the plateau is being accreted to the Australian plate, the lower part, amounting to about 80 per cent of the crust, has been subducted to a depth of 200 km or more.[49] Whether this is typical of collisions involving oceanic plateaux remains a matter of debate, since the Ontong Java Plateau is anomalous in having formed below sea level and subsided to its present depth of around 2000 m, whereas, given a crustal thickness of over 30 km, you would expect it to have risen above sea level during construction, rather like Iceland. Hence, it may be that it is composed of unusually dense rocks that reduce its buoyancy.

The Ontong Java Plateau was just one—albeit the largest—such edifice created in the interval from 125 to 120 million years ago. The second largest plateau, with an area greater than Britain, France, and Germany combined, is the Kerguelen Plateau in the Indian Ocean, which (it is believed) formed from a similar plume, starting about 120 million years ago. There is good evidence that, unlike the Ontong Java Plateau, the Kerguelen Plateau did originally rise above sea level, forming an extensive land area that was gradually drowned as the edifice cooled and subsided, leaving just a small group of islands today. In this sense, the Kerguelen Plateau is more typical of oceanic plateau evolution expected from models of mantle plumes, this particular plume remaining beneath the Indian plate, generating a continuous 'aseismic' ridge, 5500 km long, along the 90°E meridian as India rushed headlong into Asia. As with the Ontong Java, part of the original giant plateau was subsequently

[48] Schubert and Sandwell (1989).
[49] Mann and Taira (2004).

rifted away—in this case, to form Broken Ridge on the Australian plate, about 43 million years ago,[50] after which the plume remained beneath the almost-stationary Antarctic plate. The waning hotspot remains active, producing volcanism today in the Kerguelen Islands at about 49°S, and slightly further south, at Heard Island.

Recent studies of another large oceanic plateau, the Shatsky Rise,[51] which covers more than half a million square kilometres in the north-western Pacific Ocean, demonstrate clearly that the bulk of this plateau is composed of basalt sheet flows similar to those forming the Deccan Traps and other continental LIPs. The Rise was created 145 million years ago by major basalt eruptions resulting from melting of peridotite mantle roughly 50°C hotter than usual, at depths exceeding 30 km. Taken together with evidence for initial uplift to sea level, this again points to generation by a plume of hot material from the deep mantle.[52] Other large oceanic plateaux believed to result from mantle plumes include the Agulhas Plateau (South Atlantic) and the Mozambique Ridge (Indian Ocean).

At the IBM arc, Haha-jima lies adjacent to another, smaller, oceanic plateau, the Ogasawara Plateau, formed as part of the Cretaceous interval of anomalous magmatism that also produced the much larger Ontong Java Plateau. The Ogasawara Plateau, with an area of approximately 20,000 km^2, rises 2–3 km above the surrounding sea floor, and has been colliding with the subduction zone for the past 15 million years, a state of affairs that is believed to have been the cause of the uplift of Haha-juma. The arrival of thick, buoyant, crust also explains why the trench shallows at this point to just 3 km, compared with 10 km further south in the Mariana Trench. While seismic studies show that the leading edge of the plateau has been at least partly subducted, computer modelling suggests that the plateau is presently resisting subduction,[53] and is associated with the formation of a tear in the subducting Pacific plate, separating the Izu and Bonin segments. It appears that the Ogasawara Plateau was initially carried into the subduction zone, but is now offering stiff resistance, so that the majority of the plateau

[50] Other fragments have turned up in India and Antarctica.

[51] Sager et al. (2016).

[52] The mantle plume explanation has been challenged, as we will investigate in Chapter 10.

[53] Mason et al. (2010).

will eventually become accreted to the IBM margin and, ultimately, to Eurasia.

Regardless of the outcome of the controversies concerning their susceptibility to subduction, it is certain that substantial portions of oceanic plateaux *do* survive collision, and so contribute to the long-term formation of continental crust. This can be seen in the Tibetan Plateau, where an oceanic LIP, created in the Tethys Ocean around 200 million years ago, now covers an area of 200,000 km^2 of central Tibet.[54] In addition, substantial volumes of magma accompanying continental break-up above mantle plumes also make a modest contribution in places such as the volcanic continental margins of east Greenland and north-west Scotland.

There and Back Again

It all started with tiny grains of a mineral known as coesite. Christian Chopin (no relation), a mineralogist at the National Centre for Scientific Research (CNRS) in Paris, discovered the grains as inclusions in garnets extracted from metamorphic rocks sampled in the Italian Alps.[55] Since coesite is a form of quartz (silica) formed at pressures above 2.8 GPa (gigapascals), this suggested that the rocks had, at some stage, been buried to a depth of 90 km or so, that is, within the upper mantle. Other geochemical evidence pointed to a maximum temperature of 700°C, somewhat lower than expected for these depths, consistent with a subduction zone setting. Christian drew the startling conclusion that, contrary to contemporary dogma, continents *could* be subducted, and, just as startling, they could also be returned to the surface. If he was correct, then a central tenet of plate tectonics theory would have to change.

Soon after, a similar discovery of coesite within a pyroxene crystal from an eclogite collected in Norway[56] confirmed Christian's conclusions, demonstrating that continental crust had been subducted to a depth of around 100 km, more than 400 million years ago. A few years later, microscopic diamonds were reported from within a zircon extracted from metamorphosed crustal rocks in northern Kazakhstan,[57]

[54] Zhang et al. (2014).
[55] Chopin (1984).
[56] Smith (1984).
[57] Sobolev and Shatsky (1990).

indicating pressures of 4 GPa or more, and temperatures of around 1000°C, in turn suggesting depths of over 130 km around 530 million years ago (during the Cambrian). These discoveries led to an explosion of interest in such 'ultra-high-pressure' (UHP) crustal rocks and their implications for the subduction of continental crust. Soon, more than 20 coesite and microdiamond terranes[58] had been recognized, mostly within Phanerozoic (younger than 541 million years) continental collision belts. Perhaps the most likely place to find such UHP terranes is in the Himalayas. There, rocks in the Kaghan Valley experienced mantle pressures 46 million years ago, but were exhumed to mid-crustal depths just two million years later. Rapid exhumation appears to be a common feature of UHP terranes, although lower rates have also been found in some examples.

The obvious explanation for these occurrences was that passive continental margins had arrived at subduction zones during ocean closure and been dragged down to depths of 100 km or more,[59] where their felsic rocks were metamorphosed into high-pressure forms involving the transformation of silica to coesite and of carbon to diamond. Following a loss of downward slab-pull, perhaps owing to the breaking-off of the oceanic slab to which the subducting continent was attached, subduction was reversed and large slices of altered, yet still buoyant, crust returned to the surface along the subduction channel—a process termed 'eduction'. The viability of such an explanation was confirmed by laboratory and computer modelling, and this remains the favoured hypothesis. As Christian pointed out in 2003, it is unknown what proportion of subducted continental crust is regurgitated in this way, but it seems certain that at least some is swallowed to become incorporated in mantle circulation. Calculations suggest that, at depths greater than 120 km, mineral phase changes may increase the density of crustal rocks, causing them to become less buoyant than surrounding mantle, implying a 'depth of no return'. However, since a large volume of subducted continental crust clearly does return, this is an area that requires more attention. On this subject, Bradley Hacker and Taras Gerya have noted[60]: 'Why UHP material is exhumed to mid–upper crustal levels within

[58] Unfortunately, this term has refused to die, even though its definition is vague.

[59] Although buoyant, continental crust is firmly attached to the underlying mantle and is thus carried down by the subducting slab.

[60] Hacker and Gerya (2013).

continental settings may have less to do with buoyancy than surface tractions, pressure gradients, or local tectonic plate motions.'[61]

The Reckoning

A few brave souls have attempted to quantify the current rates of additions and losses of continental crust. Peter Clift (Louisiana State University), David Scholl (University of Alaska), Robert Stern (University of Texas at Dallas), and Roland Von Huene (USGS) have all tried to estimate the volumes of crust added and removed each year. Robert Stern and David Scholl very imaginatively used the Chinese concept of yin and yang to represent crustal creation and destruction in the subduction factory. They estimated[62] the total additions to the crust from the global 14,400 km of intra-oceanic arcs to be 1.5 km^3 per year, with another 1.0 km^3 per year from the 27,500 km of continental arcs. Including further additions from continental collisions and LIPS, they arrived at a total of 3.2 km^3 per year, which does not sound all that much.[63] Yet, over the lifetime of the Earth, this is more than enough to manufacture all the continental crust in existence today. On the debit side, taking into account losses from sediment subduction and erosion of volcanic arcs, together with losses in continental collision zones, Bob and Dave estimated a total rate of loss (not including delamination, the rate for which is unknown) of 3.2 km^3, giving an estimated average rate of continental growth of… nil. Today, the total stock of continental crust amounts to one cubic kilometre for every person on the planet,[64] that is, 7.5 billion km^3. However, given the rather sweeping assumptions built into such figures, it is impossible to say with certainty whether these assets are growing, shrinking, or remaining more-or-less constant.

In any case, average rates must hide large secular variations like, for example, that between the very high initial magma production rates of juvenile oceanic arcs and the much lower, long-term, values. Likewise, losses also vary through time, with the volume of sediment subducted

[61] Hacker and Gerya (2013).

[62] Stern and Scholl (2010).

[63] In fact, it is sufficient to fill Loch Lomond in less than a year, Lake Geneva in 30 years, and Lake Michigan in 1500 years.

[64] The total volume of continental crust was estimated by Schubert and Sandwell (1989), while the world's human population in February 2017 was estimated at 7.5 billion (United Nations, 2017).

increasing during periods of enhanced erosion on land caused by mountain uplift or glaciation. It has also been noted that the total length of collisional margins (such as the Alpine–Himalayan belt) at which passive margins may subduct will be much greater during the assembly of a supercontinent than during its fragmentation, varying from perhaps 1000 km to over 20,000 km.

To sum up, it now seems that, during the past 200 million years, the great majority (perhaps as much as 92 per cent) of continental crust was formed at oceanic and continental margin volcanic arcs. The primary raw material of the subduction factory is mafic magma from the mantle, and its major product is andesitic (intermediate between mafic and felsic) crust. Other important ingredients are oceanic plateaux and seamounts[65] (around 8 per cent) constructed by mantle plumes. Waste products include subducted sediments and eroded fore-arc crust (both of which may be recycled beneath arcs). A proportion of the continental crust produced is redistributed by erosion and subsequent deposition as sediment in the oceans, some of which is then accreted to arcs.

Afterthought

While it took a little longer than ten or twenty years, Teddy Bullard's tongue-in-cheek prophesy had, by 1990, been fulfilled admirably: the simple, elegant, theory of plate tectonics was now well-and-truly buggered up. Part of the problem for the geologists was that they persisted in viewing plate tectonics as a mainly oceanic phenomenon concerned with the creation and destruction of oceanic plates, but having little relevance to the complexity of rock structures they mapped on land. Over the decades, they had built up what seems to us now a rather ludicrous lexicon, revolving around imaginary concepts like geosynclines and tectogenes. Like all religious zealots, they did not want their beliefs disturbed by evidence, and, as a result, there was a tendency to throw out the baby with the bath water by assuming that continental plates do not conform to basic plate tectonics tenets. The rigid-plate assumption may have been less useful to geologists studying the deformation of continental rocks, but that was not the point: the importance of plate tectonics lies in the light it sheds on the mechanisms leading to such deformation, not in making the field geologist's life easier. For

[65] Hawkesworth and Kemp (2006).

example, when plates carrying continents converge, there is generally an almighty jolt as the two continents come into contact, but plate convergence may continue for millions of years afterwards, during which the margins of the continental plates undergo severe deformation like the crumple zones of modern cars. Yet this style of deformation in no way invalidates rigid-plate tectonics, which was responsible for the collision in the first place, any more than a rear-end shunt invalidates the internal combustion engine.

In a sentence that may surprise non-geologists, Bob Stern began a 2010 paper on intra-oceanic arcs with this: 'Continental crust is basically a mosaic of orogenic belts, and these in turn are largely nests of island arcs, welded and melded together.'[66] As we have seen, continents also incorporate significant contributions from oceanic plateaux, hotspot trails, and terrestrial LIPs, yet it is clear that volcanic arcs are the primary sites for the manufacture of continental crust, and have been for the past billion years at least.[67] Subduction zones, the locus of most of the world's most destructive earthquakes, are also the factories in which volcanic arcs and, ultimately, continents, are forged. The raw material for the manufacture of continental crust is mafic (basaltic) magma, generated in the mantle foundry and supplied to holding tanks in the crust. Cooling of this magma results in large plutons that constitute the bulk of arc crust, topped by volcanic lavas and underlain by a deep cumulate layer just above the Moho. Subduction zones are also the places at which island arcs and oceanic plateaux collide with continental margins, becoming accreted terranes, adding significantly to the bulk of the continents. Furthermore, subduction zones form the conduits by which the surface communicates with the deep mantle and core of the planet. Kiyoo Wadati and Hugo Benioff could never have imagined the full significance of the sloping sheets of earthquake foci that bear their names: the Earth is, indeed, what Bob Stern[68] has christened 'the subduction planet'.

[66] Stern (2010).

[67] Hawkesworth and Kemp (2006) estimated that around 92% of continental crust is formed at oceanic and continental margin volcanic arcs, while only around 8% originates from hotspots.

[68] Stern (2004).

7

Continents and Supercontinents

Rocks, in short, get around.
BILL BRYSON (2003)

Keith Runcorn's boys showed beyond any doubt that the continents had been touring the globe since the Precambrian, long before they converged on Pangea. Their precise pre-Pangea itineraries were, however, uncertain, for while the ancient latitude of each block could be determined from the magnetic dip of suitable rock samples of the correct age (assuming such could be found), the ancient longitude was a different matter. The magnetic compass in ancient rocks recorded the direction to the ancient pole, but did not allow continents to be placed in their correct relative longitudes. Worse still, the best evidence of relative plate motions—marine magnetic anomalies and fracture zones—had all been shredded by subduction during the assembly of Pangea. And, since the old continents had been around the block a bit since the Paleozoic, there were no continental margins that you could fit together, as had been done for the Atlantic continents by Teddy Bullard and colleagues. Geologists were left only with indirect evidence from rock outcrops to guess the relative positions of these earlier continents. On the bright side, however, the lack of constraints freed them to give full reign to their imaginations, and they quickly began postulating lost supercontinents of all kinds.

*

Collision Course

As we have seen, geologists were initially reluctant to embrace plate tectonics. This left the field clear for geophysicists to explain how continents fitted into the new theory. Dan McKenzie, as so often, was first on the battlefield, applying the newly established principles to the continental boundary between the African and Eurasian plates—that is,

The Tectonic Plates Are Moving! Roy Livermore, Oxford University Press (2018). © Roy Livermore.
DOI: 10.1093/oso/9780198717867.001.0001

the Alps and Mediterranean.[1] The Mediterranean was thought to be a remnant of the Tethys Ocean that had been subducted during the northward movement of Africa. This meant that there was, in Dan's words, 'some hope that the present tectonics of the area would demonstrate what happens when a trench attempts to consume a continent'. He presented a summary of earthquakes in the Mediterranean region, pointing out the striking difference between the width of the zone of deformation between the continental African, Eurasian, and Arabian plates and that of the boundaries between oceanic plates. While the latter were generally less than a few tens of kilometres in width, plate boundaries in continental regions such as the Alpine–Mediterranean zone were much broader and more complex. Dan also noted that focal mechanisms[2] of earthquakes in continental boundary zones do not reflect the motion of the adjacent plates in any simple way. West of Gibraltar, where the oceanic part of the Eurasian plate slid eastward past the oceanic part of the African plate, earthquake mechanisms were straightforward and consistent with movement parallel to the Azores–Gibraltar Ridge. Some very large earthquakes had occurred on this section of plate boundary, including the devastating 1755 Lisbon quake that predated modern seismometers. East of Gibraltar, however, things became much more complicated, with a variety of focal mechanisms suggesting overall convergence. This could be explained in part by the location of the Euler pole for the motion of Africa with respect to Eurasia, which predicted a change from strike–slip to convergence near the Straits of Gibraltar.

According to Dan, the reasons for the difference between the simplicity of oceanic plate boundaries and the complexity of continental boundaries were, firstly, the existence of faults in the generally older continental crust, which might act as lines of weakness, and, secondly, the difference between the silica-poor rocks of the oceanic crust and the silica-rich rocks of the continents. High-silica (quartz-rich) rocks like granite tend to be weaker, and hence deform more easily than low-silica basalts, making plates carrying continents less rigid. The pattern of epicentres in the Mediterranean region suggested the presence of a number of small plates sandwiched between southern

[1] McKenzie (1970, 1972).

[2] The technique applied by Lynn Sykes in his confirmation of Tuzo Wilson's notion of transform faults.

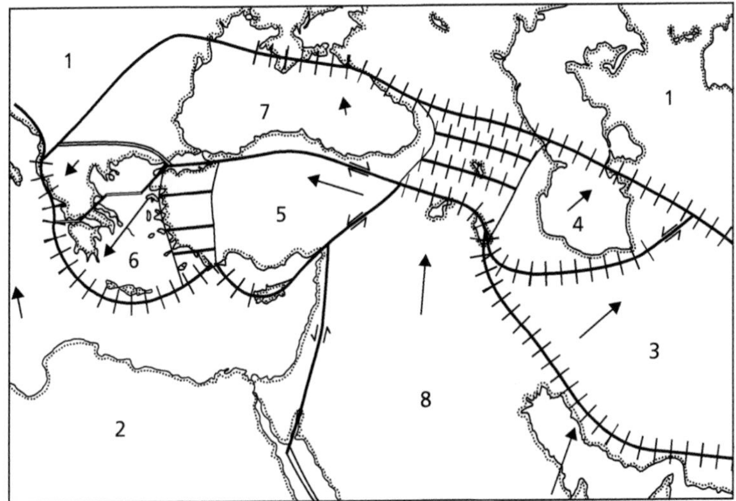

Figure 7.1 Dan McKenzie's sketch of plate boundaries and motions in the Mediterranean region. The arrows show the directions of motion of the plates relative to Eurasia and their lengths are approximately proportional their relative velocity. The plates are numbered as follows: 1, Eurasian; 2, African; 3, Iranian; 4, South Caspian; 5, Turkish; 6, Aegean; 7, Black Sea; 8, Arabian.

Europe and northern Africa, notably to the north of the active Hellenic arc, at which the sea floor of the Mediterranean was being consumed (Figure 7.1). Italy and the Adriatic formed a 'prong' of Africa, penetrating into southern Europe and driving up the Alps, while a head-on collision was proceeding between Arabia and Eurasia in the Caucasus Mountains, involving what Dan dubbed the 'Turkish' and 'Iranian' plates moving west and east, respectively. Application of rigid-plate theory, albeit with rather diminutive plates, thus permitted the interpretation of earthquake mechanisms to derive a basic understanding of the region. However, as Dan pointed out, if it turned out that a very large number of small plates was required, then the rigid-plate model ceased to be useful in the context of continental deformation, and 'continuum theory'[3] might be required.

The Alps form the western part of a continuous belt of continental deformation—the largest such belt on Earth—representing the ongoing

[3] Continuum theory will be familiar to physicists. It treats the crust and mantle as continuously deforming media (Schubert et al., 2001).

collision of Africa, Arabia, and India with Eurasia. This is the Alpine–Himalayan orogenic belt. Since India arrived somewhat earlier and was moving faster, its collision is further advanced, the intervening ocean having long since been crushed within the Himalayas and Tibet. The fun began around 50 million years ago, when the passive margin of northern India first came into contact with a subduction zone along the southern margin of Eurasia, which already sported a range of mountains similar to the present Andes. Plate reconstructions imply an abrupt drop in convergence rate at that time, from around 150 mm per year to less than 100 mm per year. This then declined gradually to around 50 mm per year by 35 million years ago, since when it has slowed only slightly. The leading edge of the Indian plate was sliced off in the collision and stacked up forming the jagged peaks of the Himalayas, but the majority of the shortening was taken up by the Eurasian plate in what is now Tibet.

The idea that India might take tens of millions of years to come to a halt (relative to Eurasia) following initial collision seems reasonable, since intuition tells us that it is going to take a very long time to stop something as big and heavy as a tectonic plate! Intuition in this case is, however, completely wrong, for the momentum (i.e. the product of mass times velocity) of a plate, given the very low velocities of plates, is very small in comparison with the forces that drive or retard it. The result is that plates respond almost instantaneously (within a million years) to changes in their driving forces, such that India's rate of convergence was reduced abruptly on contact with Eurasia[4] and, in the absence of continuing driving forces, suturing could, in theory, have occurred very rapidly. Recent work suggests that the slab of Indian Ocean lithosphere that had been providing the main driving force for northward motion became detached at about the time of first contact (about 53 million years ago[5]), causing the sudden slowing of India. Even after this, however, some driving forces remained, generated by continuing northward 'ridge-push' from the Central Indian Ridge. A consequence of the reduction in northward motion of India during the interval following initial collision was that relative motion (and

[4] Further evidence can be seen in the sharp bend in the Hawaii–Emperor seamount chain in the Pacific, caused by an abrupt change in boundary forces around 50 million years ago.

[5] Zhu (2015). Recent work suggests an even earlier date of 59 million years, based on changes in sedimentation (Hu, 2015).

sea-floor spreading) between India and Australia ceased and the two plates merged into a single Indo-Australian plate. This meant that India was now subject to northward driving forces from subduction at the Indonesian Trench (the trench visited by the 'diving Dutchman', Felix Vening Meinesz) as far east as New Guinea. In addition, it has been suggested that, if the buoyant upper crust was detached, the lower continental crust of India would be dense enough to sink into the mantle, providing a 'slab-pull' of its own. As a consequence of all this, the old, cold, crust of northern India was driven into the warm, thickened, crust of an Andean-type volcanic arc defining the southern boundary of Asia, generating a 50-million-year car crash involving further convergence of between 2000 and 3000 km (depending on where you measure it) between India and Tibet, severely telescoping the crust. But working out precisely how and where this shortening has been taken up has kept geologists busy for many years.

Early interpretations, such as that by Arthur Holmes,[6] assumed that India had simply been 'drawn down' beneath the Indo-Gangetic Trough, 'to continue its unfinished journey beneath the area that has become Tibet'. Earlier still, the Swiss geologist Émile Argand, an early disciple of Alfred Wegener, had drawn a south-to north cross section across the Himalayas and Tibet, showing the crust of Asia being underthrust by what he called 'Lemuria'. In plate tectonics terms, these models suggested that the continental crust of India had been partially subducted, producing a double-thickness crust beneath Tibet, which then rose isostatically to its present elevation of around 5000 m. A double-thickness crust implies that the mantle part of the Asian plate must somehow have been lost. This idea was elaborated by later workers,[7] some of whom suggested that the Indian upper crust may have been crushed against the Asian lithosphere, while only the lower crust and mantle lithosphere was subducted.

While Dan McKenzie had described the Mediterranean section of the Alpine–Himalayan orogenic belt in terms of a complex pattern of small, rigid, platelets, the large-scale and prolonged deformation seen in the Himalayas and Tibet suggested plastic behaviour that, to a geologist, failed to fit the theory of rigid plates. This led a young associate lecturer at the Massachusetts Institute of Technology (MIT) to reflect,

6 Holmes (1965).
7 Powell and Conighan (1973); Ni and Barazangi (1984); Zhao and Morgan (1987).

in 1973, that 'Plate tectonics, however, has not enjoyed the same success in explaining the tectonics of continental regions, and it has not been clear to many geologists that the concept of rigid plates of lithosphere would be particularly useful in understanding continental geology'.[8] A couple of years later, this same young academic, Peter Molnar, published a *Science* article co-authored with Paul Tapponnier,[9] a visiting French scientist at MIT, in which they presented satellite imagery showing deformation in the Himalayas and Tibet. They suggested that, following initial contact between India and Asia in the Eocene (50 million years ago), plate convergence of at least 1500 km was accomplished by 300 km of underthrusting of India beneath the Himalayas, crustal thickening in mountain belts such as the Pamir, Tien Shan, Altai, and Nan Shan, and by movement on several major strike–slip faults that enabled large chunks of continental crust, including Indochina, to be shifted eastward, out of the way of the advancing Indian continent. This last style of deformation, christened 'escape tectonics' by Peter and Paul, assumed that India behaved as a 'rigid indenter', penetrating the weak lithosphere of Tibet.

Soon after, Paul and his colleagues at the Institut de Physique du Globe de Paris, published the results of intriguing laboratory experiments using Plasticine to represent Asia, into which a solid metal indenter representing India was driven (Figure 7.2). This showed a surprising, if superficial, resemblance to the situation in southern Asia, suggesting that escape tectonics could be the primary mechanism by which convergence was taken up (thickening of the Tibetan crust was not reproduced in their model). For a time, this model garnered much attention, and mapping of strike–slip faults by Paul and his students seemed to support the idea. However, their explanation implied movement of thousands of kilometres on faults extending right through the crust and mantle lithosphere, and the evidence for such faults was weak.

In 1982, Dan McKenzie produced yet another important paper, this time with Cambridge postdoc Philip England, attempting to explain the deformation between India and Asia using a continuum (i.e. continuously deforming) numerical model of the continental lithosphere. They noted some obvious limitations in Molnar and Tapponnier's previous

[8] Molnar (1973).
[9] Molnar and Tapponnier (1975).

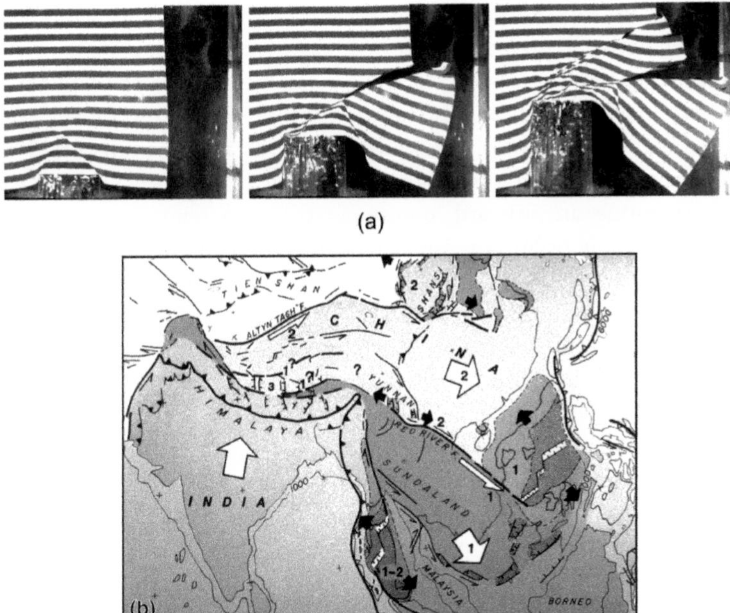

Figure 7.2 (a) Laboratory simulation of the rigid-indenter model using Plasticine. [Photographs courtesy of Professor Gilles Peltzer, UCLA.] (b) Extrusion tectonics model of the India–Asia collision. [From Tapponnier *et al.* (1982).]

modelling, chief among which was the fact that their model assumed a constant-thickness crust in a region where collision had more than doubled it to over 80 km. In their own modelling, Phil and Dan treated the crust as if it were a thin sheet of very viscous fluid material that was rucked up like a tablecloth. They made the simplifying assumption that the crust moved with a uniform velocity from top to bottom, flowing slowly over a layer with no resistance to flow (an 'inviscid substratum'). This may not sound much more realistic than simply breaking the continent into small pieces, but it provided a better approximation to the style of deformation observed in metamorphic rocks exposed on the Tibetan Plateau. As noted by Mike Searle of Oxford University, 'It might seem strange to relate the crust of Tibet to a fluid-like medium, but to any geologist working on lower crustal rocks who has seen the wildly flowing folds and ductile shear zones in metamorphic gneisses, it comes as no surprise at all'.[10] Philip went on to develop his computer models with Greg Houseman of Monash University in Melbourne, concluding that the existence of the Tibetan Plateau could be explained by two main effects. The first was the compressional stress induced by the convergence of the plates, which led to crustal thickening and high elevation. The second was the gravitational effect of the thickened crust in generating an outward 'buoyancy stress'—a tendency to spread out—opposing uplift. The balance of the two would explain the present elevation of Tibet, close to 5000 m above sea level.

During their computer modelling, Greg, Dan, and Peter found that the thickening of a plate during continental collision would result in the cold lower part (composed of mantle rocks) being depressed into the warm asthenosphere beneath. As a result, the cold bottom layer, or possibly the entire mantle part of the plate, might become detached and sink into the mantle. The overlying mantle and crust would then be warmed rapidly, leading to alteration of their minerals, producing metamorphic rocks and possibly melting to form granites, as found in the Himalayas. This delamination[11] would result in a reduction in overall density and consequently an increase in buoyancy, leading to uplift.

At the time, it was difficult to test these suggestions on the ground. China had invaded and annexed Tibet in 1951, and access to westerners was prohibited.[12] Knowledge of the Tibetan Plateau was therefore limited to descriptions of a vast desolate plateau provided by surveyors or 'pundits'[13] employed by the British Empire a century before, supplemented by satellite imagery such as that employed by Peter Molnar and Paul Tapponnier. All this changed within a decade of Richard Nixon's famous visit in 1972. A proposal to the Royal Society led to a two-month reconnaissance expedition in the summer of 1985, crossing the Tibetan Plateau along the newly-built highway from Lhasa to Golmud, more than 1000 km to the north. The two-month expedition, known as the Tibetan Geotraverse, was led by Robert Shackleton of the Open University and Chang Chenssa of the Academia Sinica in Beijing, and

[10] Searle (2013).

[11] An idea originally suggested by Peter Bird at UCLA.

[12] Anyone under forty will find it difficult to imagine the closed and mysterious land that was China during the Cultural Revolution.

[13] From the Sanskrit word for 'learned', in contrast to present usage referring to know-nothing football commentators.

included John Dewey and Peter Molnar among its participants. The results of this and subsequent field seasons transformed knowledge of the structure and evolution of the Tibetan Plateau and opened the way to new investigations and the application of modern techniques.

Tibet, it turned out, was comprised of three separate microcontinents, each with a different history, and each sutured successively onto Eurasia prior to the Indian collision. The northern limit of India was marked by the Indus suture (now known as the Yarlung–Tsangpo suture, since it follows the Yarlung–Tsangpo river valley) in southern Tibet (Figure 7.3), north of which lay the Lhasa block and, beyond this, the Qiangtang block, both ancient continental fragments of Gondwana that had made their way northward across the Tethys Ocean in advance of India. Further north, the Songpan–Ganzi block was sutured to Kunlun in North China as long ago as the Permian (260 million years ago). Each suture represented the site of a former ocean eliminated by continental collision. The Tibetan Plateau had risen by 2000 m within the past 5 million years and had also been extended in a west–east direction. Unfortunately for Peter Molnar and Paul Tapponnier, Dewey and colleagues concluded that 'Little, if any, of the India–Eurasia convergence has been accommodated by eastward lateral extrusion'.[14] Soon after, Peter Molnar abandoned this idea and went over to Philip England's view of deformation involving a viscous sheet.

There were thus three mechanisms suggested for the deformation and thickening of continental plates in the 3000 km-wide India–Asia collision zone. The first was subduction of the Indian continental plate beneath Tibet, producing a double thickness of crust. The second was crumpling of the crust beneath Tibet and the Himalayas. The third was the escape of crustal blocks eastward along numerous strike–slip faults. A study in 1989 by John Dewey, abetted by Steve Cande and Walter Pitman of the Lamont-Doherty Geological Observatory, concluded (somewhat prematurely) that: 'underthrusting is trivial, lateral extrusion is small and [...] the bulk of India/Eurasia convergence has been accommodated by the viscous, roughly north-south, shortening of the Asian lithosphere'. Fortunately, with access improving and technology at hand to measure displacements directly, the question could now be decided once and for all. In the following years, many of the newly

[14] Dewey et al. (1989).

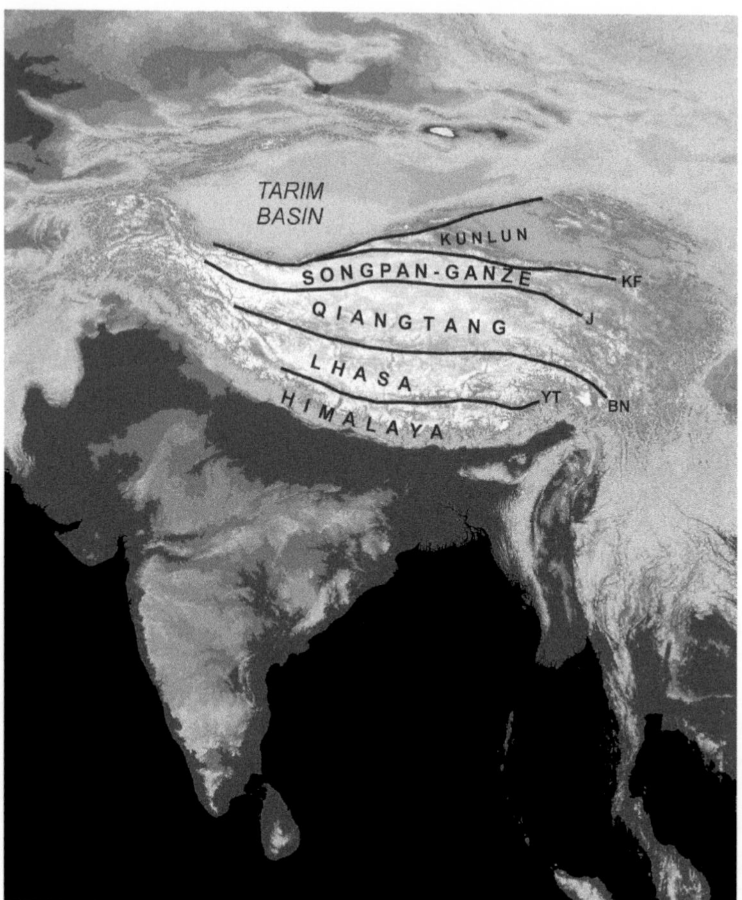

Figure 7.3 Major terranes of the Tibetan Plateau: YN, Yarlung–Tsangpo suture; BN, Bangong–Nujiang suture; J, Jinsho suture; KF, Kunlun Fault.

developed techniques described in Chapter 8 were applied to Tibet, eventually leading to a wealth of new data published by countless Chinese scientists. In fact, so great has this embarrassment of riches become that it is now very hard work to review the burgeoning literature in order to establish the current state of knowledge.

In 1991, an international programme of GPS measurements was instigated to measure crustal movements in the Tibetan region directly. A few years later, the Crustal Movement Observation Network of China (giving us the memorable acronym, CMONOC) was installed,

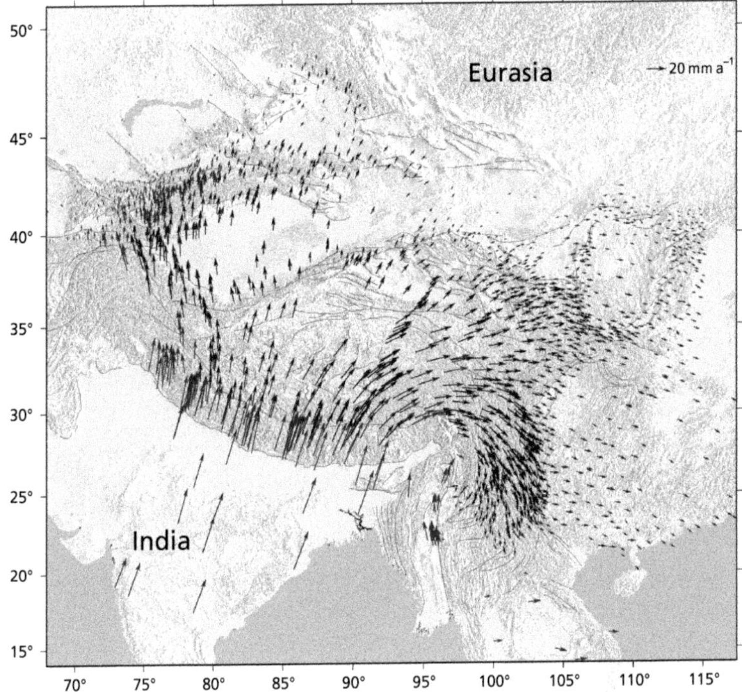

Figure 7.4 Compilation of plate motion vectors derived from GPS studies in the Himalayas and Tibet. Note that the arrows show motion in a reference frame fixed to Eurasia. [From Wang et al. (2017).]

followed by numerous other permanent and temporary installations. Within a decade, spectacular results were being obtained,[15] showing northward movement relative to Eurasia in southern and western Tibet. The latest compilations (Figure 7.4) show crust here is being squashed in response to India's continuing northward push, at rates as high as 30 mm per year, rather as modelled by England and Houseman. In the eastern part of the plateau and beyond, Tibetan crust is escaping eastward and southward, but rather than moving as discrete blocks bounded by faults as Peter and Paul had suggested, it appears to 'flow' around the eastern syntaxis[16] of the Himalayas at rates of around 10 mm per year, then turn southward towards Burma and southern China.

15 For example, see Wang et al. (2001).

16 That is, the point where the Himalayan ranges converge in the east.

And, unlike the rotation of a rigid plate, rates of movement do not increase with distance from an Euler pole of rotation, and are not constant along 'flow lines', pointing to 'deformation of a continuous medium at depth'.[17]

Chinese, US, German, and Canadian research institutions carried out another collaborative project, known as INDEPTH,[18] between 1992 and 1995, to collect a range of geophysical data along a 300 km transect from the High Himalaya to the centre of the Lhasa block. A major component was the shooting of a deep crustal profile using seismic reflection techniques developed by the oil industry, involving artificial sources of seismic waves. In 1996, a multi-author article summarizing the results was published in *Science*, with the somewhat unsettling title 'Partially molten middle crust beneath southern Tibet'.[19] This referred to a bright horizontal reflection seen on seismic records at depths of 15–20 km along much of the north–south profile, with characteristics suggesting the presence of magma, much like the magma lenses imaged beneath mid-ocean ridges, although in this case the magma was thought to be granitic (felsic). This horizon marked the top of a layer of low seismic velocity, which the INDEPTH party interpreted as hot, weak, lower crust. Their findings confirmed a suggestion made nearly a quarter of a century earlier by John Dewey and Kevin Burke,[20] who predicted such melting as a consequence of heating caused by the blanketing effect of thick crust.

Later work within the INDEPTH programme aimed to establish whether an intact Indian plate lay beneath the Tibetan Plateau or whether crustal thickening was due entirely to squashing of the Asian crust. Probably as a result of limited coverage, coupled with the complexity of mantle structure, the inconclusive results suggested that India had penetrated beneath virtually all, some, or virtually none, of Tibet. Taking a different approach, Dan McKenzie and Keith Priestley at Cambridge University used earthquake recordings (rather than artificial seismic sources) to image the Indian slab, concluding that it was likely that India had indeed underthrust most of the Tibetan Plateau. Priestley and colleagues[21] later imaged the cold Indian slab descending

[17] Zhang et al. (2004).

[18] Let's not worry about what it stands for.

[19] Nelson et al. (1996).

[20] Dewey and Burke (1973).

[21] Priestley et al. (2008).

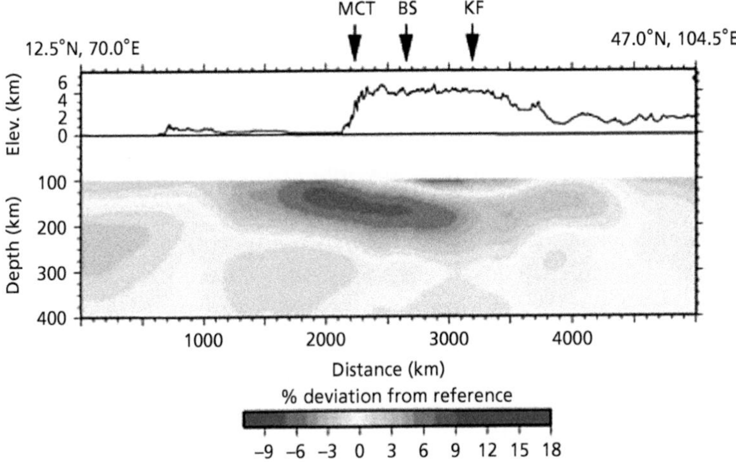

Figure 7.5 Seismic tomography from surface waves showing anomalously 'fast' mantle dipping northward beneath northern India and the Himalaya. Arrows indicate the Main Central Thrust (MCT), the Bangong Suture (BS), and the Kunlun Fault (KF). [From Priestley et al. (2008).]

to depths of 225–250 km beneath much of Tibet (Figure 7.5). Hence, the current view is not all that far from that of Émile Argand and Arthur Holmes.

Everything Flows

Whether the crust deforms by jerks or creeps depends on its strength (strictly, its rigidity) which, in turn, depends on its composition. Felsic (upper) crust, dominated by feldspar and quartz minerals, is intrinsically weaker than mafic (lower) crust, dominated by olivine. A second factor is temperature: as rocks become hotter with depth, so they become weaker and capable of flow by creep. Ultimately, the crust may begin to melt, weakening it still further. Hence, at some depth, there is a boundary between the cold, upper crust, within which earthquakes occur, and the warm lower crust, which reacts to stresses by flowing like a very, very, viscous fluid.[22]

[22] Not to be confused with the asthenosphere, which is the weak upper mantle beneath the plates.

In 1983, Peter Molnar and Wang-Ping Chen wrote an article[23] that set the scene for many of the arguments that would occupy geophysicists working on deformation of the continents for the next two decades. They noted that the maximum depth of the small and infrequent earthquakes that occur within continents depends on the age of the crust. Cold Precambrian (typically more than 800 million years old) cratons experience jerks from depths as great as 25 km, while elsewhere (except where a slab is sinking) the maximum depth is around 15 km. Moreover, while the lower crust is aseismic, quakes do occur in the uppermost mantle that forms the base of plates, particularly in regions of continental convergence like Tibet, suggesting that a strong layer of mantle rock lies beneath the Moho. It seemed that, above a temperature of 350°C, lower crustal rocks became weak enough to flow by creep, while olivine-dominated mantle rocks beneath deformed in a brittle fashion (producing earthquakes) up to a temperature of 700°C. Continental lithosphere was thus like a sandwich (Figure 7.6), with the weak lower crust as the jam-like filling. This became known as the 'jelly sandwich model' as a result of misnaming by American scientists (who refer to proper jelly by the trade-name 'Jell-o', apparently).

The 'jelly sandwich' model was challenged in a series of papers by James Jackson, Dan McKenzie, and Keith Priestley at Cambridge University.[24] While earthquakes in the oceanic mantle were common, improved data showed no clear example of hypocentres beneath the

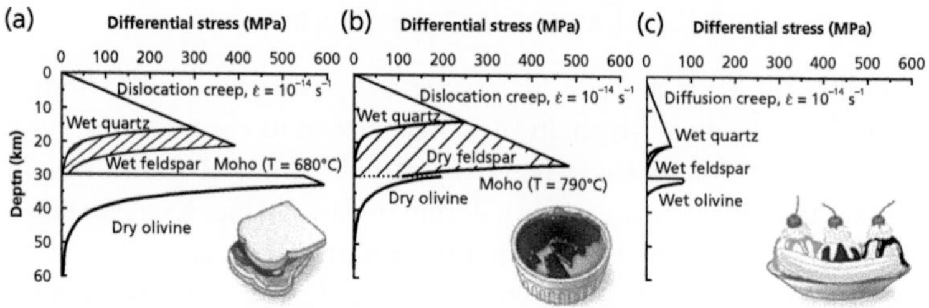

Figure 7.6 In the jelly sandwich model, plate strength is concentrated in the uppermost mantle forming the base of the continental plate, while in the crème brûlée model, it resides mainly in the crust. The banana split model is a (somewhat obscure) reference to crustal weakening around faults and shear zones. [Reproduced from Bürgmann and Dresen (2008).]

[23] Chen and Molnar (1983).
[24] See Jackson (2002).

continental Moho, although many shocks now appeared to originate in the lower crust. Dan and Keith therefore suggested that the uppermost mantle forming the bottom of a continental plate must be weak rather than strong, a result of higher temperatures, which they believed had previously been underestimated, as well as the presence of fluids. The strength of a plate was therefore concentrated in the crust rather than the mantle immediately beneath. This alternative configuration was challenged, in turn, by Tony Watts (at Oxford University) and Evgueni Burov (at the University of Paris),[25] who, following the epicurean theme, christened it the 'crème brûlée' model (Figure 7.6). In their view, such a model would not allow the persistence of mountains for more than a few million years, nor would it permit sinking of intact slabs of continental lithosphere, such as the Indian plate beneath Tibet, into the asthenosphere.

You may (or may not) think it surprising that an inter-varsity squabble could break out in the twenty-first century over something so fundamental as the physical properties of the Earth's crust, yet this simply illustrates our very limited understanding of the subject at present. As explained in a 2008 review,[26] 'Assessing the mechanical properties of rocks for the broad range of thermodynamic boundary conditions prevalent in Earth's interior remains a daunting task'.

Breaking-Up Is Hard to Do

A simple-minded view might be that, since colliding continents, such as India and Eurasia, eventually become sutured together, moving plates must inevitably assemble all continents into a single pangea. Once assembled, something else then must cause the supercontinent to explode, scattering the fragments around the globe once more. This 'something' could be what systems engineers and climatologists call a 'negative feedback': that is, a break-up mechanism set in motion by the creation of the supercontinent itself. A plausible suggestion for what this might be was advanced in 1982 by Caltech seismologist, Don Anderson, who invoked what he called 'continental insulation' of the mantle by the supercontinent. Don proposed[27] that the presence of

[25] Watts and Burov (2006).
[26] Bürgmann (2008).
[27] Anderson (1982).

thick and extensive continental lithosphere led to a rise in temperature of the underlying mantle, sufficient to cause uplift, mantle melting, and, eventually, continental rifting.[28] Hence, the formation of a supercontinent inevitably sowed the seeds of its own destruction.

Don's notion of 'thermal blanketing' was by no means new. In 1925, the famous Irish physicist John Joly, at Trinity College, Dublin, published a book entitled *The Surface-History of the Earth*,[29] in which he offered his theory of 'thermal cycles'. This was summed up by Arthur Holmes thus: 'In Joly's well-known hypothesis of thermal cycles an alternating accumulation and discharge of heat is visualised [...]. Accumulation leads to fusion of the substratum (regarded as of basaltic composition); crustal extension; volcanic activity; and marine transgressions. Westerly tidal drift of the crust slowly draws the ocean floors over the magma zone, and discharge of heat is brought about by thinning of the ocean crust. Solidification sets in by crystallization and the sinking of blocks, accompanied by crustal compression and marine recessions.'[30] Predictably, John's big idea met with a similar fate to those of his contemporaries and was soon rubbished by Harold Jeffreys on spurious mathematical grounds.

Joly's theory was certainly imperfect. His biggest error was invoking puny tidal forces to move continents, rather than continental drift. Another error pointed out by Arthur Holmes was that the mantle was more likely to be peridotite, which, when heated to the temperatures envisaged, did not melt, but did begin to convect instead. Arthur proposed that heat produced by radioactive decay in the mantle would build up beneath the continental 'blanket', creating a rising convection current. Reaching the base of a continent, this current would be deflected horizontally to either side, exerting tractions on the crust that rifted it apart and carried the two halves in opposite directions, the divergence being facilitated by subduction (he called it foundering) of the surrounding oceanic crust beneath the advancing continental fragments. The resistance offered by the sinking slabs would lead to the generation of mountain ranges at the leading edges of continents, as

[28] This may be the explanation for the existence of the Central Atlantic Magmatic Province, a large area of basaltic lavas covering parts of the Americas, Africa, and Europe, erupted around 201 million years ago, just prior to the rifting of North America from Africa. See Coltice et al. (2007).

[29] Joly (1925).

[30] Holmes (1931).

observed along the Pacific margins of the Americas. Following Arthur's logic, it required little knowledge of geology to conclude that the history of the Earth's surface must consist of a series of cycles of supercontinent assembly and fragmentation. A similar, but not identical, idea was, of course, implicit in Tuzo Wilson's later suggestion of ocean cycles generated by plate tectonics:[31] 'If the concept of life-cycles of ocean basins is correct then one can visualize that throughout much of geological time continental blocks have been coming together along island arcs, piling up mountains when they collided, reversing directions and drifting apart again.'

Even before Tuzo's insight, the seed of this idea had been sowed by John Sutton, Head of Geology at Imperial College, London. In 1963, having learned of the breakthrough achieved by Keith Runcorn's group (by the simple expedient of working in the same building), John presented his theory of geological cycles to the world in a *Nature* article, published just six months before Vine and Matthews' classic. Following the fashion, he gave his cycle a name that would be Greek to most readers: the 'chelogenic cycle' ('chelona' is Greek for 'tortoise'[32]). He suggested that four such cycles had been completed, each lasting around a billion years and ending with the creation of mountain belts worldwide. The youngest of these mountain-building episodes occurred about 1 billion years ago and had previously been given the name 'Grenville Orogeny'[33]. This, John proposed, was associated with the assembly of continental fragments, which he linked to convective cycles in the Earth's mantle, the mountain belts marking the locations of down-going convection currents. He summarized his theory as follows: 'The cycle postulated here consists of a sequence of events [...] which leads to the displacement and disruption of the continents as they existed at the start of the cycle and later to the re-grouping of these disrupted masses of continental crust.'

In 1970, Jim Valentine and Eldridge Moores, both at the University of California at Davis, published another important *Nature* article, in

[31] Wilson (1966, 1968). Note that Tuzo was, however, concerned with individual ocean basins rather than supercontinents, a point we will return to.

[32] The use of this term actually referred not to the speed of the cycle, but to the shell of the tortoise, which Sutton likened to a shield. Geologists seem to have an obsession with shields, perhaps because they live in the middle ages, or possibly it reflects a little paranoia.

[33] Named after the Canadian county in which it was first identified.

which they approached the problem from a slightly different angle. Noting previous speculation that plate tectonics had been operating for at least 3 billion years, they suggested that the resulting continental drift provided an explanation for the widespread extinctions and diversifications seen in the fossil record of marine animals and plants.[34] Speculating that these events reflected the construction and fragmentation of supercontinents, they proposed that an earlier pangea, which they referred to as 'Pangaea I', had existed in the early Cambrian (around 540 million years ago). This fragmented during the Paleozoic, concomitant with the formation of new mid-ocean ridges producing young, hot, and therefore relatively shallow, ocean floor. As a consequence, the volume of the ocean basins was reduced, leading to flooding of the continental shelves, in turn affecting climate seasonality and creating a fertile habitat for marine life. Increased isolation of marine organisms on the separating continents led to speciation and an increase in diversity. Later, during the Silurian and Devonian (roughly 425 to 400 million years ago), the continental fragments were sutured in the Appalachian–Caledonian orogeny and subsequent mountain-building episodes to form 'Pangaea II' (Wegener's Pangæa) by the Permo-Triassic (252 million years ago). During supercontinent assembly, the number of mid-ocean ridges would decline as intervening oceans were eliminated, and the remaining ocean floor, being now older on average, would thus be deeper, thereby increasing the volume of the ocean basins and hence lowering global sea level. This would expose large areas of the continental shelves as dry land, leading to an increase in climate continentality (hotter summers, cooler winters), and also reduce living space for marine life, causing faunal extinctions. The newly constructed single continent would now favour monsoon-type climates and cosmopolitan species. Jim and Eldridge's short paper thus pointed the way for much of the research that was to follow in the next decades.

Pangea Puzzles

In 1989, after participating in a trip to Antarctica, Eldridge was browsing in his university library, when he came across a paper suggesting a Precambrian link between the Pacific margin of Canada and eastern Australia.[35] Wondering whether a similar correlation might exist

[34] Valentine and Moores (1970).

[35] Bell and Jefferson (1987).

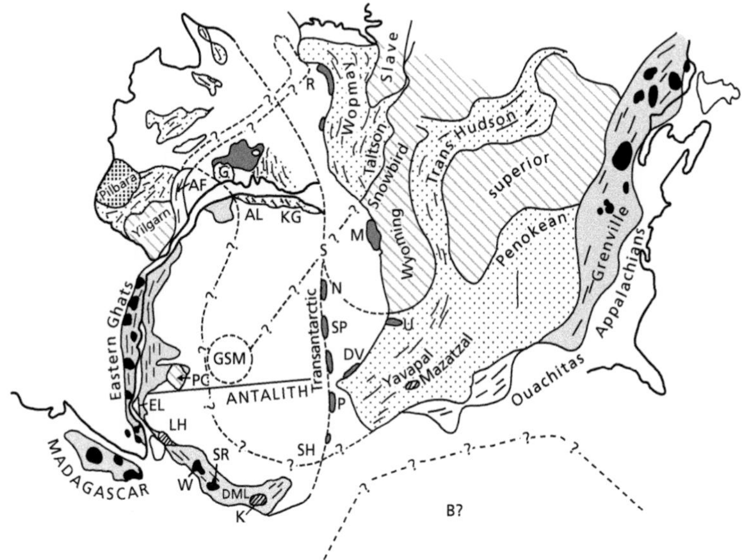

Figure 7.7 Eldridge Moore's 'crazy' SWEAT hypothesis. [From Moores (1991).]

between the US Pacific margin and Antarctica,[36] he drew a map showing the resemblance of structural features between the two and sent it to the organizer of the field trip, Ian Dalziel, a seasoned Antarctic geologist based in Texas, asking 'Is this crazy?' The following year, Eldridge submitted a short article[37] to the Geological Society of America (GSA) journal, *Geology*, presenting what he described as a 'provocative hypothesis' correlating the Precambrian rocks of the Canadian and US cordillera (representing the former western passive margin of North America) with similar rocks in the Transantarctic Mountains and eastern Australia. Since both West Antarctica and eastern Australia consist of crust accreted much later, they were eliminated, along with the present Pacific coastal regions of North America, in order for Eldridge to make his provocative reassembly of the three Precambrian continental cores (Figure 7.7). He gave his reconstruction the rather unfortunate title of 'the SWEAT connection', his clumsy acronym referring to the SW United States and East Antarctica (Australia got ignored).

36 Antarctica was adjacent to Australia in Gondwana.

37 Moores (1991).

Evidently, Dalziel did not think the idea crazy, and quickly prepared a manuscript of his own on the subject,[38] which he dropped through the letterbox of the GSA a month or two later. In it, he noted that there was other evidence in support of the SWEAT hypothesis. For a start, the total lengths of the presumed passive margins of western North America and eastern Australia–Antarctica were comparable—around 4000 km each. Then, he suggested, the Grenville Front (the boundary between rocks deformed 1 billion years ago in the Grenville orogeny and the older, undeformed rocks to the northwest) that cuts the North American margin at a high angle could be correlated with a similar feature close to the Shackleton Range of East Antarctica, which intersects the Transantarctic Mountains at a similarly high angle.[39] Furthermore, Ian now proposed that the separation of Laurentia (i.e. the Precambrian core of North America and Greenland) from Antarctica–Australia around 700 million years ago was followed by rifting of its Atlantic margin to create the Iapetus Ocean, heralding a somewhat elaborate dance of the continents that ended with the suturing of East Gondwana (Antarctica, India, and Australia) to West Gondwana (Africa and South America) during the Cambrian (around 500 million years ago), followed by collision with Eurasia 100 million years later, forming the Appalachians.

The startling implications of these articles were recognized by Paul Hoffman, a geologist at the Geological Survey of Canada, who had refereed both papers. In a third article, published just a few months after Ian's, Paul pursued the methodology of correlating orogenic belts truncated by the rifting of a proposed supercontinent, now incorporating the major continental blocks of Baltica and Siberia into the proposed reconstruction, with Laurentia as the 'core', much as Antarctica formed the core of Gondwana. A surprising feature was the juxtaposition of the Iapetus (proto-Atlantic) margin of North America, not with Britain and Europe, but with 'Amazonia', the ancient core of South America. He noted that 'rifts originating in the interior of the Late Proterozoic supercontinent became the external margins of Paleozoic Gondwanaland; exterior margins of the Late Proterozoic supercontinent

[38] Dalziel (1991).

[39] It is interesting to note that limestone samples collected by Scott's party in the Transantarctic Mountains on their return from the pole showed that some other continent had separated from Antarctica during the early Cambrian, leading to a marine transgression.

became landlocked within the interior of Gondwanaland', and asked, in an attention-grabbing title, 'Did the Breakout of Laurentia Turn Gondwanaland Inside-Out?'[40] He also highlighted the fact that the interval spanning the latest Proterozoic and early Paleozoic was one of drastic changes in global climate and sea level, accompanied by the evolution of the first animals. Might these changes be related to the fragmentation of the Precambrian supercontinent, as Valentine and Moores had suggested?

Paleomagnetism could, theoretically, provide a quantitative test of such reconstructions, but, in practice, paleopoles of the required age were too sparse and too uncertain to confirm or deny the existence of a Precambrian pangea. In fact, an agglomeration of virtually all Proterozoic (i.e. Late Precambrian) continental crust had been mooted as early as 1974 on (shaky) paleomagnetic grounds by John Piper at Liverpool University, who, after constructing tortuous paths of apparent polar wander, concluded that 'Paleopangea' had existed for more than 2 billion years prior to its break-up, 600 million years ago. Despite manning the barricades for decades in a determined defence of this model, John never gained much support for his idea, however. By the turn of the century, the configuration of the postulated supercontinent seemed to be converging on something not too dissimilar to Paul Hoffman's conception (Figure 7.8), which had been re-christened with the rather unevocative title of 'Rodinia' (derived from the Russian word *rodit*, meaning 'to beget'[41]). Reconstructing Rodinia was rather like trying to solve a jigsaw puzzle in which many of the pieces had been lost and those remaining had been chewed by the dog. Unsurprisingly, geologists soon came up with alternative ground plans for the putative supercontinent. In one suggestion, western Laurentia was matched with Siberia rather than Antarctica–Australia, while, in another, it was fitted against Australia but not Antarctica. Further modifications were required by improved paleomagnetic data[42] and geological observations, with some authors questioning Rodinia's very existence. The

[40] Hoffman (1991).

[41] According to McMenamin and McMenamin (1990). Ironically, this distinctly forgettable name was originally applied by these authors to a reconstruction entirely different from that of Hoffman and friends, with the American Pacific margin placed far from Australia–Antarctica. The McMenamins also suggested a Russian-derived name for the accompanying large ocean: 'Mirovia'.

[42] For example, Wingate et al. (2002).

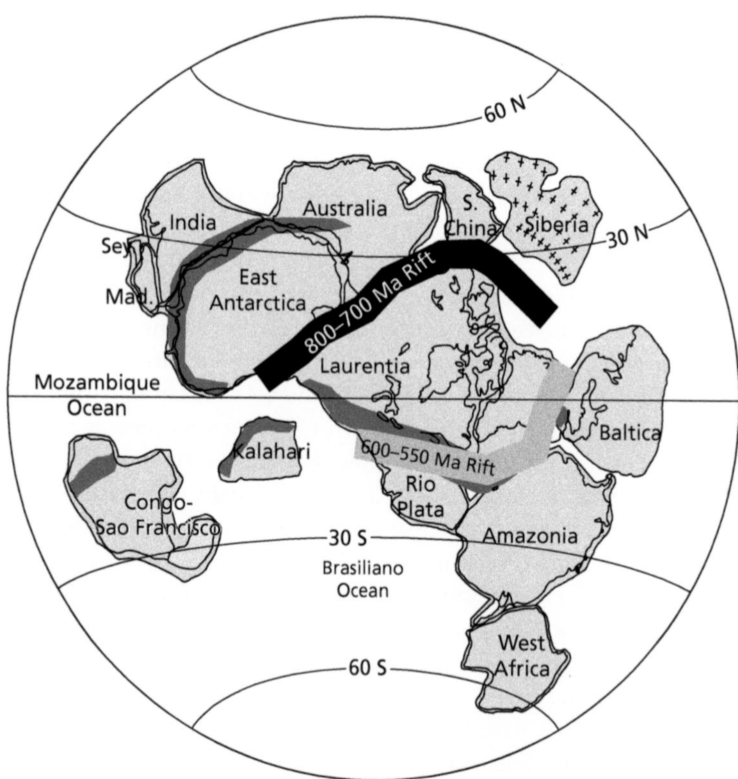

Figure 7.8 Rodinia reconstruction. [From Meert and Torsvik (2003).]

latest position seems to be that a possible pangea did exist between 1000 and 860 million years ago,[43] but its configuration remains uncertain, with some authors putting North America adjacent to Mexico and others inserting southern China between western North America and Australia/Antarctica (thereby invalidating the SWEAT connection).

Given all the uncertainties and controversies associated with the reconstruction of the billion-year-old supercontinent, anybody could see that it would be foolhardy to attempt still earlier reconstructions of the continents. Undismayed, geologists proceeded to assemble pre-Rodinia supercontinents on the flimsiest of evidence, and then arguing over what names to give them. In a riot of speculation, we had continents

[43] Li et al. (2008).

clustering at various times back to the Archean (prior to 2.5 billion years ago). One such was Nuna/Columbia/Hudsonland, a supercontinent composed of numerous continental splinters, and imagined to have existed somewhere between 1.8 and 1.3 billion years ago. Proposals for even earlier pangeas were ultimately limited only by the paltry fragments of older crust still in existence. A supercontinent as old as 3 billion years was seriously mooted. If it existed, this was likely the first pangea of all, and was named 'Ur' by John Rogers at North Carolina University[44] (though 'Er' might perhaps be more appropriate). Yet, if piecing together fragments of ancient continents seems about as productive a use of one's time as studying the *Watchtower*, then it is only fair to point out that, while the exact solutions to Precambrian continental jigsaws may be of little significance, the very existence of long-term cycles of supercontinent formation and fragmentation could be of fundamental importance to the global environment and the evolution of life. The first to realize this were, as we have seen, Valentine and Moores, who, in 1970, published their famous *Nature* paper linking the evolution of supercontinents to changes in sea level and biological evolution. The best records of such changes are to be found in deep-sea sediment cores, but these are limited to the past 180 million years or so. On the other hand, the continental crust is the repository of almost the entire history of the planet, extending back as far as 4.4 billion years, just 160 million years after its birth. Extracting that history, however, has not been easy.

A decade after Valentine and Moores' inspired insights,[45] Alfred Fischer at Princeton University had some insights of his own.[46] Alfred identified what he called 'supercycles' (comparable to John Sutton's chelogenic cycles), spanning the past 600 million years, and proposed that the formation and fragmentation both of a 'Late Precambrian Paleopangea' (Rodinia had not then been named) and of the late Paleozoic Pangea were driven by mantle convection. He linked these events with cycles of volcanism, sea-level change, and glaciations through the Phanerozoic (the past 541 million years, according to the most recent timescale). Volcanism, of course, reflected plate tectonic activity for the most part. During supercontinent assembly, mid-ocean

[44] See Rogers and Santosh (2003).

[45] See also Valentine and Moores (1972).

[46] Fischer (1981).

ridges and subduction zones would be eliminated and the number of plates reduced, resulting in a global reduction in volcanism. During break-up, however, new ridges and subduction zones would be created, resulting in greatly enhanced volcanism. Sea levels were important, as Valentine and Moores had established, because the elimination of mid-ocean ridges and young oceanic crust by continental collisions increased the average age—and therefore the average depth—of the ocean basins, leading to a fall in global sea level. Simultaneously, the total area of continental crust would be reduced by the loss of continental shelves and the formation of mountains during collisions. On the other hand, during supercontinent break-up, the continental crust would be stretched during rifting and new (shallow) mid-ocean ridges created, both leading to an overall reduction in global ocean basin volume, and thus a rise in sea-level. A record of global sea-level changes through the Phanerozoic had been compiled quite recently,[47] showing just the pattern of variation predicted by Alfred's supercycles (Figure 7.9).

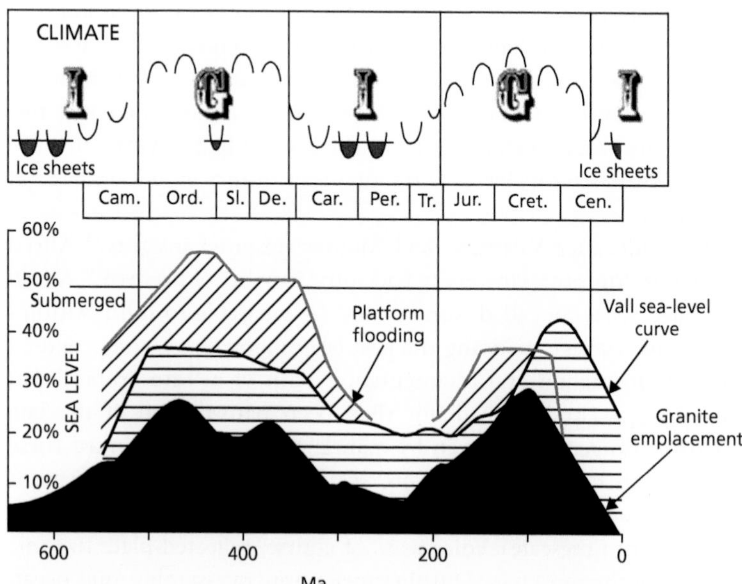

Figure 7.9 Alfred Fischer's 'supercycles'. Climate is inferred to be in either an 'icehouse' (I) or a 'greenhouse' (G) state. [After Fischer (1984).]

[47] Vail (1977).

Alfred's proposed link with climate was particularly interesting. Continental collisions producing high mountains in humid regions such as the Alps and Himalayas would result in accelerated rock weathering, a process that effectively removes CO_2 from the atmosphere.[48] At the same time, since volcanism is the primary source of CO_2, a decline in sea-floor spreading would lead to a reduction in the supply of this major greenhouse gas. The combined effect, then, was a reduction in atmospheric CO_2 concentration, leading to global cooling and, ultimately, widespread glaciation, as recorded in the 'icehouse worlds' of the late Precambrian and Permian. Conversely, supercontinent break-up would increase the concentration of CO_2 in the atmosphere, leading to the 'greenhouse worlds' of the early Paleozoic and late Mesozoic.

A little later, Thomas Worsley and Damian Nance, at Ohio University, took the argument further,[49] presenting a model of 'periodic pangeas' based on a literal interpretation of the Wilson Cycle. In essence, their model was the same as Fischer's, except that they imagined all of the past 2 billion years of Earth history to be dominated by regular supercontinent cycles, each lasting 400–500 million years, the evidence for earlier supercontinents being derived from the ages of the eroded remnants of Precambrian mountain belts. Each cycle involved fragmentation of a supercontinent into dispersing continents, between which new oceans formed by sea-floor spreading. The establishment of subduction within these oceans eventually led to closure, creating a new supercontinent by continent–continent collisions. Like the earlier researchers, Thomas and Damian identified changing sea level as both a primary influence on climate and biodiversity and a record of supercontinent evolution. In their view, each pangea would become static, because its size would tend to cancel any tractions from the asthenosphere beneath, and hence the thermal blanketing effect would be maximized, causing uplift and eventually rifting and volcanism. Their model possessed admirable simplicity, an essential property of any successful theory, but suffered from a number of problems. Firstly, it implied that continental margins external to supercontinents are fundamentally different to internal margins between separating continents. The former would always have faced a large, hemispherical, ocean, and experienced long-term subduction, whilst the latter experienced, alternately, separation and collision. The

[48] See Chapter 8.

[49] Worsley et al. (1984).

identification of ancient passive margins bounding the North American cordillera suggested, as Clark Burchfiel and Greg Davis had noted, that the Pacific Ocean had been created by the rifting of continents around 850 million years ago, potentially scuppering this idea. Secondly, they assumed that, by about 200 million years after rifting, oceanic lithosphere formed by sea-floor spreading between the separating continents becomes negatively buoyant[50] and thus subduction is inevitable. However, far from being inevitable, subduction initiation is complicated by the strength of old plates and remains a problem for geophysicists, as we will see in Chapter 11. Nevertheless, Thomas and Damian had raised awareness of the possible controls on Earth evolution, leading to further insights into what Johannes Umbgrove once called 'the pulse of the Earth'.[51]

Outside-In

In 1988, Mike Gurnis at Caltech provided a theoretical basis for the supercontinent cycle by running a much-simplified computer model of mantle convection in which he incorporated a large continental plate above a rising mantle current. Using parameters chosen to scale with those of the Earth, Mike's model developed mantle overheating of several hundred degrees, leading to uplift and rifting of the continental plate. The two halves moved away from the elevated region in opposite directions, eventually re-joining 'round the back', such that the 'outside' margins of the original continent were now sutured together to form a new 'supercontinent' that came to rest above a site of downwelling, more or less as predicted by Arthur Holmes. Insulation by the new supercontinent heated the mantle beneath, reversing the sense of convection and creating a new rising current in place of the downwelling, so that the whole cycle repeated. Based on reasonable estimates of mantle properties (notably viscosity), rates of continental drift during break-up of around 50 mm per year resulted, a good match to typical values observed for real continental plates. The time taken for the complete cycle, from one supercontinent to the next, was variable, depending on the precise values chosen for the model, but lay somewhere in the range 100 million to 2 billion years. Later, more sophisticated,

[50] In fact, the best estimates suggest that oceanic plates become negatively buoyant at an age of just 26 million years (Turcotte and Oxburgh, 1972).

[51] Umbgrove (1947).

computer models have largely confirmed Mike's findings, and suggested a typical cycle time of around 500 million years (give or take a few hundred million years).

Such a model implies that the external margins of a fragmenting pangea become the interior mountain belts and, ultimately, the sutures, within the next pangea, turning the supercontinent 'inside-out', as Don Anderson had described in his *Science* article. The term 'supercontinental extraversion' was coined by Chris Hartnady of the University of Cape Town, in his review of the SWEAT trilogy of articles in 1991, to describe this sequence of events, and later adopted by Brendan Murphy (at St. Francis Xavier University in Canada) and Damian Nance.[52] The logical alternative to 'extraversion' was, of course, 'intraversion', in which subduction was initiated within the oceans created by supercontinent dispersal, leading to the closure of those oceans and the re-formation of a supercontinent in which successive orogenic belts overprinted each other, as proposed by Worsley and Nance and, previously, by Tuzo Wilson.[53] Mike Gurnis' modelling suggested extraversion as the primary mode of supercontinent cycling, while the closure of Iapetus seemed to be the consequence of 'intraversion', a paradox that led Brendan and Damian to ask, in the title of their 2003 article, 'Do supercontinents introvert or extrovert?' However, since the two alternatives were purely theoretical 'end-members', the inevitable answer was 'a bit of both'.

Modern three-dimensional computer models (involving simulation of convection within spherical shells rather than, say, a two-dimensional annulus) show that, while supercontinents may form and fragment readily, the duration of modelled supercontinent cycles is highly variable. In one series of simulations,[54] the introduction of mantle plumes degraded the regular 400 million-year cyclicity of a plume-free model, producing only sporadic supercontinents. Other three-dimensional simulations[55] support the idea of a supercontinent cycle as 'one of Earth's major evolutionary features', but conclude that the complexity of mantle convection and the likelihood of overlapping rifting and collision (as at

[52] Murphy and Nance actually used the spellings 'extroversion' and 'introversion' (Murphy and Nance, 2003), although, as pointed out by Kaufman (2015), the term originally proposed by Carl Jung was, indeed, 'extraversion'.

[53] Wilson (1966).

[54] Phillips and Bunge (2007).

[55] Rolf et al. (2014).

the present day) make regular periodicity as envisaged by Worsley and Nance 'unlikely to be a robust characteristic of the supercontinent cycle'. The present consensus seems to be that regular supercontinent cycles do not occur, although an underlying cyclicity may be obscured by the sheer complexity of continental drift.[56] One recent view is that just two pangeas have existed: Rodinia and Pangea. These formed by a rather messy process of break-up and convergence, representing neither pure intraversion nor pure extraversion.

Such concepts have some relevance when contemplating the future of the planet, long after we (individually and as a species) have departed. As political pollsters found in 2016, prediction is very difficult, particularly about the future.[57] Nevertheless, some foolhardy individuals, including myself, have ventured to construct possible future pangeas based on pet assumptions. The essence of my effort, which I named *Novopangea*, was that the Atlantic continues to open, while the Pacific continues to contract (an example of the extraversion model), with the inevitable result that Eurasia/Australia and the Americas collide in a few hundred million years time. Other folk assumed that major subduction will soon commence within the Atlantic, which will begin to close once more, while the Pacific, which has been shrinking for the past 200 million years, will begin to widen, resulting in reconstructions like *Pangea Proxima* that epitomise the introversion model.[58] Recently, another possibility has been suggested by Portuguese geophysicists,[59] in which both the Pacific and Atlantic oceans close by subduction, producing a supercontinental configuration that they call *Aurica* (a portmanteau of 'Australia' and 'America'). This idea is based on the belief that Pacific subduction zones are invading the Atlantic via the Scotia and Lesser Antilles arcs. Its biggest drawback, as you may have realized, is that it requires Eurasia to split in two, half becoming sutured back onto the eastern margin of North America and the other half remaining in contact with the western margin. In the present state of the art, it remains a possibility that, while future supercontinents will certainly be assembled, a new pangea may not emerge at all. Furthermore, the concepts of extraversion and intraversion, as highly simplistic end-members, may not

[56] Phillips and Bunge (2007).
[57] A quote often attributed to the physicist Niels Bohr.
[58] See Nield (2007) for details.
[59] Duarte et al. (2016).

add a great deal to our understanding of the complexities of Earth evolution.

It is easy to conflate the Wilson Cycle with the supercontinent cycle, but there are good reasons why they should be kept separate. For a start, the Wilson Cycle is concerned only with the opening and closing of *ocean basins* within an *idealized* cycle (Tuzo himself noted that a given ocean need not go through every stage), so that contemporaneous oceans may be at different stages, as we see today, some just opening, like the Red Sea, while others, such as the Mediterranean, are in their death throws. On the other hand, the supercontinent cycle hypothesis requires that continents came together more or less simultaneously on four or five occasions, each time replacing oceans with mountain ranges. It is therefore intrinsically more speculative than Tuzo's eponymous cycle, and presupposes that plate tectonics was operating from very early in the Earth's history. Some leading geophysicists, notably Bob Stern and Warren Hamilton, dispute this, as we will discover in Chapter 11. Should they turn out to be correct, then there may have been no pre-Rodinia pangeas. Likewise, Kevin Burke has confessed[60] that he was not persuaded that there was 'a case for the existence of ancient supercontinents (other than Pangea) that contained all of the Earth's continental crust—let alone evidence of cycles of supercontinent assembly'.

Growing Pains

In Chapter 6, we visited a subduction factory and inspected the various production lines involved in the manufacture of continental crust. The question now arises as to whether methods of production have always been much as we see them today or whether primitive processes differed to any great extent. If the estimates of Bob Stern and David Scholl are close to the truth and crustal additions roughly balance crustal losses today, then the answer to the latter must be 'yes'. At some stage in earlier geological history, things must have been very different in order that continental crust could accumulate in the first place: either the early rates of production were much higher or the rates of loss were much less (or both). On the other hand, if continents have been growing steadily, then it may be that manufacturing technology, even in the Archean (4 to 2.5 billion years ago), was similar to today. This, of course,

[60] In his review of Ted Nield's book (Burke, 2007).

would carry the profound implication that plate tectonics has operated on the Earth since the earliest times.

During the 1980s, the latter argument (that the continental crust has grown steadily over the past 4 billion years) was taken up by a number of researchers, the most prominent of whom was the Oxford geochronologist Stephen Moorbath.[61] Ranged against Stephen and his supporters were those who believed that the crust formed early on, within the first half billion years of the Earth's existence, and remained at a more or less constant volume ever since. This point of view was spearheaded by Dick Armstrong at the University of British Columbia,[62] who argued that 'The only measure of Archean continental volume is crustal thickness and position with respect to sea level [or "freeboard", as it is known]', both of which appeared reasonably constant since ancient times. Given that both sides accepted that crust was being manufactured continually at volcanic arcs, Dick's proposition could work only if, following early creation, crust had been removed at a rate matching that of the creation of new crust. On the other hand, Stephen claimed that, since continental crust is unsubductable, it therefore must have grown continuously as juvenile crust was added each year. This, Dick asserted, was 'beside the point', since recycling of continental crust into the mantle was not achieved by wholesale subduction of continents, but by other means, such as sediment subduction and tectonic erosion of volcanic arcs. To resolve the argument, geologists needed to investigate the rock record for the ancient Earth, a record that becomes extremely sparse prior to 3 billion years ago, and disappears completely beyond about 4 billion years, leaving them groping for clues. Before proceeding further, we need to acquaint ourselves with the geological timescale for the oldest rocks.

The 'Precambrian' is really just an umbrella term representing geologists' ignorance of the first 88 per cent of Earth history before Cambrian fossils gave them a means of dating and correlating rocks worldwide. As interest in this remote and mysterious span of time has grown, it has been divided informally into three eons (Figure 7.10), named the Hadean ('Hell-like',[63] pre 4 billion years), Archean ('ancient' or 'earliest',

[61] Moorbath (1975).

[62] Armstrong (1981).

[63] The earlier vision of red-hot oceans of magma surrounding fiery volcanoes and subjected to violent collisions from meteors and comets has been replaced by a more tranquil image of blue oceans, snow-capped peaks, and even primitive life, although,

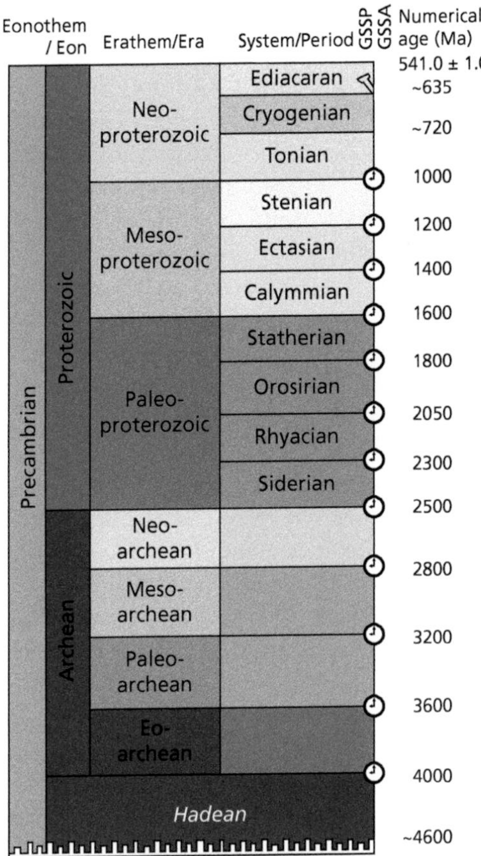

Figure 7.10 Geological timescale for the Precambrian (International Commission on Stratigraphy, version 2017/02). Note that ages of boundaries are uncertain or arbitrary.

4 to 2.5 billion years) and Proterozoic ('former life', 2.5 billion to 541 million years). The Hadean is unique in the geological timescale, since no rocks of this age are known, and it seems a fair bet that none ever will.[64] Nevertheless, evidence from Archean rocks (that we are coming

given the absence of oxygen in the atmosphere, you still would not fancy it as a holiday destination. See Arndt and Nisbet (2012).

[64] In fact the Hadean–Archean boundary was originally defined as the age of the oldest reliably dated rocks, leading Van Kranendonk et al. (2012) to propose an age of 4.03 billion years, in order to take account of the Acasta gneiss.

to) suggests that some form of felsic crust had already formed in the Hadean by processes that, given the total absence of rock outcrops, remain uncertain. This deficiency has in no way deterred geologists from proposing a variety of more or less plausible mechanisms, with a consensus that such processes, whatever they were, were unlike those operating today.[65]

Things improve as we move into the early Archean (i.e. post 4 billion years). We now have a couple of fragments of crust in Canada (Acasta gneiss) and Greenland (Isua greenstone), including some pillow basalts testifying to the presence of oceans. The record for the oldest rock, as of 2017, is 4.03 billion years, held by the Acasta gneiss in the Slave Province of north-western Canada. Although, strictly speaking, this places it just within the Hadean, the arbitrary nature of the numerical boundary means that the rocks are more reasonably regarded as early Archean.[66] Both the Isua (Greenland) and Acasta (Canada) greenstone belts contain rocks that appear to be altered boninites, suggesting an origin associated with the initiation of subduction, pointing to something like plate tectonics in the early Archean. Younger outcrops in Australia and South Africa provide a clearer picture of crust-forming processes by 3.5 billion years ago. The average composition of these crustal rocks is intermediate between mafic and felsic, with a silica content of around 60 per cent. On the face of it, this is strong support for the operation of the subduction factory throughout the 1.5 billion years of the Archean. Closer inspection reveals that, in fact, very few Archean igneous rocks actually have an intermediate composition, and the average composition is the result of combining two quite different rock types: mafic greenstones (with SiO_2 around 50 per cent) and felsic granites (with a composition of around 70 per cent SiO_2). This dichotomy is very different to anything found in younger rocks.

The boundary between the Archean and Proterozoic at 2.5 billion years corresponds roughly to a change observed by Canadian geologists from highly deformed metamorphic rocks (gneisses) to less deformed types, recording the evolution from a hot young planet to a less turbulent world, a little more like the Earth we know. Lavas known as 'komatiites',[67] with low silica and high magnesium, reflecting high

[65] Roberts et al. (2015).

[66] A claim of 4.28 billion years for the unpronounceable Nuvvuagittuq greenstone belt on the eastern shore of Hudson Bay is disputed.

[67] Named after the Komati Valley in South Africa, where they were identified.

temperatures of mantle melting, are common within Archean outcrops, but become rare in the Proterozoic and later geological record. Similar transitions are seen on other continents, but not at exactly the same time,[68] so that the boundary age of 2.5 billion years must be regarded as somewhat arbitrary. Moreover, given the fragmentary nature of the rock record, it was hard to be sure that the volume of continental rocks of any given age found by geologists was, in fact, representative of the volume of crust formed at that time. What was needed was a more objective sampling method. Once again, Gaia provided a helping hand.

Zircons Are Forever

Zirconium is a greyish transition metal with atomic number 40. Its name was derived from the mineral in which it is found, zircon, composed of the compound zirconium silicate ($ZrSiO_4$). Zircons provide, in addition to fake diamonds, very resilient time-capsules within which various trace elements are preserved. Following crystallization, unstable uranium isotopes trapped inside zircon crystals decay slowly, but steadily, to lead (half-life 4.47 billion years), providing a high-resolution chronometer first applied by our old friend Robert Strutt,[69] who determined a minimum age of 565 million years for a zircon from Ontario. More sophisticated analysis, together with the use of lutetium/hafnium decay (half-life 36 billion years), gives an age of 4.374 billion years for the oldest zircons found so far (from the Yilgarn craton of western Australia[70]).

Their remarkable resilience enables zircons to survive erosion of the rocks in which they crystallized, subsequent transport by rivers, and finally, deposition within sediments, where they become 'detrital zircons'.[71] All this relieves geologists of the arduous task of sampling individual rock outcrops, since large rivers, like the Brahmaputra, Yangtze, or Amazon, do the work for them, transporting rock particles from

[68] Rogers and Santosh (2003).

[69] Strutt (1910).

[70] Valley et al. (2014).

[71] Like other rock fragments, zircons become rounded in shape during erosion and transport, but remain largely intact. Zircons are also recovered directly from the granites, kimberlites and other igneous rocks in which they formed, and are then referred to as 'igneous zircons'. Once formed, zircons can even resist high-temperature alteration and thus end up in metamorphic rocks.

wide catchment areas within continents and then depositing them in neat layers when they reach the ocean. By sampling and dating zircon crystals within such sediments, the age of crystallization of the parent rocks from which they were eroded can be estimated, even if the zircons have been through several cycles of erosion and deposition. What makes them particularly useful is that, being silicate minerals, they form preferentially from magmas with a surplus of silica (i.e. granites) and thus are found mainly in quartz-rich sediments—such as quartzites and sandstones—formed by the erosion of granitic rocks making up the continental crust. Hence, the more continental crust created at any stage in the history of the Earth, the more zircons of that age you would expect to find deposited in later sediments, even if the original rocks were completely destroyed by erosion. On the other hand, if there was a time before which continental crust did not exist (say, because subduction had not got under way), then you would not expect to find zircons older than this. Zircons could thus be used to address the major outstanding questions that had kept Stephen Moorbath and Dick Armstrong at each other's throats for years: when and how was the continental crust formed?

A plot of the numbers of detrital zircons against age (Figure 7.11) reveals several very interesting things. Firstly, only a few zircons give uranium–lead ages older than 3 billion years, suggesting a minimal

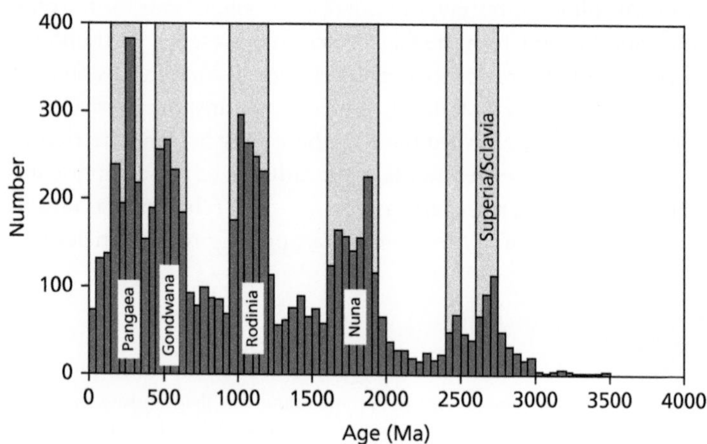

Figure 7.11 Uranium–lead ages of 5246 detrital zircons from 40 major rivers. [From Campbell and Allen (2008).]

volume of felsic crust before that time. But perhaps the most surprising thing about this plot is that there are half a dozen clear peaks when zircons are two to three times more abundant. The same pattern is seen on all continents, suggesting a global phenomenon, perhaps reflecting primary processes at work on the Earth. For example, the peaks could represent times during which continental crust production was particularly high because of enhanced volcanic arc or mantle plume activity. In other words, the growth of continents was episodic. Other data, such as the ages of metamorphism and mineralization, the abundance of continental margins formed by rifting, and the composition of seawater, all support the idea of an episodic geological history, showing well-defined peaks and troughs. To explain these, geologists have suggested, variously, that episodes of rapid continental growth may reflect short bursts of mantle plume activity producing LIPS, variations in the extent of subduction, and even intermittent plate tectonics.

On the other hand, some researchers, notably a group of British geochemists based at Bristol University,[72] have challenged such interpretations, arguing instead that the zircon record is not a primary record of crustal formation, but instead reflects variations in the *preservation potential* of felsic crust at different times. Clearly, the number of detrital zircons of any given age depends on the amount of erosion globally, which, the group point out, must vary according to the phase of the supercontinent cycle. As supercontinents assemble by the subduction of oceanic plates, mountains are formed by continental collisions. The uplift and erosion associated with these processes provides plentiful sediments containing plentiful detrital zircons. During supercontinent break-up phases, there will be much less erosion and so fewer zircons will be deposited. A strong relationship between the number of detrital zircons and the existence of supercontinents might therefore be expected as a consequence of this preservation bias, even if new continental crust is generated continuously. Chris Hawkesworth and colleagues thus argue that the bias in preservation of igneous rocks and zircons results from the integration of three separate phases of the supercontinent cycle: subduction during the closure of ocean basins, continental collisions, and rifting/break-up of supercontinents. The volume of new magma produced from the mantle is highest during the closure phase when volcanic arcs are extensive, while preservation is

[72] Hawkesworth et al. (2009).

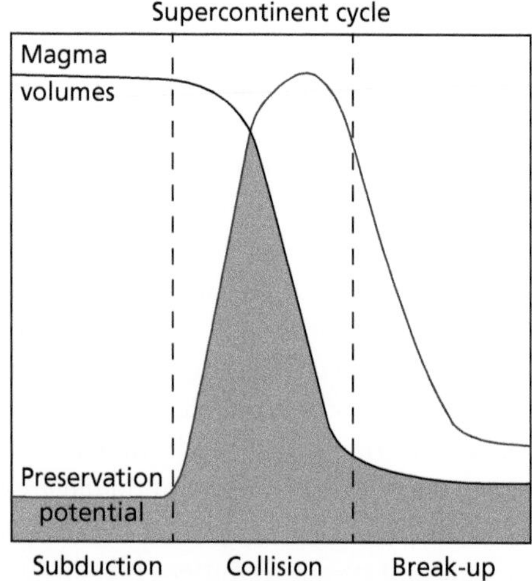

Figure 7.12 Schematic showing the volumes of magma generated and the preservation potential of rocks crystallized from magmas during the three phases of the 'supercontinent cycle'. The overlap of the two curves defines the preserved record, which is thus episodic. [From Cawood et al. (2013).]

most likely following collision and suturing of fragments to form the new supercontinent (Figure 7.12). Superimposing the two curves produces a peak of preserved zircons during the collision phase. In other words, the collisional phase tends to preserve zircons and associated granitic rocks produced during the preceding ocean closure phase. This explanation is not universally accepted, however, and some geologists still cling to the idea that there were periods of accelerated crustal growth, perhaps linked to mantle convection.[73] Regardless of whether the record reflects original contributions or preservation potential, there is a strong, though irregular, cyclicity in zircon ages resulting from supercontinent formation and break-up, a cyclicity that could be said to reflect the 'pulse of the Earth'.

Industrial Revolution

Back in 1973, Kevin Burke and John Dewey introduced the concept of the 'permobile Earth'. Noting that sea-floor spreading, followed by

[73] For example, Arndt and Davaille (2013).

cooling of plates, is the planet's primary means of heat loss, they argued that, since heat production by radioactive decay within the Earth was twice as great in the Archean (prior to 2.5 billion years ago), plate tectonics must have involved much faster spreading and smaller plates, with a much greater total length of mid-ocean ridges—plate tectonics on steroids, you might say.

Reviewing the situation in 1981, John Dewey and Brian Windley concluded that continental growth rates had also been high during the Archean (prior to 2.5 billion years ago), with an estimated average rate of growth of 4.06 km^3 per year. After 2.5 billion years ago, in the Proterozoic, the average rate fell to 1.46 km^3 per year, before dropping to just 0.6 km^3 per year in the Phanerozoic. Current estimates based on modelling the uranium–lead, hafnium and oxygen isotope ratios of detrital zircons imply that an extensive continental crust with a volume of 60–70 per cent that of today was already in existence by 3 billion years ago,[74] a date that seems to mark a fundamental change in the net rate of crustal production, from around 3 km^3 per year before that time to just 0.8 km^3 per year afterwards.[75] This leads to a major conundrum, since less than about 10 per cent of exposed continental crust today is older than 3 billion years, implying recycling on an industrial scale. Hence, if the rate of addition of new crust from mantle melting has been fairly continuous, destructive processes such as subduction and delamination must have become much more effective at about this time, to the point where they more or less matched the rate of new crust production. This occurred at about the time that geologists see a transition in continental crust composition, from the bimodal 'granite–greenstone' rock types of Archean cratons, to the modern andesitic composition.

This all seems to point to a major shift in the style of plate tectonics 3 billion years ago, involving the mode of subduction. It has been proposed that subduction during the Archean, if it occurred, involved thicker, warmer, and therefore more buoyant, oceanic crust, which would simply underthrust the overriding plate rather than sinking at a steep angle. Something like this is observed in the central Andes, where the young Nazca oceanic plate is subducting today. In that case, there would be no 'mantle wedge' within which partial melting could be

[74] Hawkesworth et al. (2013).
[75] Dhuime et al. (2012).

induced by fluids released from the slab, but, being hotter, the slab itself would be liable to begin melting, producing a magnesium-rich magma known as adakite. Geochemists have shown that the best fit to the trace-element composition of felsic rocks (tonalities) is obtained from a model in which a thick (35–50 km) mafic crust is subjected to melting. It thus seems more likely that we are dealing with extensive oceanic plateaux (like the Ontong Java Plateau) rather than normal oceanic crust. Such plateaux probably formed in the early Archean above hot mantle plumes. Partial melting of these plateaux then generated felsic magmas that either erupted or cooled within the crust, producing the tonalites that form the Archean cores of many continents.

Nevertheless, exactly how and when the continental crust formed remains a major unsolved problem today. Growth curves showing the volume of crust existing through time have been suggested by many authors and include just about every possibility, from creation very early in Earth history,[76] to rapid growth within the past billion years[77] (Figure 7.13). In a recent review,[78] the Bristol group proposed a history of the crust in five stages that provides a good working basis for fresh research, although, as they themselves observed, the subject remains highly controversial. Stage 1 occurred in the Hadean, beginning with the accretion of the Earth around 4.567 billion years ago, and involved the separation of the iron-alloy core from the silicate mantle, and the formation of a deep magma ocean sporting a mafic crust[79] that gradually thickened until, on reaching a thickness of around 20 km, this crust itself began to melt at depth, producing the first felsic magmas. Stage 2 began in the Archean, when plumes of hot mantle peridotite from deep within the planet ascended into the crust, where they melted, creating large igneous provinces. These magmas not only thickened, but weakened the early lithosphere, inhibiting plate-like behaviour and preventing

[76] Armstrong (1981).

[77] Hurley and Rand (1969).

[78] Hawkesworth et al. (2016).

[79] Recent work (e.g. Reimink et al., 2014, 2016) has tended to contradict the earlier conclusions regarding the composition of the earliest crust, suggesting that the crust that formed following the Moon-forming impact around 4.5 billion years ago was predominantly mafic (basaltic), possibly formed at the surface of a magma ocean. Re-melting of this mafic crust produced small quantities of more felsic magma (O'Neil and Carlson, 2017). Hence, it now seems that no extensive continental crust existed prior to 4 billion years ago.

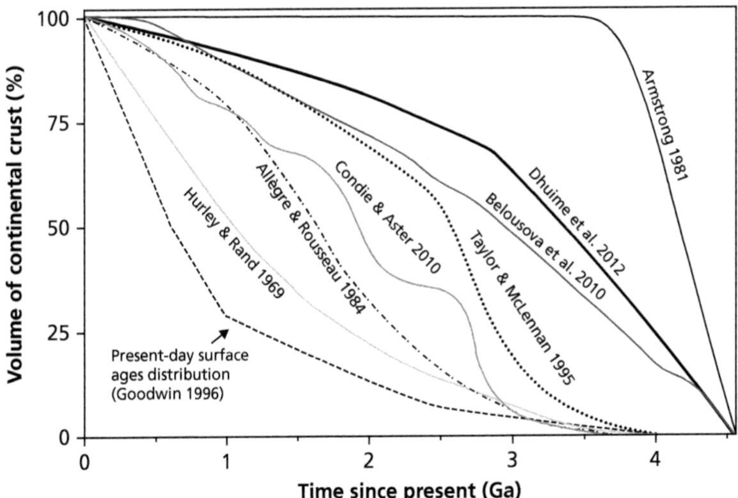

Figure 7.13 The wide variety of growth curves proposed for the continental crust. [After Hawkesworth et al. (2013).]

subduction from occurring. Nevertheless, crust was returned to the mantle by delamination and as a result of meteor impacts.

Stage 3 commenced around 3 billion years ago, by which time the thickness of felsic continental crust had reached 30–40 km, and its volume was around 70 per cent of that seen today. Following the onset of an early style of plate tectonics, the net rate of continental growth slackened as a type of recycling known as 'hot subduction' began. During this stage, the movement of plates inevitably resulted in the earliest continental collisions and the construction of the first supercontinents. Stage 4, widely referred to as the 'boring billion', began around 1.7 billion years ago in the Proterozoic, and appears to be a period of stability during which few passive margins were created by rifting, implying limited reorganization of continental fragments prior to the assembly of Rodinia. The mantle was still hot enough that the base of the thickened crust was subject to melting, producing anorthosite magmas. This 'middle age' of the Earth was terminated by the break-up of Rodinia, 750 million years ago, heralding the transition to stage 5 and the onset of 'cold subduction', that is, processes like those observed in modern subduction factories. Limited subduction of continental crust now produced ultra high-pressure terranes, while blueschists formed at oceanic subduction zones.

The continents began a dance that culminated in the assembly of Pangea, followed by rifting and dispersal of the fragments to create the present-day map of the world.

Afterthought

Teddy Bullard showed in 1965 that the continents around the Atlantic had remained rigid for nearly 200 million years, allowing them to be fitted neatly back together. Slightly less satisfactorily, the 'stable' cores of continents could be assembled in unlimited ways to create putative supercontinents of earlier times. Cycles of deposition and mountain-building recorded by continental rocks can be placed in the context of supercontinent fragmentation and assembly, the processes that determine first-order geological cycles and so may regulate the pulse of the planet. Given the similarities between current thinking and early ideas on continental drift, geological cycles, and supercontinents, you could be forgiven for concluding that, despite all the technical advances that have occurred, our understanding has not come all that far since the work of Umbgrove, Holmes, and Sutton. Certainly, the insights gained by these pioneers from limited data, in the absence of supercomputers and sophisticated laboratory techniques (other than radiometric dating), was truly remarkable, and deserves much wider appreciation.

Continents, the product of plate tectonics, are what make the Earth unique among the planets and moons of the Solar System. They have been carted around the globe for billions of years, assembling and disassembling supercontinents along the way. Present-day Eurasia, for example, has formed by the collision and suturing of a set of ancient continental blocks including Baltica, Siberia, Armorica, North China, and South China. North America has also been in contact with Eurasia for millions of years, the plate boundary proving somewhat cryptic as a result of very slow relative motion, although it is known that a large region of north-east Russia actually lies on the North American plate. Today, Africa and India are in the process of colliding with Eurasia, and Australia is about join the party. The next pangea is forming right beneath our feet.

8

All at Sea

> Each time certain properties of the sea floor were postulated on theoretical grounds, investigations in situ have upset the picture.
>
> Ph. Kuenen (1958)

According to first-generation plate tectonics, sea-floor spreading was nice and simple. Plates were pulled apart at mid-ocean ridges, and weak mantle rocks rose to fill the gap and began to melt. The resulting basaltic magma ascended into the crust, where it ponded to form linear 'infinite onion' magma chambers beneath the mid-ocean tennis-ball seam. At frequent intervals, vertical sheets of magma rose from these chambers to the surface, where they erupted to form new ocean floor or solidified to form dykes, in the process acquiring a magnetization corresponding to the geomagnetic field at the time. Mid-ocean ridge axes were defined by rifted valleys and divided into segments by transform faults with offsets of tens to hundreds of kilometres, resulting in the staircase pattern seen on maps of the ocean floor. All mid-ocean ridges were thus essentially identical. Such a neat and elegant theory was bound to be undermined as new data were acquired in the oceans.

*

Slumbering Giant

As with Pearl Harbor, the Americans were very slow to recognize the threat to the status quo. A major revolution in marine science had taken place right under their noses, yet some US scientists scarcely noticed, while others followed Harold Jeffreys' example and obstinately opposed the new ideas. Their intransigence was all the more surprising, since the US Navy, as we have already seen, had played a central role in developing the technology that underpinned the new theory. By 1966, however, magnetic profiles acquired over mid-ocean ridges provided evidence that finally convinced many American scientists that all this

The Tectonic Plates Are Moving! Roy Livermore, Oxford University Press (2018). © Roy Livermore.
DOI: 10.1093/oso/9780198717867.001.0001

rubbish about sea-floor spreading was, in fact, true. Following the general acceptance of plate tectonics, the US military continued to develop new technologies, including submersibles, satellites, and sonars, that were to be at the forefront of a new wave of global exploration beginning in the 1980s. The only problem was their reluctance to release the results of their good work to the academic community, preferring to 'classify' much of the newly acquired data, at least initially. As Fred Vine has pointed out, perhaps if the high-resolution bathymetry collected alongside Raff and Mason's magnetic profiles had been released at the same time, the symmetry of the anomalies over the Juan de Fuca Ridge might have been spotted earlier and history would have taken a different path.

Twenty years after the birth of plate tectonics, we entered the era of Big Science in the oceans. Integrated studies and multi-author papers were now *de rigueur*. Six years after the discovery of hydrothermal vent communities on the Galapagos Rise, a programme was set up by the US National Oceanographic and Atmospheric Administration (NOAA) in 1983 to 'systematically explore, discover, and, ultimately, characterize the environmental impacts of submarine volcanism and its associated hydrothermal activity'. Wisely eschewing acronyms, this initiative called itself, simply, the 'Vents Program', and concentrated its efforts, at least initially, on the Juan de Fuca and Gorda ridges, parts of which lay within the US Exclusive Economic Zone. Later, its compass was extended to include vents discovered in the western Pacific.

Some of the biggest marine science was conducted within an intensive programme of mid-ocean ridge exploration under the ingenious acronym 'RIDGE'.[1] Established by the US National Science Foundation in 1989, RIDGE aimed to 'integrate exploration, experimentation, and theoretical modeling into a major research effort to understand the geophysical, geochemical, and geobiological causes and consequences of the energy transfer in the global rift system through time'.[2] In effect, it provided a large fund for marine scientists to go and explore various bits of the 60,000 km mid-ocean ridge system using a range of newly developed technologies. The idea was soon imported to the UK by Joe Cann (now at the University of Leeds) and Roger Searle (at Durham University), who established a consortium of British research groups

1 Standing for 'Ridge Interdisciplinary Global Experiments'.

2 Ocean Studies Board (1999).

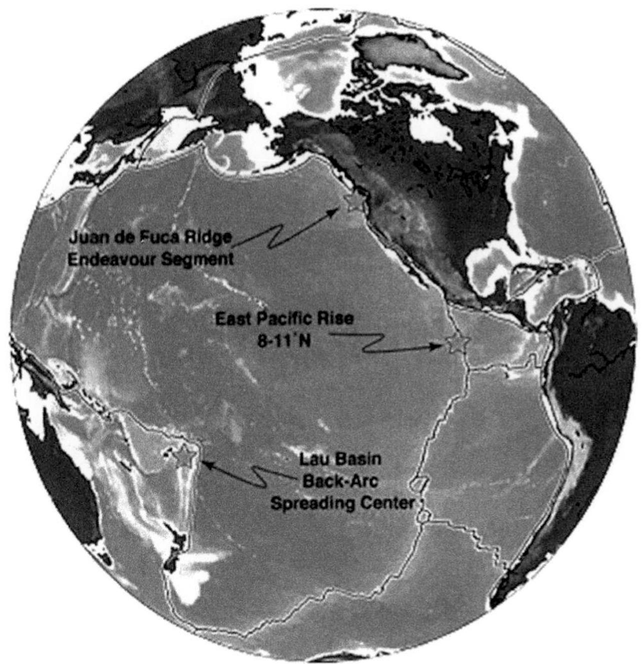

Figure 8.1 Locations of the Ridge 2000 integrated study sites in the Pacific Ocean: Endeavour segment of the Juan de Fuca Ridge, the East Pacific Rise at 9.5°N, and the Lau Basin.

that became known as 'BRIDGE', for which they successfully obtained Government funding via the UK's Natural Environment Research Council in 1993. Other nations instigated their own versions of these initiatives with more or less unimaginative titles, such as the German 'D-Ridge' and French 'Dorsales' programmes. Eventually, an international body had to be established to bring some order to proceedings and to avoid the embarrassment of research ships from different nations turning up on the same bit of ridge at the same time. This body was 'InterRIDGE', which alone remains in existence now that the Big Science juggernaut has moved on.

A second phase of RIDGE exploration, known as Ridge2000, was focused on three selected locations at which long-term observations would be made (Figure 8.1). These were, firstly, a site at 9.5°N on the East Pacific Rise (a fast-spreading ridge); secondly, a segment of the Juan de Fuca Ridge (an intermediate-rate ridge); and, thirdly, a spreading centre in the Lau Basin in the western Pacific (a back-arc ridge). The idea was to study these sites in detail 'from mantle to microbe', that is, from the initial melting of mantle rocks and the creation of oceanic plates to hydrothermal vents and the organisms that lived there. Eventually, after shedding much light on the processes associated with sea-floor spreading, Ridge2000 suffered a fate not unlike that of the ephemeral vent ecosystems it had studied, and, in 2012, it was extirpated as the funding was withdrawn. BRIDGE funding had ended in 1999, having enhanced the UK's reputation for global exploration in places as far-flung as the Lau Basin and Scotia Sea, as well as supporting concerted efforts in the North Atlantic. The Vents Program was subjected to 'restructuring' in 2013, and continues as the 'Earth–Ocean Interactions Program'. The ending of this phase of intensive exploration,[3] during which the axial zone of half the global mid-ocean ridge system was mapped, did not, however, slow the rapid progress of plate tectonics in the new century any more than Harry Hess' ill-conceived pessimism thirty years before.

Sound Science

Perhaps the best example of the 'swords to ploughshares' theme in plate tectonics research is the development of sonar systems for imaging and mapping the sea floor. Indeed, the British towed sonar GLORIA had anti-submarine warfare embedded in its very name.[4] Developed in the 1960s by the National Institute of Oceanography,[5] GLORIA consisted of a mighty 15 m towfish, containing arrays of transducers (effectively, lines of loudspeakers) transmitting sound at frequencies between 6 and 7 kHz, powered through its tow-cable. Resembling a giant torpedo, this behemoth was so cumbersome that it required its own davit and launch system, restricting its use to a single ship. Deployment required calm, shallow, waters in which divers from the mother ship could safely launch the thing, which was then towed at a depth of 300 m at a speed of a few knots. Once underway, it insonified an area of sea floor extending up to 20 km on one side with beams of sound directed at a low

[3] Perhaps not quite the end: a programme funded by the US National Science Foundation, focused on continental margins and known as 'MARGINS', commenced in 1998 and continues today as the 'GeoPRISMS Program'.

[4] The 'A' stood for 'ASDIC', or 'Anti-submarine Division –ic'.

[5] Subsequently the Institute of Oceanographic Sciences, subsequently the Southampton Oceanography Centre, and currently the National Oceanography Centre.

angle—hence the name, 'sidescan sonar'. Backscattered energy was detected by the transducer array and used to construct what was, in essence, a photograph of the sea bed. This was very useful for distinguishing young volcanic ridges devoid of sediment cover, which backscattered brightly, from areas covered with sediment, which appeared dark on images. The system had the great advantage that large areas of the sea bed could be imaged relatively rapidly on a reconnaissance basis. With so much of the oceans unknown before the 1990s, this was a valuable trait that has become less significant in the twenty-first century. A somewhat more manageable version, GLORIA II, was developed by 1977, and used to image large areas of the Atlantic and adjacent oceans, subsequently being employed by the USGS to obtain complete coverage of the US Exclusive Economic Zone. Its effectiveness can be gauged from the fact that the entire West Coast zone—an area of 650,000 km^2—was imaged in just 112 days using an old Hull trawler to tow GLORIA back and forth. Higher-frequency sonars were also developed, designed to be towed just a few hundred metres above the sea floor for high-resolution imaging, particularly of mid-ocean ridges. Two such deep-towed systems are Sea MARC I and TOBI, developed in the USA and UK, respectively. Later sidescans employed two rows of transducers, using the difference in arrival times at each to compute depths, thus providing both sidescan imagery and depth measurements across a wide swath of sea floor, all in one package. Examples of this type of towed sonar are Sea MARC II and HAWAII-MR1 (both developed at the Hawaii Institute for Geophysics). GLORIA II was also developed to provide rudimentary bathymetry, and re-christened, humorously, 'GLORI-B'.

A second type of sonar used in the deep ocean is the echo-sounder. The difference between sidescan sonar and an echo-sounder is rather like the difference between taking a photograph of a landscape and surveying it with a theodolite: one produces a picture while the other provides quantitative measurements from which a map can be made. As mentioned in Chapter 1, echo-sounders had been developed by the US Navy during the 1950s to provide many simultaneous depth measurements across a swath of sea floor by means of multiple 'beams' of sound transmitted into the water via arrays of low-frequency (typically 12 kHz)[6] transducers built into the ship's hull. As you will remember from

[6] This frequency gives an optimal compromise between range and resolution (lower frequencies travel further, but higher frequencies provide better resolution).

school physics lessons, sound waves from adjacent sources interfere with each other—where a peak coincides with a trough, the pulse is cancelled, but where two peaks or two troughs coincide, the pulse is magnified—resulting in a pattern of 'beams' across (or 'athwart' to old seadogs) the ship. A second array, aligned with the ship's hull, receives echoes from the sea bed, which it uses to make depth estimates at numerous points at ranges up to three times the water depth. The first successful 'civilian' multibeam echo-sounder was a cut-down version of the SASS (Sonar Array Sounding System) developed by General Instrument Inc. for the US Navy. Known as 'SeaBeam', this provided the capability of mapping a swath of deep-sea floor 2 km wide with its meagre sixteen beams (SASS had sixty-one beams). Perhaps hamstrung by the military obsession with secrecy, the US lost its lead in sonar technology after the introduction of SeaBeam. First, the Germans eclipsed SeaBeam with their 'Hydrosveep' system, which was, in turn, supplanted by the Norwegian Simrad systems, each advance providing more beams, higher precision, and wider swaths. By the time that SeaBeam 2000 appeared in 1990, it was too late: Simrad had cleaned up in the market for hull-mounted multibeam sonars. The latest oceanographic research vessels are typically fitted with the Simrad EM122 system with 288 beams for deep-sea work, and a range of higher-frequency systems for work on continental shelves and other shallow regions.

Of Ships Sailing the Seas

The critical Indian Ocean magnetic anomaly profiles of Vine and Matthews had been collected by HMS *Owen*, a ship originally designed as an anti-submarine frigate, but completed in 1949 as a survey ship at Chatham Dockyard. It was only on completion that she was given the name *Owen*, thereby saving her and her crew from the Navy's intended appellation, *Loch Muck*. She then went on to work mainly in the Indian Ocean where, as we saw in Chapter 1, she collected the data that kicked off the plate tectonics revolution.

The first purpose-built research ship was launched nearly fifty years earlier, in 1901. This was the Royal Research Ship (RRS) *Discovery*, built in Dundee as the last three-masted, wooden-hulled, ocean-going vessel constructed in the UK. Her initial mission was to support the British National Antarctic Expedition of Ernest Shackleton and Robert Scott, but, just four years later, she was sold to the Hudson Bay Company

Figure 8.2 RRS *Discovery*, launched in 2013, is operated by the UK's Natural Environment Research Council.

and, ironically, converted to a cargo ship. Even greater ignominy befell her in 1922, when she was demoted to a scout hut for the 12th Stepney Sea Scouts. Happily, she was soon reassigned to research duties in the Antarctic, where she remained for several years, before being moored alongside the Victoria Embankment in London, serving as a training ship. Finally, in 1985, she returned to Dundee to become a tourist attraction. Three later British research ships have also carried the name *Discovery*, the last being launched in 2013 (Figure 8.2). A typical twenty-first century research ship, the new *Discovery* is almost 100 m stem to stern, and sports the low aft deck and A-frame characteristic of oceanographic vessels. She is fitted with hull-mounted Simrad EM122 and EM710 multibeam echo-sounders, dynamic positioning (enabling ROVs and AUVs[7] to be deployed), and also has capabilities for seismic surveys, sediment coring, and trawling, amongst other worthwhile activities.

In the USA, the first dedicated oceanographic research ship was the 142-foot steel-hulled sailing ship *Atlantis*, which entered service with the Woods Hole Oceanographic Institution in 1936. Incredibly, this ship is still in service today, having been sold to Argentina in 1966 and refurbished

[7] ROVs are 'remotely operated vehicles'; AUVs are 'autonomous undersea vehicles'.

in 2009, making her the world's oldest research vessel. Her name has been applied to other vessels, including the ship that currently operates the submersible *Alvin*, and, like HMS *Challenger*, to a Space Shuttle, in this case the vehicle that carried out the final Shuttle mission in 2011. As we shall see, the name has also been used for undersea features, including a major Atlantic fracture zone.

Most of the marine data upon which first-generation plate tectonics was erected were, however, collected by research ships that, like HMS *Owen*, had been converted from military or cargo vessels, and even, as in the case of the *Vema*, a luxury yacht. *Vema* was a three-masted, iron-hulled, racing schooner that, when built in 1923, had sported 'teak decks, Louis XV bedroom, marble-rimmed fireplace, oriental rugs, stained-glass windows and gold-fauceted bathrooms'.[8] After war work in the Second World War, the *Vema* was purchased, sadly shorn of her luxury fittings, by the Lamont Observatory and used for oceanographic research worldwide, collecting over a million nautical miles of under-way magnetic and bathymetric data, as well as countless sediment cores.[9] In 1962, she was joined at Lamont by a new, purpose-built, research vessel, the *Conrad*, which also contributed mightily to the data acquisition effort. In the same year, a third ship, the *Eltanin*, a double-hulled, ice-capable, cargo ship, joined the Lamont fleet, providing access to polar regions. Magnetic profiles acquired by *Eltanin* over the Pacific–Antarctic Ridge in 1965 were used by Fred Vine to demonstrate the validity of the Vine–Matthews Hypothesis, and convince sceptics that sea-floor spreading was a reality.[10] Ships of other nations have also made important contributions to plate tectonics research, notably the German vessels *Sonne*, *Meteor*, and *Polarstern*, the French vessels *Jean Charcot*, *Marion Defresne*, *Atalante*, and *Pourquoi Pas?*, and the Spanish vessel *Hesperides*.

Fabric of the Oceans

While the Apollo programme provided us with the first views of the Earth from space, the images gave little away about the structure of the

[8] See the Lamont-Doherty website: http://www.ldeo.columbia.edu/research/office-of-marine-operations/history/vema.

[9] Today, restored to something like her former glory, she serves as a tourist vessel cruising the Caribbean under the name *Mandalay*.

[10] Pitman and Heirtzler (1966); Vine (1966).

Figure 8.3 GLORIA sidescan images of abyssal hills near the East Pacific Rise. (Searle, 1984).

planet beneath. In the same way that the surface of Venus is masked by dense clouds, so most of the Earth's continents are obscured by vegetation or ice, and the ocean basins are almost totally hidden from view by the oceans they contain. Seeing through several miles of water has proven much more challenging than cutting through Venus' atmosphere, however. Sidescan sonar provided an impression of what the ocean basins might look like if the water could be removed temporarily, early surveys showing that, unlike the continents, much of the sea floor has a remarkably geometrical, not to say rectilinear, appearance of bright and dark lineaments more or less parallel to the mid-ocean ridges (Figure 8.3).

Echo-sounders revealed major topographic features and provided base-maps to guide further exploration. Even low-resolution surveys with early single-beam echo-sounders showed that large areas of the sea floor were covered with linear features.[11] These 'abyssal hills', as they are called, are typically a few hundred metres high, around 5 km wide, and several tens of kilometres long. In effect, they represent the

[11] For example, see Menard and Mammerickx (1967).

surface of the basaltic layer—layer 2—of the oceanic crust and, since they cover the floor of the majority of the oceans (and probably exist beneath the sediments that cover the remainder), they are widely regarded as 'the most common physiographic form on the face of the earth'.[12] Clearly, they are related to the spreading process, but precisely how was not clear until detailed sonar and submersible studies were undertaken in the 1990s.[13]

On the flanks of the East Pacific Rise (see Figure 8.4 for location), abyssal hills appeared to be dominated by long, parallel, 'normal faults' (that is, dip–slip faults[14] in the sea floor caused by extension) formed between 2 and 5 km from the axis. Some faults faced towards the axis of the mid-ocean ridge while others faced away, but while the former tended to be straight and steep, dipping towards the ridge at around 45°, the latter were more irregular and frequently covered with pillow lavas. It seemed that two quite different processes were at play in producing these hills: tectonic stresses leading to the ridge-facing faults, and volcanic constructions resulting in the more gentle back slopes. More detailed mapping of slow-spreading ridges confirmed this idea of abyssal hill topography being the result of a kind of dual between magmatism (intrusion of gabbros and eruption of basalts) and brittle deformation (faulting). Magmatism was clearly the fundamental process by which spreading occurred and new plates were constructed. If, however, magma supply was not constant, then, during times of ample supply, a thicker, smoother, volcanic crust would form, but, when magma supply was reduced, spreading would require faulting of existing crust to achieve the necessary amount of extension. If magmatism and faulting alternated, then an abyssal hill fabric like that observed could be produced.[15] Sea-floor morphology thus recorded a history of regular magmatic pulses interspersed with episodes of faulting. On fast-spreading ridges, faults were more closely spaced and their offsets were less, giving a less rugged appearance to the sea floor.

Roger Buck at the Lamont-Doherty Earth Observatory has spent many years studying abyssal hills, trying to establish the mechanisms behind the growth of ocean floor fabric at mid-ocean ridges (Figure 8.5).

[12] Luyendyk (1970).
[13] Macdonald et al. (1996).
[14] See Chapter 4.
[15] See Thatcher and Hill (1995).

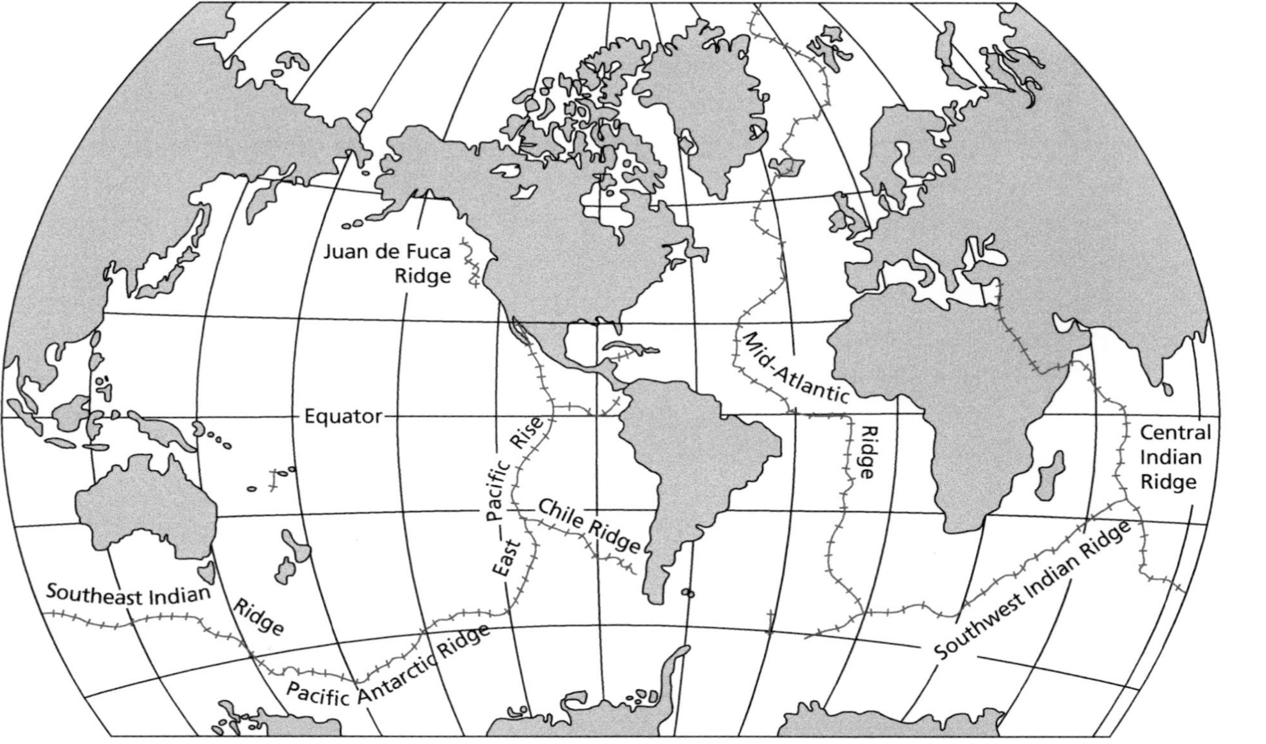

Figure 8.4 The main mid-ocean ridges of the world. The Carlsberg Ridge is the northern part of the Central Indian Ridge, the Gakkel Ridge is the extension of the Mid-Atlantic Ridge into the Arctic Ocean, and the East Scotia Ridge is the short, isolated, section in the South Atlantic. [Source: USGS.]

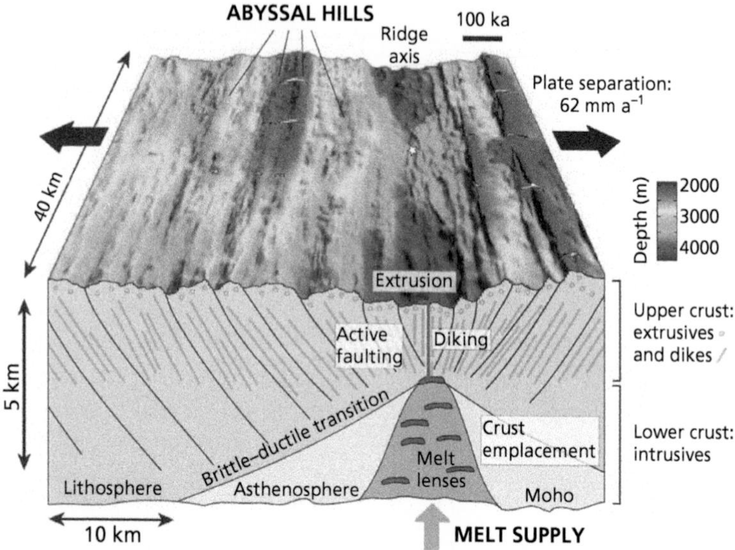

Figure 8.5 Abyssal hill formation at an intermediate-rate spreading centre (the Chile Rise). [From Olive et al. (2015).]

Being a theoretician, he does this by creating computer models of seafloor spreading to see if he can reproduce the patterns of abyssal hills observed. As he once observed, where there are few examples of a particular feature, the best route is to study these individually, but, where there are a very large number of examples, it is easier to study them statistically. And this is exactly what he has been doing in the case of abyssal hills. Together with colleagues, Roger has created a numerical model of faulting at mid-ocean ridges into which he has incorporated 'dyking', the injection of vertical sheets of molten basalt.[16] Faulting behaviour is still too complex to model from first principles, so that simplifying assumptions are necessary even with modern computer power. By varying the proportions of spreading due to magmatism and faulting, Roger and his colleagues have been able to reproduce the different abyssal hill patterns observed at the world's mid-ocean ridges. At slow-spreading ridges such as the Mid-Atlantic Ridge or Central Indian Ridge (Figure 8.4), where magmatism (dyking) contributes 50 per cent of spreading, inward-facing normal faults form close to the axial rift and grow as they migrate away,

[16] Buck et al. (2005).

soon becoming 'locked in' as the young plate cools and thickens (Figure 8.5). At fast-spreading ridges, where magmatism contributes about 95 per cent of spreading, young oceanic crust forms at an axial high and then subsides as it moves away. In the process, Roger and his colleagues suggest, it has to 'unbend' as the sea floor flattens out, creating linear faults that dip away from the axis as well as towards it.

Recently, some researchers have suggested that abyssal hills may represent more than simply the result of steady sea-floor spreading.[17] If melting beneath mid-ocean ridges is a consequence of decompression, they argue, then any factor that reduces pressure on subsurface magma chambers is likely to promote enhanced melting. The factor they have in mind is glaciation, or, rather, deglaciation. Since glaciation involves the transfer of large amounts of water (estimated at around 50,000 trillion tonnes) from the oceans to ice sheets on land, the lowered sea level will noticeably reduce the pressure on mid-ocean ridges, increasing the extent of melting in the mantle beneath. Conversely, melting of glaciers at the end of ice ages will reduce the pressure on subglacial volcanoes, including the section of Mid-Atlantic Ridge forming Iceland, resulting in a burst of volcanic activity. For the past few million years, ice sheets have been waxing and waning on very rigid timescales paced by variations in the Earth's orbit, resulting in cycles at 23,000, 41,000, and 100,000 years—the so-called 'Milankovitch cycles',[18] as revealed in ice cores from Antarctica and Greenland. During periods of glaciation, sea level naturally falls, typically by around 100 m, thereby reducing pressure on the sea bed and, specifically, on mid-ocean ridges, which could promote enhanced melting, calculated at up to twice the long-term average.[19] You might therefore expect to find the cycles of glaciations and melting imprinted on the sea floor as a series of highs (produced during glaciations when sea levels are low and magma supply is increased) and lows (produced during interglacials when sea levels, and hence pressure, are high). The topographic highs would form linear hills parallel to mid-ocean ridges—in other words, the abyssal hill fabric.

The idea that an extra 100 m of water covering a ridge could result in the production of linear hills hundreds of metres high may seem far-fetched.

[17] Crowley et al. (2015); Tolstoy (2015).

[18] Named after their discoverer, the Serbian geophysicist, Milutin Milankovitch (see Chapter 9).

[19] Lund and Asimow (2011).

Yet the journal *Science* has devoted many pages to the idea, and it has been taken seriously by numerous scientists. If correct, it would mean that abyssal hill topography represents a 'proxy' for glacial cycles. So far, researchers have presented only a limited amount of evidence from maps of abyssal hills on the flanks of the Southeast Indian Ridge and East Pacific Rise, suggesting that the 100,000 year Milankovitch orbital cycle might be recorded, and, in the former case, possibly the shorter cycles too. The idea runs into difficulty, however, when the nature of the hills is considered. They are clearly not simply volcanic constructions, being defined by obvious faults that represent brittle deformation rather than volcanism. Furthermore, such a hypothesis predicts that the spacing between abyssal hills should become greater as spreading rate increases, whereas the opposite is observed.[20] In addition, it was argued that the oceanic crust, even at fast-spreading ridges, would not reflect such short-wavelength variations owing to its finite strength and the smoothing effect of the sea-floor spreading mechanism, recalling the arguments over isostatic compensation during the twentieth century. On the basis of their modelling, Jean-Arthur Olive and colleagues expected 'the tectonic fabric of the Australian–Antarctic Ridge [i.e. the Southeast Indian Ridge (Figure 8.4)] to be insensitive even to extreme fluctuations in melt supply on Milankovitch frequencies'. Debate on this subject has continued in the pages of *Science*,[21] so far without resolution.

Fascinatingly, if speculatively, David Lund and colleagues have taken the argument one step further,[22] suggesting that, if melt production at mid-ocean ridges is truly modulated by glacial–interglacial cycles, then there may be a negative-feedback mechanism between glaciation and spreading. Accepting that the lowering of sea level leads to an increase in mid-ocean ridge volcanism, they point out that this would result in an increase in the amount of CO_2 delivered to the oceans and atmosphere (as well as in the direct input of heat to the oceans), which, in turn, would raise global mean temperatures by the greenhouse effect and hence encourage melting of ice sheets. For the time being, however, it has to be said that the question of whether abyssal hills provide a link between plate tectonics and glacial cycles remains unanswered.

[20] Olive et al. (2015); Goff (2015).

[21] For example, Huybers et al. (2016).

[22] Lund et al. (2016).

Faultless

While the transform fault concept was the crucial step in establishing first-generation plate tectonics, and fracture zone traces left by transforms provided the trajectories of plates separating at mid-ocean ridges, the phenomenon was by no means universal. Transforms had been observed on the Mid-Atlantic Ridge and Carlsberg Ridge, at which plates separate at 40 mm per year or less, but in the Pacific Ocean, where rates of 100 mm per year or more occur, the ridge-transform model began to break down. As we noted in Chapter 1, the topography of mid-ocean ridges seemed to be a reflection of the rate of separation, and detailed sonar mapping soon showed that the entire structure of a ridge, together with that of the crust and mantle beneath, depended in surprising ways on this rate.

Marine geophysicists therefore classified mid-ocean ridges according to their spreading rates: those spreading at rates of 20–50 mm per year were classed as 'slow', while those spreading at rates above 80 mm per year were 'fast'. Unsurprisingly, ridges with rates in the range 50–80 mm per year were dubbed 'intermediate'.[23] It was soon realized, however, that the southern part of the East Pacific Rise, spreading at rates above 120 mm per year, exhibited features not seen on more tardy ridges, and a new 'superfast' class was born. Likewise, parts of the Southwest Indian Ridge between Africa and Antarctica (Figure 8.4), had unique characteristics that appeared to be related to their 'ultraslow' spreading rates of less than 20 mm per year. In reality, there is a continuum of ridge styles related to spreading rate and, ultimately, to magma supply. Many of these styles can be observed on a trip down the East Pacific Rise, such as that made in the summer of 1982 by the Scripps research ship *Thomas Washington*, which steamed 2200 km from the Gulf of California at 20°N as far south as 2°N,[24] and later to 20°S.[25] The ship had recently been fitted with SeaBeam, the capabilities of which were sufficient to change ideas about mid-ocean ridges completely.

Where it emerges from the Gulf of California (Figure 8.6), the East Pacific Rise spreads at a rate of around 50 mm per year and has a form

[23] These boundaries are not rigidly defined: different researchers will use slightly different rates.

[24] Macdonald and Fox (1983).

[25] Lonsdale (1983).

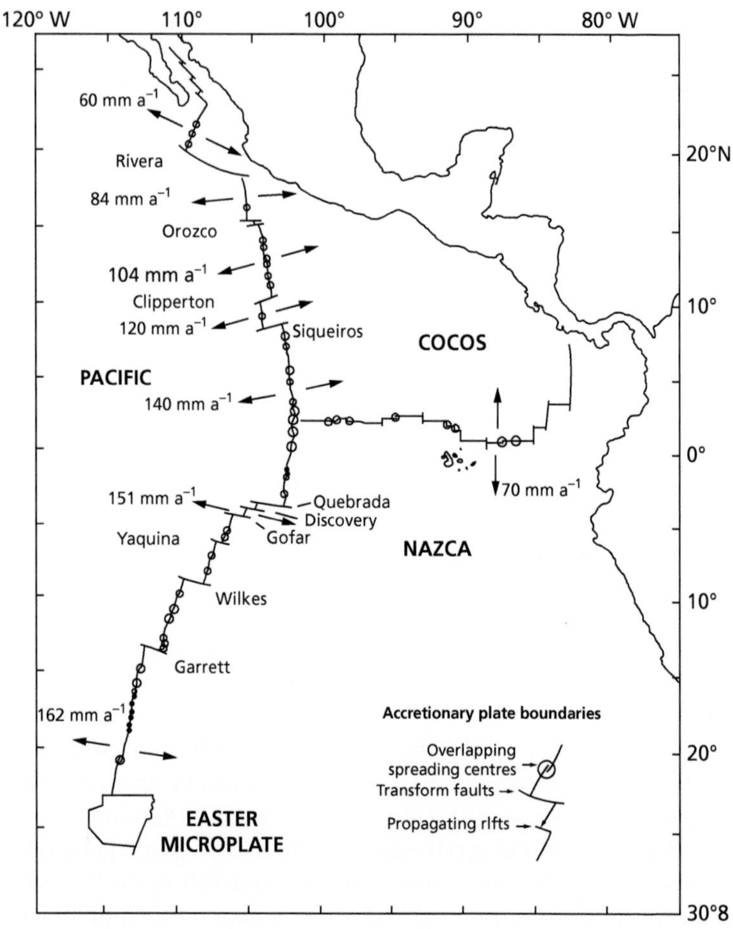

Figure 8.6 The East Pacific Rise. [After Macdonald (1986).]

similar to the Atlantic ridges, with short segments offset by transform faults. The rates of spreading on the East Pacific Rise reflect the westward migration of the Pacific plate away from a series of plates forming the eastern flank of the Rise, starting with the North American plate within the Gulf of California.[26] The left-stepping Tamayo transform

[26] In fact, GPS measurements suggest that Baja California is moving at a slightly different rate from the Pacific plate, and may therefore be regarded as a semi-independent microplate (Plattner et al., 2007).

connects the sedimented ridge within the Gulf to a 300 km ridge segment running southwest to the curved Rivera transform. The plate to the east of this segment is a small oceanic plate known as the Rivera plate, but, further south, this is replaced by the larger Cocos plate, the eastward motion of which is driven by its subduction beneath Central America.[27] As you head south, rates of Pacific–Cocos spreading increase from 84 mm per year to well over 120 mm per year, and the East Pacific Rise here is consequently classed as a fast-spreading ridge.

The zone of plate accretion, that is, the region within which new crust is added, was known to be extremely narrow here—less than 1 km—which, given the enormous width of the plates, is truly remarkable. As mentioned in Chapter 1, it is the narrowness of this zone that provides the very high resolution of magnetic anomalies formed at the ridge—a wider zone would tend to smear out the reversal record, erasing the shortest anomalies. In detail, however, the axis was found to be much less uniform than previously supposed. In some places, it was marked by a narrow peaked ridge at typical depths of 2500 m, in others by a shallow axial trough at depths of 2800 m or so, superimposed on a broader swell.[28] An along-axis profile undulated continuously from shallower 'inflated' axial highs near the centres of ridge segments to the deeper 'rifted' profiles near segment ends. It was as though the supply of new basaltic magma to the ridge was robust near segment centres, but declined gradually towards segment ends. Three major transform offsets had previously been mapped along this part of East Pacific Rise (Figure 8.6): the Orozco, Clipperton, and Siqueiros fracture zones, the first and last of which were left-stepping offsets. However, the most surprising discovery was that, between these transforms, the Rise was not continuous, but was interrupted by an entirely new kind of ridge axis geometry.

The discovery was made by Ken Macdonald, a Californian geophysicist, together with colleague Paul Fox of Rhode Island.[29] At many points

[27] Surveys off-axis have shown that, in contrast to the Mid-Atlantic Ridge, the East Pacific has undergone many major reorganizations and, as a result, does not reflect its original form following continental break-up. The plates presently forming the eastern flank of the Rise were once part of a single large plate, the Farallon plate, which was disrupted by interactions with the Americas.

[28] It later emerged that the summit rift was nearly ubiquitous, representing a kind of elongated volcanic caldera.

[29] Macdonald and Fox (1983).

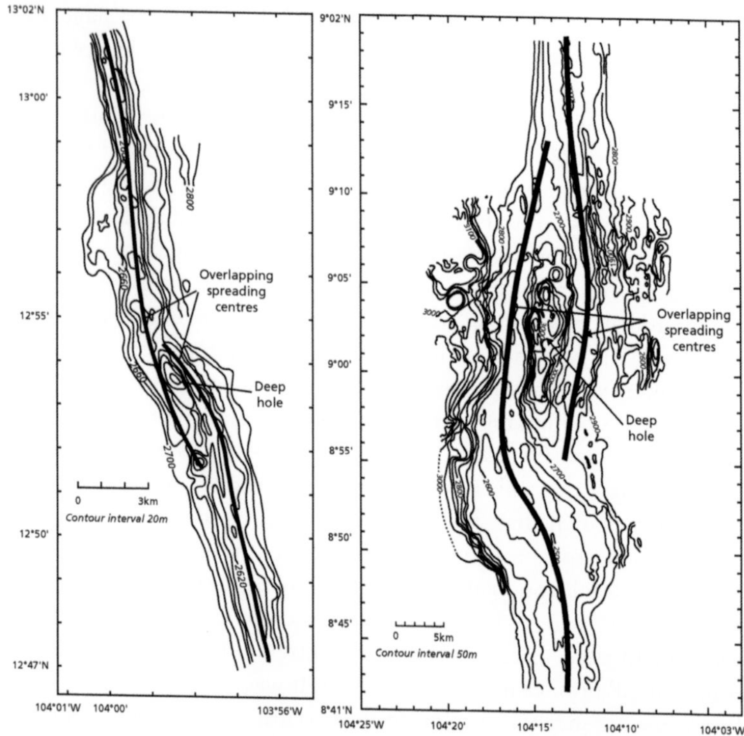

Figure 8.7 Examples of overlapping spreading centres on the East Pacific Rise at 12°54´N (left) and 9°N (right).

along the rise axis, there were gaps in the axial topographic high. In these cases, the tips of the adjacent highs overlapped in broad curves, resembling a handshake (Figure 8.7), enclosing a small 'overlap basin'. These 'overlapping spreading centres' (or OSCs as they became known in the trade) were of variable dimensions, some with offsets of just a few kilometres, while others were separated by 10 km or more and overlapped by 30 km or so. Like segment ends, the OSCs appeared to reflect a restricted magma supply, being at slightly greater depth with narrower volcanic ridges. Larger OSCs appeared to mark segment boundaries in the same way as transform faults. These large OSCs also generated 'disturbed zones' away from the axis, consisting of cuspate volcanic ridges and depressions. Typically, the overlapping ridges were at different depths, suggesting that their magma sources were distinct from one another, each representing the distal part of its respective magmatic segment.

A couple of degrees north of the equator, the intermediate-rate Galapagos Rise—the boundary between the oceanic Cocos and Nazca plates—theoretically meets the East Pacific Rise in a ridge-ridge-ridge triple junction. 'Theoretically', because true spreading on the Galapagos Rise only begins about 60 km east of the East Pacific Rise, the region in between being an area of complex deformation. The Galapagos Rise, of course, was the ridge at which hydrothermal vents and associated vent communities were first observed from *Alvin* in 1977.[30] It was also the place at which another spreading-related phenomenon was first observed. Three years before the *Alvin* discovery, Dick Hey at the Hawaii Institute of Geophysics and Peter Vogt at the US Naval Oceanographic Office collected magnetic anomaly and bathymetry data showing that, about 400 km west of the Galapagos Islands, the ridge axis jumped 27 km to the north.[31] The jump, however, was not a discrete event, but represented a process of ridge migration that had begun 3 million years earlier and had resulted in the offset shifting 150 km to the west. The Galapagos Islands, like Hawaii and Iceland, sit above a mantle hotspot, resulting in abundant volcanism. It appeared that a large pulse of magmatism had propagated away from the islands along this ridge segment, causing the tip of the ridge to penetrate into existing sea floor, creating a new rift (Figure 8.8). By this means, the ridge segment had migrated west to 95.5°W as the new rift propagated, leaving a V-shaped wake in the bathymetry. Similar propagating rifts were later discovered on fast and intermediate ridges elsewhere.

Continuing south, the East Pacific Rise spreads at still faster rates, owing to the pull exerted by subduction of the Nazca plate beneath South America. Between the Galapagos triple junction and 20°S, 6 right-stepping transform faults were mapped by SeaBeam,[32] starting with the Quebrada transform at 4°S and ending with the Garrett transform at 14°S. The ridge axis is thus divided into six major segments, the longer ones exhibiting small OSCs. A large OSC was discovered on the ridge segment between the Gofar and Yaquina transforms at 5° 30´S,

[30] High-temperature black smokers have recently been discovered a few hundred kilometres west of the original site.

[31] Hey and Vogt (1977).

[32] Lonsdale (1989).

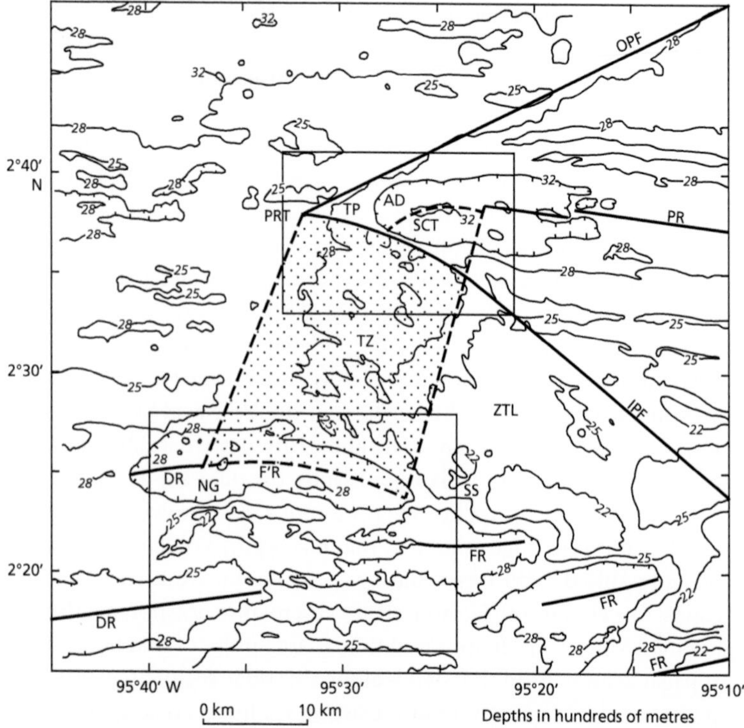

Figure 8.8 Propagating rift on the Galapagos Rise at 95.5°W. PR - propagating rift; DR - dying rift; TZ - transfer zone between the two; FR - failed rift.

where the spreading rate is around 145 mm per year. Here, two axial ridges overlapped by 20 km, and were separated by a 9 km wide basin filled with volcanoes.[33]

Beyond the Garrett, we enter the fastest-spreading section of mid-ocean ridge on the planet today, with rates exceeding 160 mm per year (although rates over 200 mm per year have occurred in the past)—too fast, apparently, for transforms to keep up, since there are none for over 1000 km. At 20°40′S on this segment, a remarkable ridge offset was mapped by Ken Macdonald and colleagues using SeaBeam.[34] It appeared to represent a pair of ridges offset by 15 km that had been propagating alternately at rates comparable to the superfast spreading rate. These

[33] Lonsdale (1983).
[34] Macdonald et al. (1988).

'duelling propagators' had created a broad, V-shaped, zone of disturbed topography on the flanks of the ridge, as first the southern rift propagated northward at the expense of the northern rift, only for the situation to be reversed. The net result was a slow migration of the ridge offset towards the south at around 20 mm per year.

In April 1983, the *Thomas Washington* surveyed the southern part of the East Pacific Rise near Easter Island at around 25°S, where the spreading rate is 155 mm per year. Here, SeaBeam and magnetic anomaly data revealed a quite spectacular new mode of spreading. Previously, it had been reported on the basis of earthquake epicentres and focal mechanisms that this region could be acting as an independent 'microplate'.[35] The new SeaBeam data showed that much of the abyssal hill fabric had trends quite different to the normal north–south orientation of East Pacific Rise fabric, while the magnetic anomalies indicated two sets of overlapping and propagating rifts. To the east, these rifts were propagating northward, while to the west, they propagated south. Overall, the result was to isolate a region of oceanic plate around 400 km across that was behaving rather like a cam, rotating clockwise at between 10° and 20° per million years. This region became known as the 'Easter microplate', and was the first 'plate' to be imaged in its entirety (its total area is 200,000 km^2) by sidescan sonar[36] (Figure 8.9).

A similar situation appeared to exist further south, near the junction of the East Pacific Rise with the Chile Rise at 33°S (Figure 8.10). Here, however, three major plates were involved: the Pacific, Nazca, and Antarctic plates,[37] making things more complicated. Nevertheless, the same pattern of rifts propagating southward on the western boundary and northward on the eastern boundary defined another microplate, christened the 'Juan Fernandez microplate',[38] that was also rotating clockwise, but at a slightly slower rate of around 10° per million years. In total, the cores of both the Easter and Juan Fernandez microplates are estimated to have rotated by about 80° in the past 3–5 million years. The best explanation for the rapid rotation of these microplates is that they are caught up like

[35] Called the 'upper plate' by Ellen Herron (Herron, 1972). See also Forsyth (1972).

[36] Searle et al. (1989).

[37] It may come as a surprise that the Antarctic plate extends as far north as 33°S. This emphasizes the enormous difference between the size and shapes of continents and of the plates on which they are transported.

[38] Both the Easter and Juan Fernandez microplates are named after the nearest land, which, in the latter case, is over 1000 km away.

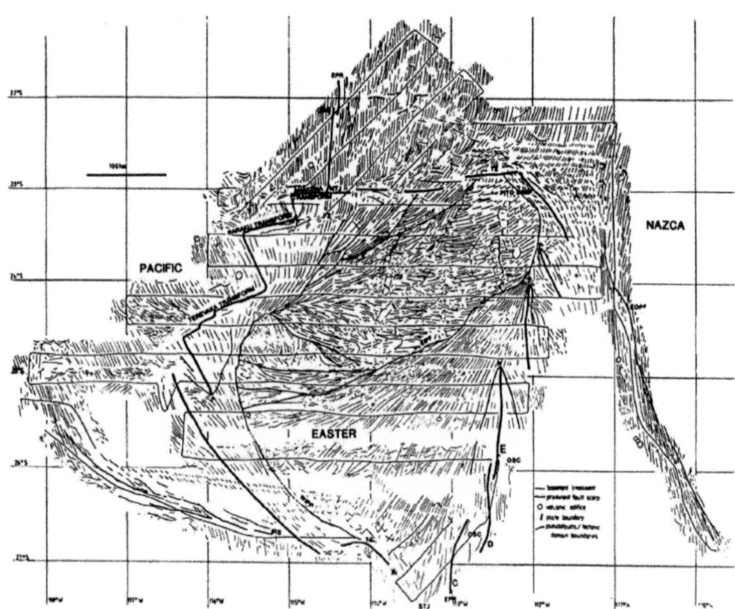

Figure 8.9 Interpretation of GLORIA and SeaMARC II sidescan images of the Easter Microplate (Rusby, 1993).

roller-bearings between major plates.[39] The Easter microplate experiences eastward drag in the north owing to contact with the eastward-moving Nazca plate, while, in the south, the Pacific plate is exerting a westward drag, with the result that the microplate spins clockwise. Likewise, the Juan Fernandez microplate is subject to a torque applied by the Nazca plate in the north and the Antarctic plate in the south, with much the same consequences. Recent work suggests that the complex region west of the Galapagos Rise may also be acting as a microplate.

On the section of East Pacific Rise between the Easter and Juan Fernandez microplates, there were no transform faults, but, at 29°S, beautiful GLORI-B sidescan and SeaBeam2000 datasets collected in 1993[40] revealed spreading centres overlapping by 120 km, making this the largest OSC on the mid-ocean ridge system. It appeared that the giant OSC had evolved from a normal ridge-transform offset about

[39] Schouten et al. (1993).
[40] Hey et al. (1995).

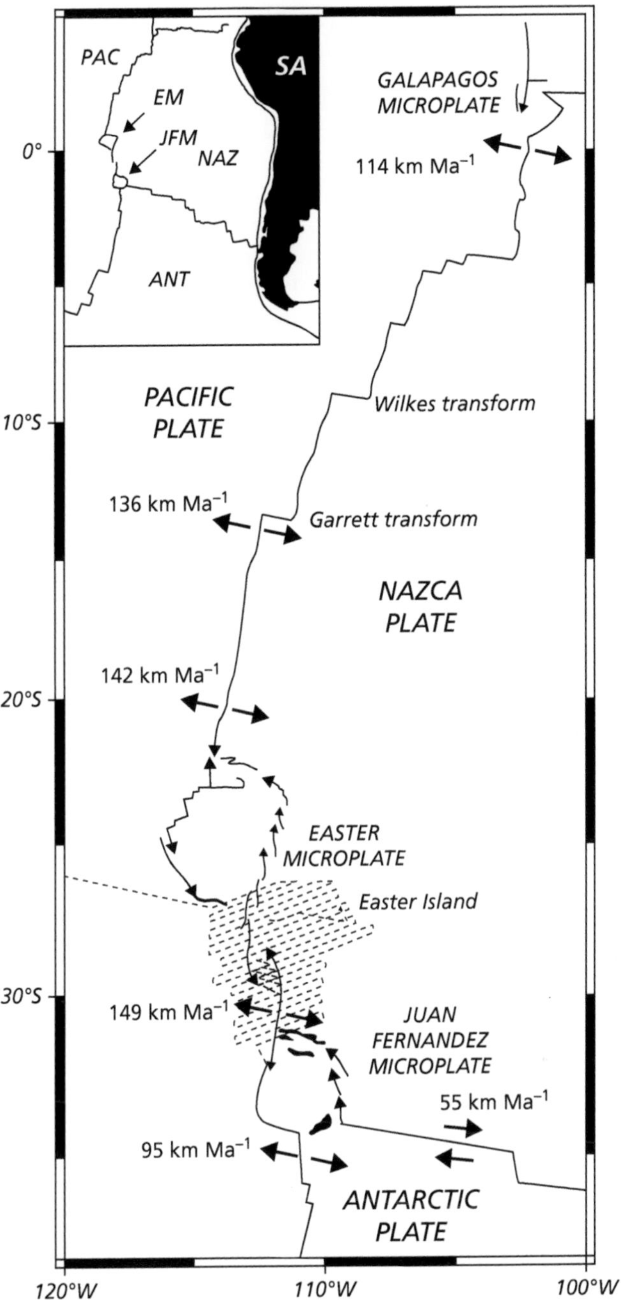

Figure 8.10 The East Pacific Rise south of the Galapagos Rise. This section incorporates two rotating microplates, separated by a large pair of overlapping rifts. It also includes the fastest-spreading section of mid-ocean ridge on Earth today. [From Hey et al. (1995).]

2 million years ago, since when the two rifts have competed for supremacy, with the eastern ridge currently propagating north at the phenomenal rate of 500 mm per year. The consequence of this duelling is an overlap zone of 120 km by 120 km that, quite possibly, could evolve into a future microplate.

Continuing southward from the Juan Fernandez microplate, the East Pacific Rise now separates the Pacific and Antarctic plates and is therefore commonly referred to as the Pacific–Antarctic Ridge.[41] This was the section of the East Pacific Rise over which the famous *Eltanin* magnetic anomaly profiles mentioned earlier were acquired. More recent exploration, notably by Peter Lonsdale of Scripps, has revealed much about the nature of this section of the Rise. The Antarctic plate is, like the Nazca plate to the north, being subducted beneath South America, but, being a far larger plate carrying a major continent, is also subject to tractions from its extensive boundaries elsewhere. The upshot is that the Antarctic plate is moving eastward more slowly than the Nazca plate, and so spreading rates south of the Chile triple junction are lower—declining from about 100 mm per year just south of the triple junction to about 84 mm per year at the Eltanin Fracture Zone. This section appears to be a single 1500 km-long ridge segment, incorporating several southward-propagating ridge offsets. Transform faults reappear at 50°S, with the Menard transform, a right-stepping offset of 200 km, followed southwards by four more transforms, all right-stepping, and all named after American oceanographers. At 54°S, the East Pacific Rise meets the world's largest transform fault system, the Eltanin Fracture Zone, which offsets the Rise by 1000 km. The Eltanin is comprised of two closely spaced offsets, the Heezen and Tharp transforms, and their inactive fracture zone extensions. Further southwest, the Pacific–Antarctic Ridge becomes an intermediate-rate spreading centre, separating New Zealand from Antarctica in one of the remotest oceanic regions on Earth.

Tubeworm Barbeque

Following the surprising discovery of warm springs at the Galapagos spreading centre[42] by *Alvin* in 1977, it was soon established from the

[41] Although Lonsdale (1994) refers to it as the 'Pacific–Antarctic East Pacific Rise', reserving 'Pacific–Antarctic Ridge' for the sections south of the Heezen–Tharp fracture zones.

[42] See Chapter 1.

chemistry of the emerging fluids that the hydrothermal vents must be a consequence of the circulation of seawater through the hot basalt crust of mid-ocean ridges. In 1979, *Alvin* discovered jets of water at 380°C shooting from mineralized chimneys on the sea floor at a segment of the East Pacific Rise just south of the Gulf of California at 21°N.[43] The chimneys were associated with sulphide mounds, and emitted black 'smoke' produced as the hot, mineral-laden, jet mixed with the ambient cold seawater and precipitated metal sulphides. Calculations showed that the power output of a single black smoker chimney in heating cold seawater to these high temperatures must be around 300 megawatts—equivalent to the total input of heat from the mantle along a 6 km stretch of mid-ocean ridge.[44] Hence, such hot springs must be very episodic, being active only for months or years before their supply of heat is exhausted.

A decade later, a group of American scientists from the University of California at Santa Barbara and the Lamont-Doherty Geological Observatory in New York towed a camera platform about 10 m above the East Pacific Rise between the Clipperton and Siqueiros transforms near 9° 50'N (Figure 8.5). Cheaper (and a lot more comfortable) to operate than a submersible, the camera platform, known as ARGOS, nonetheless provided evidence of hundreds of active vents, including high-temperature black smokers with thriving communities of vent organisms comprising masses of tubeworms, together with numerous clams, mussels, and fish.[45] Encouraged by this plenitude, the group decided to return a couple of years later, equipped with *Alvin*, to investigate this deep-sea Garden of Eden further. On reaching the site this time, however, they got a surprise. For, where a thriving ecosystem had previously bathed contentedly in the warm (if dark and toxic) waters surrounding the vents, a scene of carnage now presented itself. Dead tubeworms and mussels were strewn over the sea bed, which had acquired a white covering of microbial mats. Nearby, a jet of hot water surged from the bare basalt floor at a temperature measured at 403°C—evidence of a very recent eruption.[46] Later examination of some of the corpses in the lab revealed that the victims had, in fact, been cooked by exceptionally hot fluids emerging from the sea floor. These fluids had completely overwhelmed this section of the

[43] Spiess et al. (1980).
[44] Macdonald (1980).
[45] Haymon et al. (1991).
[46] Haymon et al. (1993).

ridge, searing the hapless invertebrates and partially burying them in lava. The location thus became known as the 'tubeworm barbeque'. No more dramatic illustration could be imagined of the risks of living next door to a fickle hydrothermal vent.

New discoveries of hydrothermal vent communities at mid-ocean ridges spreading at rates from 30 to 145 mm per year showed that, while those at fast-spreading ridges such as the East Pacific Rise were more diverse, they were also more ephemeral, since eruptions occurred more frequently. Vent communities depend on the effusion of reduced chemical compounds—such as hydrogen sulphide and methane—that can be used to release energy. The critical reactions are carried out by microbes (chemoautotrophic bacteria), which makes them highly desirable companions for multicellular animals such as shrimp and crabs, which feed on the microbes, and tubeworms, which go so far as to incorporate the food-producing bacteria into their tissues.

It seemed that such hot springs were unlikely at slow-spreading ridges where the supply of magma—and presumably of heat also—was much less. However, evidence of hydrothermal venting on the Mid-Atlantic Ridge was discovered near 26°N during the NOAA Trans-Atlantic Geotraverse (TAG) expedition in 1974, between the Kane and Atlantis transforms. Active vents, including black smokers, were discovered in 1985 located on a 50 m high mound, similar in size to the Colosseum in Rome,[47] close to one wall of the axial valley where faults and fissures acted as a conduit for cold seawater penetrating the hot crust. Other, lower-temperature, vents were also discovered nearby. Later work has shown that the TAG vent site is the largest and longest-lived of all high-temperature mid-ocean ridge hydrothermal systems (at least, so far as we know), emitting copper-rich fluids intermittently over a period of 150,000 years, and generating massive sulphide deposits reckoned to exceed 3.9 million tonnes. No large tubeworms were seen, although the vents were covered in thousands of shrimp (at densities of 40,000 m^{-2}) plus a few crabs and large fish. As elsewhere, white bacterial mats covered parts of the site.

It was becoming clear that hydrothermal circulation was 'the principal agent of energy and mass exchange between the ocean and the earth's crust'[48] and that particularly large chemical plumes detected in

[47] Karson (2015).
[48] Baker et al. (1995).

the water column were caused by fresh eruptions of lava. This began a competition to observe such eruptions while still in progress, which meant detecting mid-ocean ridge earthquakes as early as possible. Much of the effort of the NOAA Vents Program was focused on the Juan de Fuca Ridge, where, in 1986, scientists detected a 20 km-diameter 'megaplume' of anomalously warm water above the southernmost segment of the Ridge, which they attributed to 'cataclysmic hydrothermal venting'.[49] Following a brief massive injection of hydrothermal fluids equal to the output of several hundred hydrothermal chimneys, the megaplume had risen more than 1000 m above the sea floor. Two months later, there was no trace of the plume. A similar short-lived plume was discovered the following year, 45 km further north, while follow-up work showed that a volcanic eruption had occurred on this segment between 1983 and 1987. The most likely explanation for the formation of these megaplumes was therefore rifting events associated with the eruption of mounds of pillow lavas at the black smoker vent sites. These and other megaplumes subsequently detected on other ridges were therefore dubbed 'event plumes'.

In order to locate fresh event plumes on the Juan de Fuca Ridge, detection of small earthquakes associated with magma movement within the crust was required. Fortunately, permanent hydrophone arrays, known as the 'Sound Surveillance System' (SOSUS), were already installed and operated by the US Navy in order to detect Soviet submarines. In 1991, following the collapse of the USSR, the Navy agreed to make the data available to academics, and, a couple of years later, a real-time data link was established. Just a week after the link became operational, a swarm of earthquakes was detected on a segment of the Juan de Fuca Ridge. A Canadian ship was commandeered to go to the site, where it detected an event plume. After this, ROVs, such as the Canadian ROPOS vehicle, were developed for work on hydrothermal vents, incorporating still and video cameras, vent sniffers, and robot sampling arms. In what some might regard as an act of vandalism, ROPOS recovered, in 1998, 2 m of an active chimney by means of a chainsaw. When the chimney was revisited 72 hours later, it had regrown the lost 2 m, demonstrating the intensity of hydrothermal processes during the limited periods of black smoker activity.

[49] Baker et al. (1987).

Around a third of the 60,000 km global ridge crest has now been prospected for hydrothermal plumes, permitting analyses of the relationship between frequency of vent fields and spreading rate. The incidence of hot water (black smoker) venting shows a very clear positive relationship with spreading rate,[50] which has been used to predict the number and location of vent sites yet to be discovered.[51] It seems likely that a total of around 1300 vent sites could exist worldwide, of which a third have already been discovered. The majority of the undiscovered vent sites are probably in the Southern Ocean, many at slow or ultraslow ridges.

Secret Chambers

As long ago as 1976, the Lamont group shot a pair of multichannel seismic reflection profiles across the East Pacific Rise at the 9°N site.[52] After processing their data, they believed that they had detected a reflection from the top of a magma chamber, 2 km beneath the ridge axis. This discovery depended on new techniques of digital recording of artificial seismic signals produced by arrays of airguns towed behind a ship. By suitably arranging the ship's speed and shot rate so that many seismic waves were reflected from the same point, it was possible to add or 'stack' reflections and apply digital filtering to recover weak signals from the noisy recordings. Their findings were confirmed ten years later when a series of new lines was shot both along and across the axis. The new profiles showed a strong reflection extending for tens of kilometres beneath the axial ridge at depths of 1.2–2.4 km, with phase characteristics suggesting a liquid. This was interpreted as a 'lens' of basaltic melt at the top of a partially crystallized magma chamber. A year or two before, a group from Scripps had used the seismic refraction technique to establish the existence of a low-velocity zone 1.4 km thick some 2 km beneath the ridge axis at this location. Their results were interpreted as representing a magma chamber consisting largely of crystals with a small amount of melt between them—what geologists call a crystal mush. The new reflection results thus established the presence of a lens of magma, tens to hundreds of metres thick, right at the

50 Baker (2009).
51 Beaulieu et al. (2015).
52 Herron et al. (1978).

top of the mush chamber, where it collected owing to its buoyancy. While no more than a few kilometres wide, the melt lens extended for tens of kilometres north and south of the Clipperton transform. A similarly narrow (less than 1 km) magma lens was detected beneath the superfast ridge segment south of the Garrett transform.[53] In this case, the melt lens was just 1.2 km beneath the sea floor and displayed remarkable continuity along this segment of the Rise. It seemed as though the higher the spreading rate, the shallower and more continuous was the melt lens.

Similar seismic experiments at slow-spreading ridges such as the Mid-Atlantic Ridge failed to show evidence of a melt lens beneath most of the ridge axes. At the TAG hydrothermal field, the existence of black smokers and colonies of vent organisms suggested that a magma chamber existed within the crust beneath, but seismic profiling could find no melt lens, which is surprising, since the heat necessary to drive hydrothermal circulation beneath ridges can only come from the crystallization of hot magma.[54] Recent work suggests that the heat may be derived from magmatic intrusions below the ridge axis at depths of more than 7 km[55]. By contrast, at the Lucky Strike hydrothermal field at 37°N, seismic profiles shot in 2005 by the French research ship RV *l'Atalante* showed clear evidence of a melt lens at a depth of 3 km beneath the axis, extending for 60 km along the ridge.[56] It thus appears that hydrothermal activity on slow-spreading ridges corresponds to periods when magma is being supplied to crustal (or, in the case of TAG, subcrustal) magma chambers. Between these inflationary phases, vents may become inactive and their fauna homeless.

Megamullions

Transform faults, you recall, were so-called because they transformed the extension related to spreading on mid-ocean ridge segments into parallel, 'strike–slip', movement on the faults themselves. Where a transform met a spreading segment, there was a sharp corner in the boundary separating two plates. In fact, there were two corners: an

53 Kent et al. (1994).
54 Cann and Strens (1982).
55 Canales et al. (2007).
56 Singh et al. (2006).

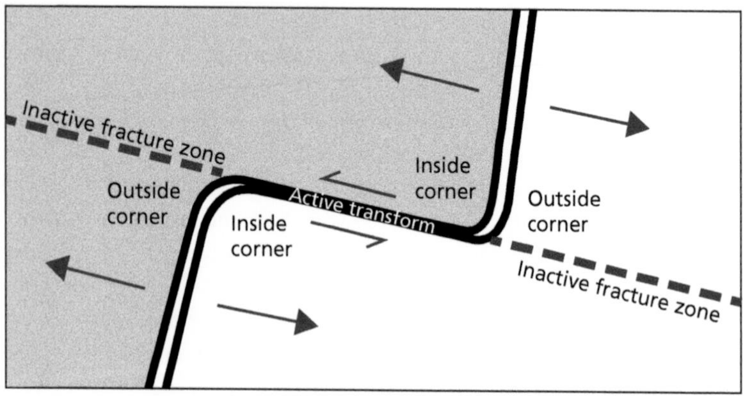

Figure 8.11 Configuration of a typical transform fault offset, showing the locations of inside and outside corners. Note that only the central section between the two spreading segments is active.

'inside corner' between the ridge segment and the active transform, and an 'outside corner' between the segment and the inactive fracture zone extension (Figure 8.11). While the crust beyond the outside corner seemed perfectly normal, with parallel abyssal hills, that on the inside corner of slow-spreading ridges such as the Mid-Atlantic Ridge was frequently anomalous. Magnetic anomalies showed that the crust here was of the age expected, yet bathymetric surveys quite often revealed a shallow region of lumpy topography lacking the usual linear hills. As well as irregular fault patterns, Brian Tucholke and Jian Lin of the Woods Hole Oceanographic Institution showed that many inside corners exhibited thinned crust, an absence of volcanic features, and, if dredge samples were anything to go by, rocks typical of the lower crust, like gabbro, rather than the usual basalts.[57] The pair suggested that what they called a 'low-angle detachment fault' had sliced off the upper crust at the inside corner, exposing the lower crust and uppermost mantle.

In 1997, Joe Cann and colleagues presented remarkable images of the bathymetry at two inside corners adjacent to the Atlantis transform near 30°N on the Mid-Atlantic Ridge, where the spreading rate is around 24 mm per year[58] (Figure 8.12). Both areas exhibited slightly

[57] Tucholke and Lin (1994).
[58] Cann et al. (1997).

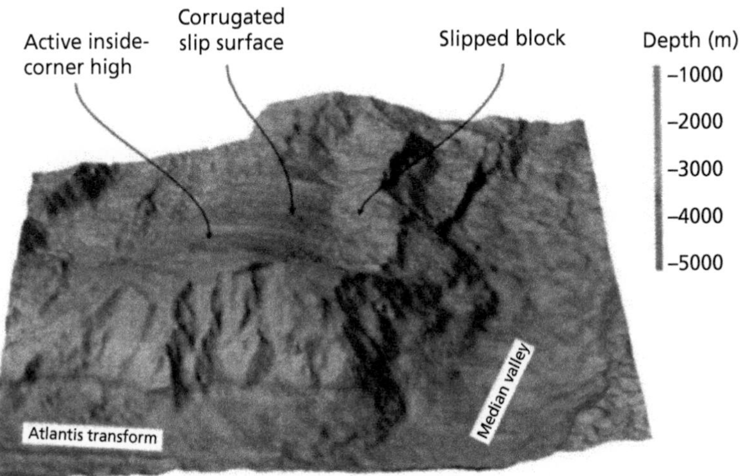

Figure 8.12 Bathymetry of the inside-corner high mapped by Cann et al. (1997), showing the corrugated surface or 'megamullions'.

domed features roughly 15 km wide, rising to shallow depths of 1000 m or less. Their surfaces had strange corrugations parallel to the direction of plate movement—that is, more or less perpendicular to the surrounding abyssal hills. These surfaces were the detachment faults proposed by Tucholke and Lin. Brian and Jian reckoned[59] that, whereas faults near the centre of such segments were supplanted within a few thousand years by newer faults formed closer to the ridge axis, the crust towards the ends of segments lacked magmatism and thus became colder and stronger, inhibiting the formation of new faults. In addition, reaction of mantle peridotite with seawater produced serpentinite, a soft metamorphic rock that could serve to lubricate the existing fault. The fault thus continued to slip for 1–2 million years, finally exposing the rocks of the lower crust and upper mantle. Brian and Jian dubbed these fault surfaces 'megamullions'[60] and suggested that they evolved from steeper normal faults, becoming shallower with time, thus forming a domed structure. Similar detachment faults in continental crust exposed rocks from the lower crust metamorphosed

[59] Tucholke et al. (1998).

[60] The term 'mullion' is applied by geologists to corrugations in the bedding surfaces of sedimentary rocks.

by the elevated pressures and temperatures to which they had been subjected. Such exposures were known to geologists as 'core complexes', and so, by analogy, Brian and Jian proposed that the megamullions were diagnostic of oceanic core complexes.

In 2000, *Alvin* visited the dome, along with an ROV, both operated from R/V *Atlantis*, and made an important discovery A new kind of hydrothermal venting was observed, 20 km from the ridge axis, producing white calcium carbonate spires and chimneys up to 60 m high[61]. These vents expelled highly alkaline fluids (pH of 9 and above), rich in hydrogen and methane, at temperatures of 40–70°C. Unlike the high-temperature vents on the axes of mid-ocean ridges, they were disappointingly depauperate in vent animals, although a thriving flora of bacteria and archaea was present. The high calcium concentrations and high alkalinity were in sharp contrast to the highly acidic (pH 3–5) and sulphide-rich fluids emanating from typical basalt-hosted black smokers, and were strong indicators of an ultramafic (i.e. peridotite) host rock. The latter, as we observed in Chapter 1, is highly unstable in the presence of seawater and breaks down to form serpentinite, a process involving the absorption of large volumes of water and the generation of enough heat to raise the rock temperature by 260°C. Some of this heat was thought to power the vent field, dubbed the 'Lost City' (i.e. Atlantis). Seven out of eleven vent sites on the northern Mid-Atlantic Ridge, including the TAG hydrothermal field,[62] are now known to be associated with detachment faults, supporting the idea that these faults serve as conduits through which heated hydrothermal fluids rise towards the sea floor.

Exposure of the lower oceanic crust and uppermost mantle presented a golden opportunity for sampling by deep drilling. Unfortunately, drilling technology has still not developed to the point where boring through the entire 6 or 7 km of oceanic crust is feasible, but, by drilling a series of holes 1–2 km deep in the megamullions, a composite section through the oceanic crust could be assembled. This was soon achieved by JOIDES *Resolution* on expeditions 304/305 of the Integrated Ocean Drilling Program (as it is now called) in 2004 and 2005. However, the rocks recovered in one 1.4 km borehole were nearly all gabbros, with hardly any of the expected fresh mantle rocks. Peridotites were, however,

[61] Kelley et al. (2001).
[62] Canales et al. (2007).

common to the south of the dome, where many samples were dredged from the ocean floor, so that drilling to date has raised more questions than it has answered.

Given the frequency with which these core complexes occur, it is likely that an appreciable proportion of Atlantic sea floor—maybe 20 per cent—was produced in this way and is therefore formed of gabbro and peridotite rather than basalt. Hence, serpentinite must form quite large areas of the sea bed adjacent to slow (or ultraslow) ridges. Some authors have taken this as a kind of vindication of Harry Hess' theory of oceanic crust formation, but serpentinization of mantle peridotite exposed by crustal faulting is entirely different to Harry's idea involving the formation of the entire oceanic crust by serpentinization rather than by magmatism.

Remote Ridges

Possibly the most surprising of the US military's many contributions to plate tectonics research came in the early 1990s. At that time, the Eurasian Basin within the Arctic Ocean was known to be an extension of the North Atlantic, but, for obvious reasons, few data existed to constrain the sub-ice bathymetry. In 1993, the US Navy invited US scientists to participate in an Arctic cruise aboard USS *Pargo*, a Sturgeon-class nuclear submarine, promising that data collected could be published openly following the expedition, which was christened 'SCICEX' (for 'Scientific Ice Experiments'). Bernie Coakley, a research scientist at the Lamont-Doherty Earth Observatory, described his participation in the expedition in biblical terms as 'forty days in the belly of the beast',[63] during which the vessel surfaced at various places, including the Pole. In 21 days of survey, he was able to acquire 10,000 km of new bathymetric and gravity data. The Navy's generosity extended to the following seasons and the SCICEX programme was able to generate high-quality data, freed from the difficulties of negotiating permanent sea ice.

In 1998 and 1999, cruises were carried out by USS *Hawkbill* (Figure 8.13), equipped with specially designed sonar pods that acquired both swath bathymetry and sidescan sonar. The results established that the Eurasian Basin had begun opening at the same time as the North Atlantic—around 60 million years ago—creating the Gakkel Ridge spreading centre.

[63] Coakley (1998).

Figure 8.13 Tent set up on the Arctic ice during SCICEX-99. The structure on the right is the conning tower of USS *Hawkbill.* [Photograph: Mark Rognstad.]

Only two plates, the North American and Eurasian, were involved, and a single set of Euler rotations describes the evolution of both ocean basins. Owing to the location of the Euler pole in northern Russia, close to the Laptev Sea, spreading rates decline along the Gakkel Ridge as it crosses the Eurasian Basin, endowing it with the status of slowest-spreading ridge on the planet, with plate separation rates that decline from around 13 mm per year near Greenland to just over 6 mm per year where it reaches the margin of Siberia. The SCICEX mapping showed that this 1800 km mid-ocean ridge had several unusual features. Despite its length, the ridge axis lacked transforms or other types of offset, and lay within an axial rift at depths that in places exceeded 5000 km, nearly twice as deep as typical mid-ocean ridges. These characteristics reflect the very slow spreading rates, which favour spreading mainly by faulting. However, three large volcanic features were mapped during the 1998 cruise, demonstrating that, in places, the supply of magma to the spreading centre was greater than expected for an ultraslow ridge.

In 2001, the US Coast Guard's new icebreaker, USCGC *Healy*, accompanied by the German polar research vessel *Polarstern*, surveyed 1100 km of the Gakkel Ridge using modern multibeam echo-sounders in order to produce high-quality bathymetric maps. Known simply as the 'Arctic Mid-ocean Ridge Expedition' (AMORE), this joint US–German initiative surveyed the ridge for hydrothermal plumes in the water column as evidence of venting on the sea floor, successfully locating at least nine separate plumes, all associated with volcanic features. It appeared that, unlike faster-spreading ridges where magma supply was more evenly distributed along-axis, the Gakkel Ridge enjoyed an enhanced magma supply focused on discrete sites, constructing circular volcanoes. Elsewhere, magma supply was restricted, and in some places, there was no crust at all, exposing mantle peridotites on the sea floor.

Six years later, a group of scientists revisited the Gakkel ridge, hoping to locate and photograph active high-temperature vents and their communities of vent animals. Dubbed, somewhat optimistically as it turned out, 'AGAVE' (for Arctic Gakkel Vents), this highly publicized expedition went well prepared. Using the Swedish icebreaker *Oden*, the expedition carried two state-of-the-art AUVs: the PUMA AUV for sniffing out active vents and the JAGUAR AUV for near-bottom geophysical and photographic surveys of high-temperature vents. They were also equipped with a wireline system for acquiring digital images and samples of black smokers and their fauna. Woods Hole engineers had spent four years configuring the AUVs specifically to survey the Gakkel Ridge, aided by a $3 million contribution from NASA, who thought they might be useful for future exploration of Jupiter's moon Europa. Embarrassingly, despite their best efforts, the AGAVE team were unable to locate any active, or even recently active, vent sites, and the hoped-for discoveries did not materialize. Nevertheless, in a creditable face-saving operation, the party published papers in *Nature* and elsewhere on explosive volcanism beneath part of the Gakkel Ridge, making the expedition feel like a success. No doubt, future expeditions will have more luck.

A series of cruises between 2008 and 2010 did manage to locate black smokers further south at 73.5°N on the Mohns Ridge, a section of Mid-Atlantic Ridge spreading at around 16 mm per year in the Norwegian–Greenland Sea, north of Jan Mayen Island. The vent fauna imaged by the team was, it has to be said, rather disappointing. There were tube worms, but not the giant, blood-red, creatures of the tropical Pacific: these were small beard worms that were not even endemic to the vent sites, and the vent shrimp characteristic of the Mid-Atlantic Ridge further south were notable by their absence. While the emerging vent fluids were as hot as 300°C, the surrounding deep sea had a temperature below zero, perhaps deterring immigrants from venturing this far north, even if they were able to navigate past Iceland.

At the other end of the planet, scientists were having more luck. In their search for hydrothermal vents at spreading centres surrounding Antarctica, one ridge in particular caught their eye. This was the East Scotia Ridge, a spreading centre that separates the small Sandwich and Scotia plates at respectable rates of 65–70 mm per year. Unlike the Gakkel Ridge, the East Scotia Ridge is not directly connected with the global mid-ocean ridge system, but is located in the East Scotia Sea, a textbook example of a back-arc basin,[64] that is, a miniature ocean basin associated with a volcanic arc—in this case, the South Sandwich arc. Using *Geosat* data as a guide, a survey was conducted onboard the British Antarctic Survey ship RRS *James Clark Ross* in January 1995 as part of the original BRIDGE programme. This succeeded in mapping the entire ridge for the first time, using the HAWAII-MR1 towed sonar system, revealing nine segments separated by a variety of migrating non-transform offsets, two of which had 'inflated' characters suggesting an enhanced magma supply (Figure 8.14). Seismic profiling showed that these shallow portions of the ridge axis were underlain by melt lenses like those beneath the East Pacific Rise, thus supporting this interpretation.[65] Focusing on these two segments, another cruise by *James Clark Ross* in 1999 used the deep-towed sonar TOBI, fitted with optical sensors with which it detected chemically enriched plumes in the water column, confirming the existence of active high-temperature venting beneath both segments.[66] The ship returned early in 2009 to locate the vents more precisely, which it not only succeeded in doing, but also managed to obtain the first camera footage of black smokers and vent fauna in the Southern Ocean. The following season, the then-new UK research ship RRS *James Cook* visited both sites identified previously and deployed an ROV (*Isis*) to investigate and sample the vents and their biota.

At the northern segment, vents were located on a major fissure near a bathymetric feature known as the Mermaid's Purse. Chimneys up to

[64] The East Scotia Sea was, in fact, one of the first back-arc basins to be identified as such.

[65] Livermore et al. (1997).

[66] German et al. (2000).

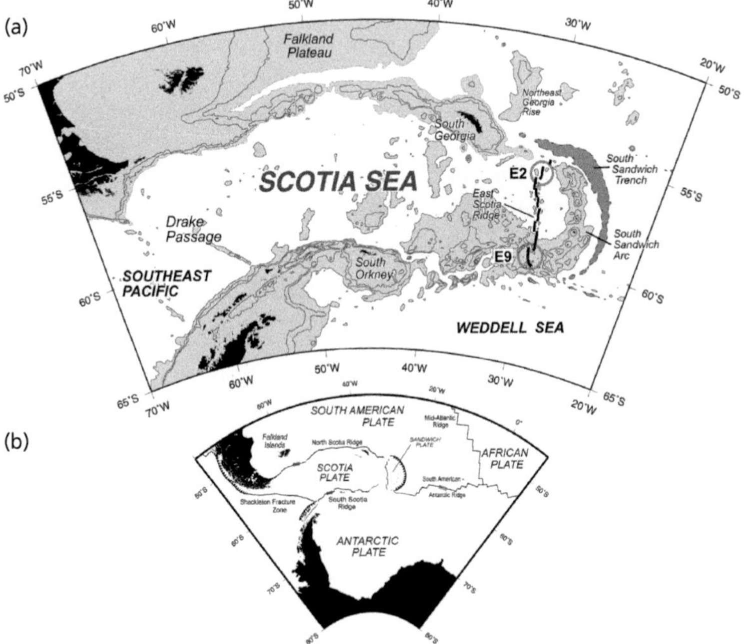

Figure 8.14 (a) Location of the East Scotia Ridge in the Southern Ocean, showing segments E2 and E9, where new chemosynthetic communities were discovered at hot vents. (b) The plate tectonic setting.

15 m high vented fluids at 353°C, producing black 'smoke' on contact with the cold seawater, and were surrounded by small volcanic cones and craters.[67] Vents on the southernmost segment were situated on fissures associated with a volcanic caldera, 1 km in diameter and 300 m deep, and emitted fluids at up to 383°C. The vent fauna resembled neither the tubeworm-dominated ecosystems of the Pacific ridges nor the shrimp-dominated fauna of the Mid-Atlantic Ridge, consisting instead of seething masses of small, white, 'hairy-chested yeti crabs' (*Kiwa tyleri*), curious decapods that farm bacteria on their chest hairs and arms, which they then devour in a crab-like manner.[68] These animals covered the chimneys at both sites at densities far greater than observed elsewhere, and were thus the primary animal species, together with stalked

[67] Rogers et al. (2012).
[68] Rogers et al. (2012); Thatje et al. (2015).

barnacles, gastropods, sea spiders, and anemones. To the biologists, these communities represented a new and unique biogeographic province, in which yeti crabs appeared to characterize Antarctic vents in the same way as tubeworms on the East Pacific Rise or shrimp on the Mid-Atlantic Ridge. Hence, rather than acting as a highway for vent fauna dispersal between the major oceans, it seemed that the Scotia Sea served as something of a barrier.

Making Space

Back in July 1975, the end of the Space Race was marked by the high-profile 'Apollo Soyuz Test Project', during which an American command/service module docked with a Soviet Soyuz spacecraft in a demonstration of détente. The scientific value of this billion-dollar project was, to put it politely, modest. By contrast, just a few months earlier, NASA had launched, almost unnoticed by the world's press, a Thor-Delta rocket from Vandenberg Air Force Base in California, carrying a diminutive cut-price satellite that heralded a new era in Earth exploration. The satellite, known as *GEOS-3*, was built from spare parts from its predecessors, and was designed to provide information on a range of geophysical parameters, including the size and shape of the Earth, the structure of the crust, the Earth's tides, and, above all, the gravitational field. Results far exceeded expectations. One sensor in particular turned out to be of immeasurable value during the satellite's four-year operation. This was a radar (microwave) altimeter that measured the altitude of the vehicle above the sea surface as it orbited the planet every 102 minutes. Since the position of the vehicle was known precisely from continuous tracking by ground stations, this allowed the height of the sea surface to be derived to high precision along the satellite's ground track. Any submarine feature, such as a seamount or submerged plateau, produces additional gravitational attraction, causing water to 'pile up' above it, creating a bulge, perhaps a few metres high, in the sea surface. Hence, from sea surface height measurements, it was fairly straightforward to calculate the gravity anomaly, which, while of lower resolution and accuracy than measurements made from ships, provided ocean-wide coverage more rapidly (only a small fraction of the world's ocean basins has been mapped by survey ships even now) and more cheaply.

Three years later, in June 1978, a second satellite was launched with the primary aim of monitoring the oceans. Known as *Seasat*, this spacecraft carried the first civilian imaging radar, which, to the consternation of the military, proved capable of detecting the wakes of their submerged nuclear submarines. It also carried a radar altimeter similar to that of *GEOS-3*, but capable of much higher precision. During the 101 days of operation prior to its catastrophic failure (which conspiracy theorists link to the security threat posed by revealing submarine locations), *Seasat* provided unprecedented coverage of the ocean basins, including remote regions never surveyed by research ships. Unfortunately, it failed before the austral summer, so that the most remote and poorly known region—the Southern Ocean—was not covered, owing to the presence of extensive winter sea ice (height trackers on satellites become confused over sea ice, producing large errors).

In 1985, the US Navy launched yet another satellite that proved even more revolutionary than the abortive *Seasat* mission. *Geosat*, carrying a further improved radar altimeter, spent 18 months completing a secret 'Geodetic Mission' for the Navy before being shifted into an orbit with ground tracks that repeated every 17 days—the Exact Repeat Mission. This permitted averaging of numerous passes and thus the removal of oceanographic 'noise' from the geophysical signal. However, spatial resolution was limited by the wide separation of the repeat ground tracks, whereas the earlier Geodetic Mission phase provided unprecedented spatial resolution combined with high precision. Sadly, the Geodetic Mission data remained 'classified' by the US Navy until 1990, when data south of 60°S were released to the scientific community, enabling NOAA scientists Dave McAdoo and Karen Marks to construct high-resolution gravity grids of the oceans between 50° and 72°S.[69] Features down to around 12 km could now be resolved, providing marine geophysicists, at least those working in the Antarctic, with a marvellous free gift. Moreover, features buried beneath sediments, and therefore invisible on bathymetric maps, could be resolved by the new gravity data. For example, in the Weddell Sea, the newly released data revealed a striking 'herringbone' pattern of ridges and troughs that was interpreted as a set of sediment-covered fracture zones created by the migration of South America away from Antarctica during the past 120 million years (Figure 8.15). Marine geophysicists were later able to arrange their ship surveys to follow these ridges and acquire magnetic

[69] McAdoo and Marks (1992).

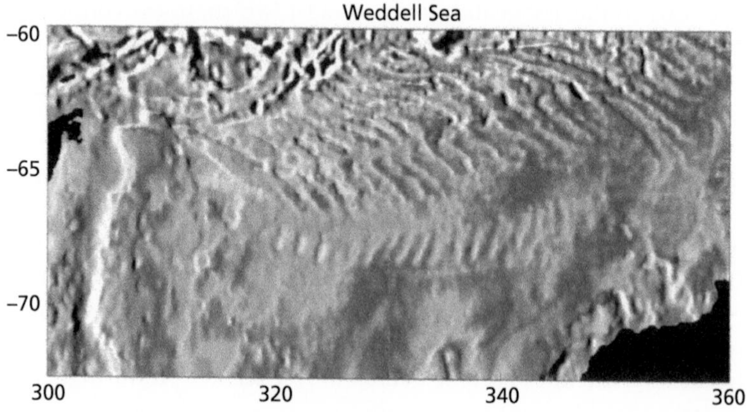

Figure 8.15 Gravity field of the Weddell Sea derived from *Geosat* Geodetic Mission data (McAdoo and Marks,1992).

anomaly profiles with which to date the history of motions between these two plates.

In 1992, the European Space Agency launched the European Remote Sensing satellite, *ERS-1*, carrying its own microwave altimeter. This satellite was programmed to follow closely spaced ground tracks similar to the *Geosat* Geodetic Mission, with the result that, a few years later, the US Navy realized that the game was up and finally released all their Geodetic Mission data, permitting geophysicists Walter Smith (at NOAA) and Dave Sandwell (at Scripps) to construct high-resolution gravity and 'predicted bathymetry' maps of the world's oceans based on combined *Geosat* and *ERS-1* data.[70] These maps, and the underlying data, were released to researchers worldwide and have since contributed mightily to the planning of research cruises and to the interpretation of new and existing data. Later gravity field models incorporated data from more recent satellite missions, such as CryoSat-2 (ESA) and Jason-1 (NASA), providing improved coverage and even higher precision, thereby permitting the identification of features as small as 6 km (Figure 8.16). Only with the latest models has the resolution improved sufficiently to show the subtle anomaly associated with the axis of the East Pacific Rise, as well as smaller features such as an extinct spreading centre in the Gulf of Mexico, buried beneath thick sediments.[71]

[70] Sandwell and Smith (1997); Sandwell and Smith (2009).

[71] Garcia et al. (2014).

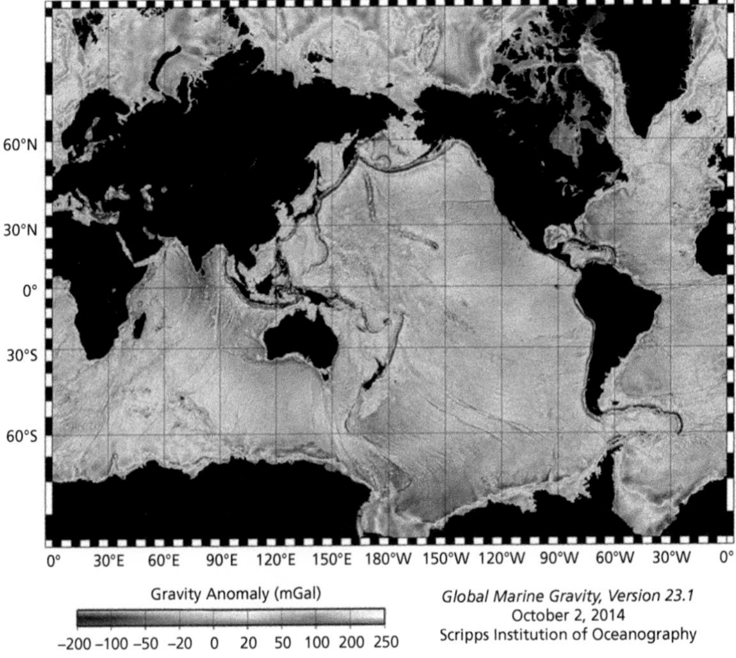

Figure 8.16 Satellite gravity field produced by Sandwell and Smith (2014).

Going Global

A major landmark in plate tectonics was reached in 1972. After discussions with John Sclater and Dan McKenzie, Clement Chase at the University of Minnesota assembled data from plate boundaries worldwide and constructed the first global numerical model of present-day plate motions,[72] using a computer to find the best-fitting set of Euler vectors describing the recent movement of eight major plates in relative motion with one another. The data he used were of three kinds: rates of spreading derived from young magnetic anomalies in all the major oceans, the trends of transform faults mapped from ship crossings, and earthquake slip-vectors from the mid-ocean ridges. The

[72] Chase (1972). Previously, rotations had been obtained by Jason Morgan and Xavier Le Pichon by fitting pairs of plates and then adding the Euler vectors. However, this was the first attempt to compute a global set of Euler vectors by simultaneously inverting all available data.

resulting Euler vectors could be used to calculate the rate and direction of relative motion at any point on a boundary between any two plates, even where this was a convergent boundary or collision zone. Clem's model thus allowed estimates of the rate at which crust was added or destroyed for each plate pair. So, for example, relative motion between the Pacific plate and the Americas (North and South America were regarded as a single plate) over the past 1 million years produced 56,000 km^2 of new crust at the East Pacific Rise, while consuming 197,000 km^2 of sea floor at the Aleutian Trench, a net loss of 141,000 km^2. However, anyone who still believed in Earth expansion or contraction would have been disappointed to learn that Clem's estimates of the *global* rates of growth and destruction balanced at 2.9 million km^2 per million years for each.

A couple of years later, Bernard Minster (Caltech), Thomas Jordan (Princeton), and Peter Molnar (University of California at San Diego) assembled their own dataset from plate boundaries worldwide and used a similar method to invert them, producing a model of relative motions for ten separate plates, which they christened 'RM1'.[73] Bernard and Co. assigned North America and South America to separate plates and were able to resolve significant motion between them, though the location of the boundary in the Atlantic was anyone's guess. Having obtained this relative plate motion model, they then included the trends of hot-spot trails from the Pacific and Atlantic oceans, in order to derive the absolute motion of each plate with respect to the deep mantle. The consistency of their results suggested that the hotspots had been fixed with respect to one another for at least 10 million years, allowing them to use their model, 'AM1', to predict the motion of any plate at any point on its surface. Within a few years, Minster and Jordan published an improved model, 'RM2', incorporating many more data collected in the interim, as well as including a Caribbean plate, making a total of eleven plates, while Clem revised his model, going one further with twelve plates.[74] Rates for RM2 were averaged over about 3 million years rather than the 5 million years of the earlier models, reducing the effects of recent changes in plate motions. In order to obtain a good fit to data in the South Atlantic and elsewhere, Bernard and Tom were

[73] Minster et al. (1974).

[74] Minster and Jordan (1978); Chase (1978).

forced to separate India from Australia, treating them as independent plates (they had formerly been combined on a single Indo-Australian plate), while incorporating the Bering Sea 'plate' into North America.

Twelve years later, a much-improved model inversion, benefitting from a great deal of new data derived from a decade of intensive exploration, including many new transform azimuths and an expanded set of magnetic anomalies over ridge crests, was published by researchers at Northwestern University, Illinois, led by Chuck DeMets.[75] This model, dubbed 'NUVEL-1', included a dozen major plates and used new sonar data to constrain the direction of motion at mid-ocean ridges, greatly improving the overall fit of the model to the data. India and Antarctica were again assigned to separate plates, based on new data suggesting an incipient boundary between the two. The results were significantly different from the earlier models, particularly in the Indian and Pacific oceans, and NUVEL-1 became, for many years, the standard model of global plate motions (NUVEL-1A was published in 1994 to take account of a revised timescale, but NUVEL-2 never appeared).

While the use of sea-floor data to estimate plate motions in this way was one of the most important steps in the evolution of plate tectonics, it had its limitations. Spreading rates could be estimated from the spacing of dated magnetic anomalies, but resolution was limited by the rate of field reversal—for example, present-day rates could only be estimated from the width of the Brunhes anomaly over mid-ocean ridges, so that any changes within the past 773,000 years could not be resolved. Fortunately, measurement techniques using space geodesy were becoming sufficiently precise that current plate motions could be estimated directly from measurements made over just a few years. Several techniques were involved, the most unlikely of which was developed during the 1970s using quasars (distant quasi-stellar objects, some of which lie billions of light-years away at the edge of the known universe), to provide a fixed reference frame in which to make high-precision estimates of the separation of locations hundreds or even thousands of kilometres apart. By tracking the same quasars at several stations, the difference in arrival time of light from these bodies could be measured and used to calculate the separation or 'baselines' between stations. Accurate measurements of the separation of two or more stations on

[75] DeMets et al. (1990). Chuck's supervisor at the time, Richard Gordon, was the prime mover behind this effort.

different plates repeated over months and years provided a direct estimate of present-day plate motion. Movements as small as 10 mm could be detected by this means, so that even slow motions between plates could be determined within a decade. Moreover, motion between converging plates could now be estimated directly, rather than by adding Euler vectors as was done with models like NUVEL-1. There was one serious drawback, however, and that was that you needed a radio telescope at each station. Since radio telescopes are expensive and hardly portable, this limited its application somewhat. Nevertheless, the technique, known as 'very long baseline interferometry' (VLBI), was applied successfully by NASA within its Crustal Dynamics Program from 1980 onwards. A network of VLBI stations was used to measure the relative motions of the Pacific and North American plates, including movement on the San Andreas Fault, as well as of plates on either side of the Pacific and Atlantic oceans, finding surprisingly good agreement with models such as NUVEL-1A, despite the latter's averaging interval being a million times longer.

At about the same time, a second space geodetic technique was developed, using lasers to determine the distance to retro-reflectors mounted on satellites in high (5900 km) stable orbits, and also on the Moon. The satellites, known as LAGEOS (Laser GEOdynamics Satellite), are 60 cm metal spheres each fitted with 426 reflectors, resembling the mirror balls you find in ballrooms and nightclubs.[76] Laser transmitters are used to bounce short pulses off these reflectors, the round-trip time for which is used to determine the range to the satellite. Estimates from many stations are combined to compute the satellite's trajectory and the coordinates of the tracking stations. Repeat measurements then provide estimates of plate movement over periods of years to a precision of less than 10 mm. Known as 'satellite laser ranging' (SLR), this technique has the advantage that ground stations are mobile and can therefore be arranged to investigate specific plate boundaries or deformation within a plate. An analysis of ten years of SLR data from five plates concluded that relative motions were very close to those predicted by

[76] Unlike most space junk, LAGEOS, with no moving parts, is expected to remain in orbit for around 8 million years. It carries a plaque designed by Carl Sagan, showing maps of the continents today, 268 million years ago (in their Pangea configuration), and 8 million years in the future.

NUVEL-1, with a suggestion that rates had slowed slightly since 3 million years ago.[77] The discrepancy was, however, largely removed by the recalibration of the magnetic reversal timescale that resulted in NUVEL-1A.

In 1985, just as Geosat was getting off the ground, other spacecraft were being launched from Vandenberg Air Base on a regular basis. These satellites, eleven in all, were put into circular repeating orbits at around 20,000 km altitude, forming the basis of a prototype navigation system known as Navstar-GPS. By October, the constellation was complete, providing fixes with a precision of less than 10 m for the US Air Force, but much poorer for civilian users, for whom the service was deliberately degraded by means of something called 'selective availability', which added errors of 50–100 m to fixes using the GPS signal. Nevertheless, for research ships, this system represented a significant improvement over an earlier satellite system, known as 'TRANSIT', and allowed much more accurate maps of sea-floor features than hitherto. A further nine improved satellites were in orbit by 1990 and, by 1997, another nineteen had been launched, by which time Bill Clinton had decided that selective availability should been turned off, permitting navigation to an accuracy of 1 m. Not wishing to have their weapons systems dependent on the enemy's technology, the Russians developed their own satellite navigation along similar lines, which they called 'GLONASS', while the French installed a characteristically idiosyncratic system known as 'DORIS', based on a global set of ground stations that transmit to a range of suitably equipped Earth observation satellites. As of early 2016, there were over thirty satellites in the GPS constellation, providing global coverage, with up to nine satellites visible (a minimum of four is required to obtain a fix, but additional satellites improve accuracy). Together, GPS and GLONASS provide the means for half the world's population to track their many 'friends' to within a metre or so via their mobile phones. More sophisticated receivers can reduce the error to just a few millimetres, sufficient to allow direct measurement of plate movements.

The low cost and convenience of GPS measurements meant that this method soon became the primary method of measuring plate velocity fields, particularly in actively deforming regions adjacent to

[77] Robbins et al. (1993).

plate boundaries, such as the San Andreas Fault, the Alpine Fault of New Zealand, and the Himalayas (see Chapter 6). In 2010, Don Argus at Caltech and colleagues published a model of the motion of 11 major plates based only on direct measurements made using GPS, VLBI, SLR, and DORIS. This model, which they dubbed 'GEODVEL', predicted plate velocities that differed slightly from NUVEL-1A for all but two plates and showed that spreading rates on eight mid-ocean ridges had slowed down over the past few million years. In the same year, Chuck

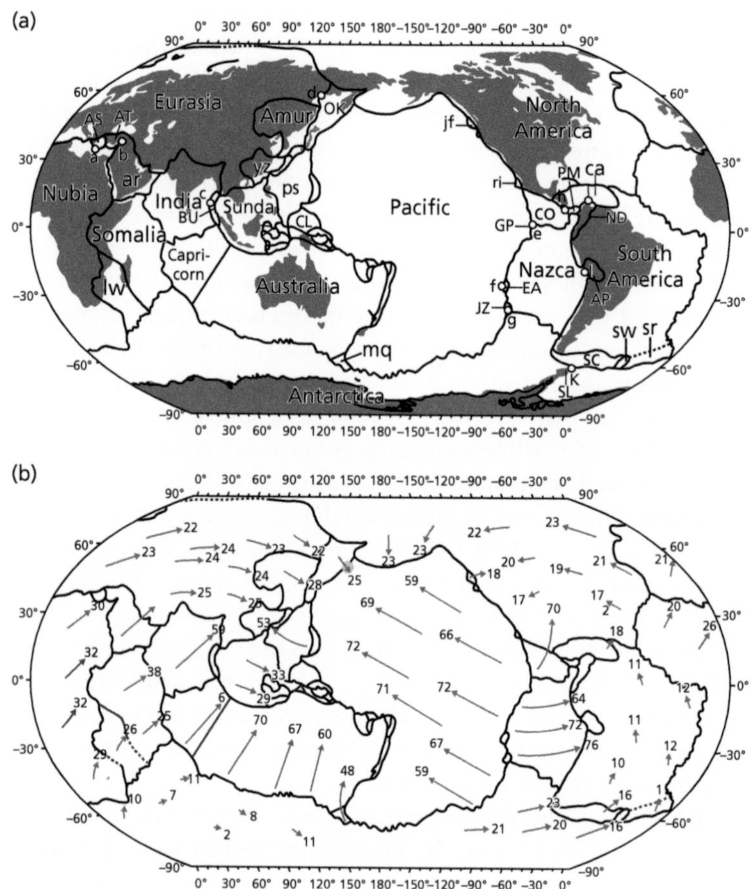

Figure 8.17 Plate outlines from MORVEL56 (a) and estimated absolute motions (in mm a^{-1}) in the no-net-rotation reference frame (b). Africa has been divided into three subplates. [After Argus (2011).]

DeMets and colleagues published a new model for the motion of nineteen major plates, based on a greatly expanded set of spreading rates and fracture zone traces (slip vectors from subduction zones were not used), but also incorporating the results of GPS studies for six minor plates. They dubbed this model 'MORVEL' (Mid-ocean Ridge VELocity). Like the earlier models, MORVEL averaged motion over 3 million years on slow-spreading ridges, but reduced the averaging window to just 773,000 years (i.e. the duration of the Brunhes normal polarity interval) for fast-spreading ridges, limiting the effect of geologically recent changes. The latest version, 'MORVEL56', was published in 2011[78] and has fifty-six plates, comprising the twenty-five from MORVEL plus thirty-one small plates defined by Peter Bird[79] (Figure 8.17). The motion of each plate is given in a 'no-net-rotation' frame of reference, an approximation to an 'absolute' reference frame obtained by adding motions over the surface of the planet and then removing the net motion from each vector. Despite minor differences, the overall agreement of models based on sea-floor spreading over the past 1–3 million years with models derived from GPS and other space geodetic measurements over the past few decades shows that plate motions are remarkably continuous. In the modern world, these motions are no longer of purely academic interest. Australia, for example, is moving northward at a rate of about 68 mm per year. This means that, in less than fifteen years, it moves more than a metre, requiring maps and navigation systems to be updated to preserve accuracy.

Afterthought

More of our world's surface has been explored in the past fifty years than in all of human history. Since the overall form of the ocean basins was revealed by satellite altimetry during the 1980s, half of the entire tennis-ball seam of mid-ocean ridges has been mapped and imaged by surface ships. Detailed mapping of mid-ocean ridges led to deeper (in both senses) insights into how plates are constructed and how sea-floor spreading processes vary with the rate of plate separation. It turns out that not only does the seam represent the defining feature of Earth, but

78 Argus et al. (2011).

79 Bird (2003).

careful examination of the stitching tells us something fundamental about processes operating within our planet's interior. The era of intensive exploration of the oceans with GEOSAT, GLORIA, SeaBeam, TOBI, and the rest thus represented a necessary step on the road towards a full understanding of plate tectonics on Earth and perhaps elsewhere.

9

Chilling Out

> Our modern climate system is a product of millions of years of plate tectonics.
>
> Mark Maslin (2013)

We humans, it seems, have lately been responsible for changing the climate through our overenthusiastic consumption of hydrocarbons. Yet the Earth's climate changes naturally on all timescales. At the short end of the spectrum—hours or days—it is affected by sudden events such as volcanic eruptions, which raise the atmospheric temperature directly, and also indirectly by the addition of greenhouse gases such as water vapour and carbon dioxide. Over years, centuries, and millennia, climate is influenced by changes in ocean currents that, ultimately, are controlled by the geography of ocean basins. On scales of thousands to hundreds of thousands of years, the orbit of the Earth around the Sun is the crucial influence, producing glaciations and interglacials, such as the one in which we live. Longer still, tectonic forces operate over millions of years to produce mountain ranges like the Himalayas and continental rifts such as that in East Africa, which profoundly affect atmospheric circulation, creating deserts and monsoons. Over tens to hundreds of millions of years, plate movements gradually rearrange the continents, creating new oceans and destroying old ones, making and breaking land and sea connections, assembling and disassembling supercontinents, resulting in fundamental changes in heat transport by ocean currents. Finally, over the very long term—billions of years—climate reflects slow changes in solar luminosity as the planet heads towards a fiery Armageddon. All but two of these controls, you notice, are direct or indirect consequences of plate tectonics.

*

The Tectonic Plates Are Moving! Roy Livermore, Oxford University Press (2018). © Roy Livermore.
DOI: 10.1093/oso/9780198717867.001.0001

A Foreign Country

These days, it seems, there is no escape from the charter-flight brigade. Once-remote and beautiful islands and coasts worldwide have been turned into wretched concrete resorts where overweight westerners lie basting their outsized carcasses on plastic sunbeds. However, should you be looking for a tropical paradise far from the madding crowd's ignoble strife, might I suggest Wilkes Land? This is a place you may not have considered for a beach holiday, particularly as it lies on the coast of Antarctica at latitudes of around 70°S. Here, you will be surprised to learn, you may find palm-fringed beaches backed by dense rainforest, within which giant baobabs, spectacular proteas, and huge tree ferns provide homes for marsupial mammals, large ungulates, and a wide variety of insects. Or, at least, this is how it was 55 million years ago (so you can put away your sunscreen and Ray-Bans).[1] Nevertheless, this tropical[2] paradise was, at the time, in much the same high latitudes as Antarctica today. Back then, in the Eocene, annual rainfall was over 1000 mm and winter temperatures never fell much below 8°C, despite the absence of sunlight for 50 days a year. High latitudes, therefore, do not guarantee cold weather or glaciers. Something else is required, something that occurred between the good times of the early Eocene (55 million years ago) and the chilly world of the late Eocene (50 million years ago), when evergreen conifers replaced the palms, followed by the downright freezing conditions of the Oligocene (33 million years ago), when everything disappeared under ice.[3]

Thus, while it is clear, as Ted Irving showed, that climate indicators in sedimentary rocks agree much better with the latitudes derived from paleomagnetism than with their present locations (thereby providing support for the theory of drifting continents),[4] closer examination reveals major discrepancies. If Brian Harland was correct and sea-level glaciers

[1] This idyllic picture is based on the results of deep ocean drilling near the continental margin of Wilkes Land during Expedition 318 of the Integrated Ocean Drilling Program (successor to DSDP). Plants were identified from their spores or pollen preserved in the sediments—see Pross et al. (2012) and Contreras et al. (2013).

[2] Strictly, this environment is labelled as 'paratropical'.

[3] This decline is often referred to as the 'greenhouse-to-icehouse' transition, but since it can be quite chilly in a greenhouse in winter and, at least in Britain, ice houses are even less common than duck houses, these are hardly useful terms. 'Hothouse-to-freezer' might have been better choices.

[4] See Chapter 3.

approached the tropics 700 million years ago, then climate regimes very different from the present are possible. The zonal hypothesis, employed by Wladimir Köppen and his son-in-law Alfred Wegener to position the latter's continental reconstructions in their likely paleolatitudes, was based on the generally accepted principle of 'uniformitarianism',[5] espoused by the father of modern geology, James Hutton, in the eighteenth century. This postulate, summed up by James as 'the present is the key to the past', did not mean to imply that nothing ever changed, but rather that the history of the Earth was, in the words of the great nineteenth-century geologist Charles Lyell, 'one uninterrupted succession of physical events, governed by the laws now in operation'. Despite its syllabic superfluity, uniformitarianism served the subject of geology well for over a century, but, in the end, like so many other widely held assumptions, eventually became an intellectual straightjacket, constricting free thought.

The limitations of the zonal hypothesis were noted by Alfred Wegener himself in the fourth (1929) edition of his *Origin*,[6] where he observed that 'All isotherms show the predominance of a zonal arrangement of climate, but there are characteristic deviations from this caused by the distribution of land and sea'. The latter idea derived from Lyell, who, in his classic tome, *Principles of Geology*,[7] presented the inspired insight that the transition from a hotter to a colder Earth, as recorded by rocks of Tertiary and younger age (i.e. rocks formed during the past 66 million years), was a consequence of changing geography. This, he imagined, was accomplished by the sinking of mountains and the rising of the ocean floor—that is, vertical rather than horizontal motions. He nevertheless drew a series of maps showing the present continents distributed in unlikely ways (see Figure 9.1), which might, he believed, lead to warm or cool climates such as those recorded in the rocks. The essence of his theory was that land surfaces were heated much more quickly by solar radiation than was the ocean. Hence, when continents were gathered in low latitudes, the land would be 'heated to an excess under the equatorial sun', causing warm currents of air to sweep towards the polar regions. With land near the poles, the climate would be cooled, resulting in refrigeration, the effect being amplified by the movement of bergs and the formation of clouds—phenomena that

[5] The word itself (along with the term 'scientist') was, apparently, coined by the Cambridge scientist William Whewell in the early nineteenth century.

[6] Translated in 1966 by John Biram.

[7] Published in twelve editions between 1830 and 1875.

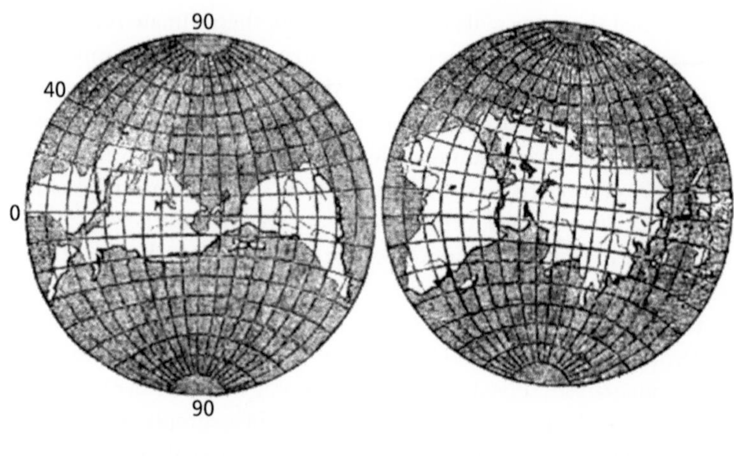

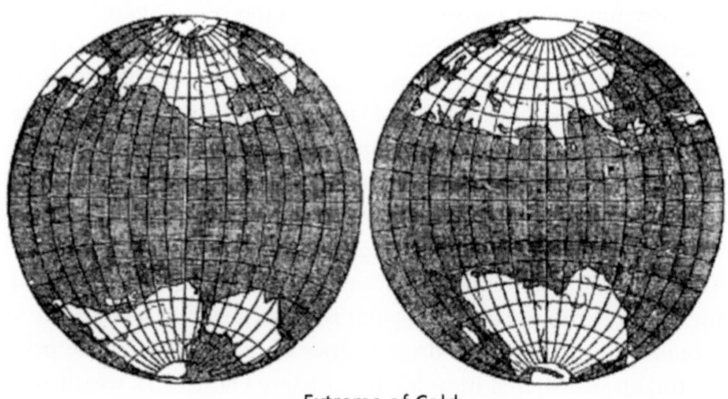

Figure 9.1 Continental configurations conducive to warm and cool climates, as constructed by Charles Lyell. [From Lyell (1837), *Principles of Geology*.]

would today be classified as 'positive feedbacks'. It thus appeared that, as far as global climate was concerned, the present was a somewhat ill-fitting key to the past. In fact, it was beginning to look as though the present cool climate was rather exceptional, and the Earth's normal disposition was somewhat sunnier than today.[8]

[8] A recent estimate by Bill Hay suggests that interglacial climates like the present represent just 2.5 per cent of the time since the Cambrian explosion, 541 million years ago.

With the arrival of plate tectonics, a perfect mechanism had been provided to accomplish the very changes envisaged by Lyell. Continents, transported around the globe on the backs of the plates, had assumed a variety of configurations, sometimes gathered in supercontinents, at other times scattered carelessly about the globe. The world map of land and sea had been redrawn time and again and the challenge now was to test the notion that global climate reflected this.

Supercomputers

In 1977, two Lamont scientists, William Donn and David Shaw, published the results of numerical modelling experiments designed to establish whether plate motions alone could account for gross changes in climate on a geological timescale.[9] They identified changes in the Earth's albedo as a prime suspect. The albedo is simply the proportion of solar energy arriving on Earth that is reflected back into space: light-coloured surfaces, notably continental ice sheets and sea ice, naturally throw back a greater proportion than darker ones, such as forests and oceans. Today, around a third of incoming solar energy ends up being returned whence it came, but this might have been different in the past. Using a thermodynamic computer model designed for long-range weather forecasting (though not, perhaps, quite as long-range as Bill and Dave had in mind), they computed the temperature at various times during the past 250 million years, based on an early set of plate reconstructions. Making the assumption that the reconstructed continents represented land areas in the past (thereby ignoring changes in sea level), they estimated the balance of incoming and outgoing solar radiation, taking into account the changing latitude distribution of land and sea. This assumption broke down during the Cretaceous when shallow seas covered much of the continents, with the result that land areas were far smaller than the continental outlines would suggest, and so they took the simple step of avoiding this period altogether. From the results of their modelling, they concluded, somewhat optimistically as it turned out, that 'the gross changes of climate in the Northern Hemisphere can be fully explained by the strong cooling in high latitudes as continents moved poleward'.

Previously, John Crowell and Lawrence Frakes, geologists at the University of California, had addressed the problem of the causes of ice

[9] Donn and Shaw (1977).

ages.[10] 'Clearly', they noted, 'changes in the amount of heat retained within the Earth system[11] are sensitive to changes in the albedo and to changes in the proportions of water vapour, CO_2, and dust in the air'. The question was: 'what were the factors that could alter the albedo and the composition of the atmosphere sufficiently to bring on ice ages?' Dismissing variations in CO_2 concentration as 'a third-order effect', John and Larry focused on changing geography as the likely driver of climate change and glaciation. The supercontinent of Gondwana had migrated over the south pole during the Late Paleozoic (359 to 252 million years ago), as revealed by paleomagnetism. Geological evidence of glacial deposits showed that extensive ice sheets had formed and the centres of glaciation had actually tracked this movement, finally disappearing from South America and Africa as this part of Gondwana moved into lower latitudes. The pair highlighted the effect of continental configuration on ocean circulation, in particular noting that the free low-latitude circulation between the Pacific and Atlantic oceans prior to 3.5 million years ago was ended by the emergence of the Isthmus of Panama, which was soon followed by the glaciation of the northern continents. The closure, they speculated somewhat counter-intuitively, may have enhanced the northward flow of warm water in the Gulf Stream and North Atlantic Drift, providing moisture for the growth of ice sheets. Thus, they concluded, 'the spread of ice seems to coincide with times when ocean currents were forced far north and south through blocking by continents. The arrangement of continental masses deflected vigorous zonal [i.e. east–west] circulation in low latitudes and carried warm waters into polar regions.'

A few years later, Eric Barron and colleagues at the Rosenthal School of Marine and Atmospheric Science in Miami carried out some simple calculations of their own, in which they averaged the proportions of land, sea, and ice in 10° latitude bands and then assigned a mean albedo to each in order to estimate the balance of incoming and outgoing radiation.[12] They were surprised to find that land in high latitudes did not seem to have been a controlling factor in global cooling after all, while changes in sea level appeared to be more significant. In 1984, Eric, now at the US

[10] Crowell and Frakes (1970).

[11] What they meant by the 'Earth system' was the solid Earth, oceans, atmosphere, and glaciers, all acting in concert.

[12] Barron et al. (1980).

National Center for Atmospheric Research (NCAR) in Colorado, reported a series of computer simulations of the Cretaceous hothouse climate[13] (the interval omitted by Donn and Shaw) that he had carried out to test the idea further. Recognizing, as had Crowell and Frakes, that a major effect of changing geography could be drastic changes in the circulation of the atmosphere and oceans, Eric applied the GENESIS[14] general circulation model, developed at NCAR, to the problem. Despite some serious limitations, notably a representation of the ocean which did not include heat storage, transport or diffusion, Eric concluded that changing geography was indeed a substantial climatic forcing factor, but insufficient on its own to fully explain the much higher global temperatures of the Cretaceous. Something else was required. By 1990, Eric had moved to Penn State University, where the 'Earth System Science Center' had been established in 1986, and continued to apply the NCAR GENESIS general circulation model to the problem of Cretaceous warmth. However, in an about-turn that warns against uncritical acceptance of numerical models, he now concluded[15] that 'Model experiments suggest that the role of geography is negligible', and favoured an alternative explanation for the dramatic change in global temperatures, namely variations in atmospheric carbon dioxide concentration.

Proxy Science

The development of techniques for recovering long cores of mud and sediment from the sea bed held out the prospect of continuous records of environmental change, with the ultimate prize of datasets derived from DSDP drilling stretching back to the Jurassic (180 million years ago). Inevitably, the new sampling techniques required more sophisticated methods for extracting the high-resolution records locked up in the cores. Analysis of stable (i.e. non-radioactive) oxygen isotopes in the calcite ($CaCO_3$) shells of microscopic single-celled animals known as foraminifera[16] preserved in the sediments provided one of the most valuable sources of information concerning past environmental conditions.

[13] Barron and Washington (1984).

[14] For aficionados of daft acronyms (ADA), this is 'Global Environmental and Ecological Simulations of Interactive Systems'.

[15] Barron et al. (1993).

[16] The name, apparently, means 'hole bearer' and refers to the existence of a hole or foramen between the walls of the chambers of the organism.

The technique is based on the ratio of heavy to light oxygen: light oxygen accounts for about 99.8 per cent of the oxygen in the air and in the oceans, and has 8 protons and 8 neutrons in its nucleus, giving it a mass of 16, while heavy oxygen makes up most of the remaining 0.2 per cent, and has 8 protons and 10 neutrons in its nucleus, giving it a mass of 18.[17] Foraminifera remove calcium (Ca^{2+}) and carbonate (CO_3^{2-}) ions from seawater in order to construct their calcite shells or 'tests', incorporating oxygen isotopes in a ratio dependent on that of seawater at the time.[18] Following their brief lives of a few months, they may become buried and their tests preserved in sea-bed sediments for millions of years, providing an extremely valuable record of sea-water temperature and glaciation (providing that it can be deciphered).

In the 1940s, University of Chicago professor Harold Urey, who in 1934 had received a Nobel Prize for his discovery of heavy hydrogen (deuterium or ^2H), and had also played an important role in the Manhattan Project, performed careful laboratory experiments demonstrating that the ratio of heavy to light oxygen in crystallizing calcium carbonate was slightly greater than that in the water from which it crystallized. As the temperature decreased, so the enrichment in heavy oxygen increased—very slightly, but enough to be measurable. In a famous quotation, Harold recalled, 'I suddenly found myself with a geological thermometer in my hands!' One of Harold's students, Cesare Emiliani, later applied the oxygen isotope thermometer to the tests (i.e. shells) of deep-living foraminifera, in order to determine the past temperature of deep bottom waters, which he did with great success. Results from shells extracted from Pacific sea-floor sediments indicated that the sea bed had experienced a period of pronounced cooling, from 10.4°C roughly 30 million years ago, to 2.2°C by about 2 million years ago.[19] After analysing tests from short sediment cores covering the past half-million years, he was able to identify 13 'marine isotope stages' corresponding to cycles of glacial advance and retreat, thereby providing support for the theory of nineteenth-century Scottish physicist James Croll, who had attributed these cycles to variations in the Earth's orbit. Cesare noted that the formation of extensive land ice would affect his

[17] For the sake of completeness, it should be noted that a very rare form of oxygen with 9 neutrons also exists in nature.

[18] There are also 'vital effects' that alter the ratio slightly, but corrections can be applied for these.

[19] See Pearson (2012).

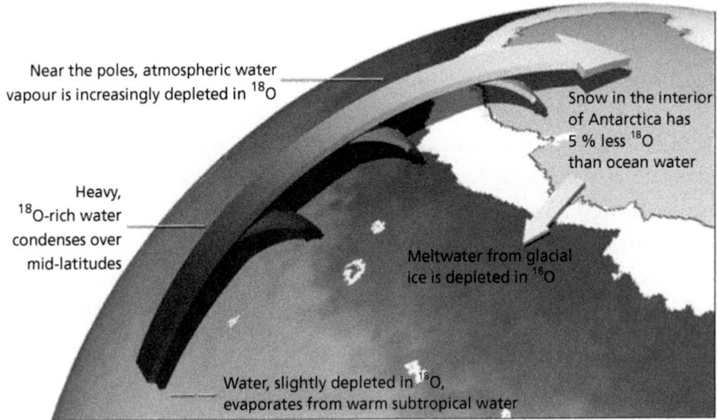

Figure 9.2 Variation in oxygen isotope composition in rain and snow in high latitudes. [Illustration by Robert Simmon, NASA GSFC.]

results, since the locking up of light oxygen in ice sheets would increase the ratio of heavy to light oxygen in the oceans.[20] The reason for this is that water containing the light isotope evaporates preferentially from the sea surface, while water vapour with the heavier isotope condenses more easily[21] and is rained out of the atmosphere sooner. The upshot is that, as moist air makes its way into higher latitudes (Figure 9.2), it becomes more and more depleted in the heavy isotope, so that snow falling in the polar regions has a slightly lower ratio of heavy to light oxygen. When global temperatures fall on entering an ice age, the heavy oxygen rains out even more rapidly, leaving the snow falling in high latitudes with even lower $^{18}O/^{16}O$ ratios than before. With all this light oxygen locked up in polar ice, the ratios in seawater are correspondingly increased. Cesare therefore needed to correct for this effect, which he estimated might be around half as large as the sea-temperature effect. Subsequently, Nicholas Shackleton[22] at Cambridge University, demonstrated that the effect of ice volume was greater than proposed by Cesare: in fact it was as great as, or even greater than, the temperature effect.[23]

20 Emiliani and Edwards (1953).

21 The difference is slight but highly significant.

22 Great grand-nephew of Ernest.

23 Shackleton (1967).

In 1967, Ian Devereux at the Victoria University of Wellington, New Zealand, applied the oxygen isotope technique to the shells of various organisms, including bryozoans, oysters, and planktonic foraminifera of Eocene and Oligocene age. He was surprised to find a rapid jump in the oxygen isotope ratios at the Eocene–Oligocene boundary, around 33.5 million years ago. Ignoring possible changes in ice volume, this jump indicated a sudden drop in sea-surface temperature on the New Zealand shelf from 21°C to 12°C within just a few million years, followed by a partial recovery. If this was a widespread, or even global, phenomenon, then something dramatic had occurred at that time.

The Floodgates Open

During the 'Decade of Dreams', as Tjeerd Van Andel described it,[24] *Glomar Challenger*, the drilling ship that had so spectacularly confirmed Fred Vine's predicted ages for Atlantic Ocean crust a few years earlier, visited the south-west Pacific in 1973, where it drilled a series of holes at sites on the Campbell Plateau and South Tasman Rise.[25] Nick Shackleton and James Kennett, a marine geologist at the University of Rhode Island, constructed a temperature curve for the past 35 million years derived from oxygen isotope analysis of bottom-living foraminifera in sediment cores recovered at DSDP Site 277 on the Campbell Plateau. Using a relationship developed previously by Nick, they calculated that surface temperatures in the region had fallen from a balmy 20°C in the early Eocene to a chilly 11°C by the Late Eocene, and to just 7°C in the early Oligocene (Figure 9.3). At about 33 million years ago, a spectacular drop in deep-sea temperatures from 10°C to 5°C had occurred in just 100,000 years, confirming the remarkable step found by Ian Devereux. Since bottom water is produced by the sinking of surface water around Antarctica, this pointed to a dramatic cooling of these surface waters to freezing point and hence the growth of sea ice. It was not clear what had caused the rapid cooling, but they noted that 'The Cainozoic paleoclimatic development of Antarctica is marked by rather rapid temperature decreases superimposed on a steady climate cooling after the early Eocene. The Eocene–Oligocene boundary event therefore probably represents a critical threshold level in Antarctic climatic development, which in itself is related to more gradual

[24] See Chapter 1.
[25] DSDP Leg 29.

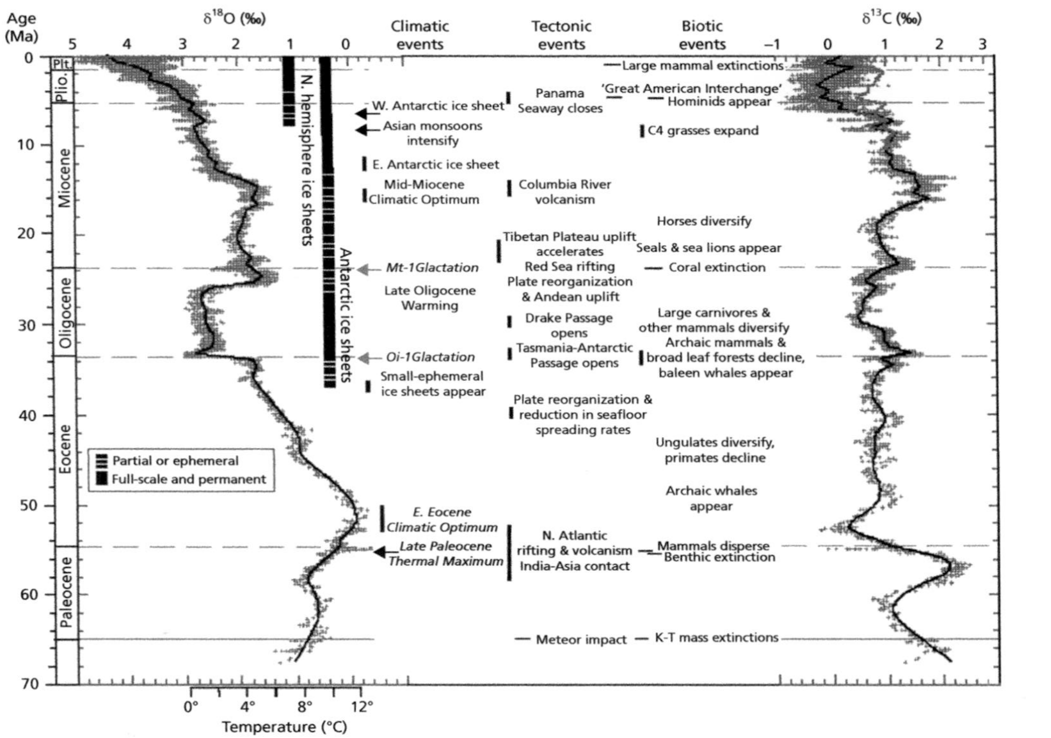

Figure 9.3 Global deep-sea oxygen and carbon isotope records based on data compiled from more than 40 DSDP and ODP sites by Zachos et al. (2001).

isolation of the continent, resulting from the northward spreading of Australia and the opening of the Drake Passage.'[26]

They also pointed out that other gradual changes in global oceanic circulation had taken place during the past 50 million years, involving the initiation of circumpolar circulation in the Southern Ocean, matched by a decline in equatorial circulation. A third major change was the development of a worldwide system of deep (more than 1000 m) ocean circulation. However, Jim and Nick dismissed the idea of major Antarctic glaciation at the time of the abrupt step, believing that extensive ice sheets had formed much later, during the Miocene (around 14 million years ago), and thus 'If an ice sheet were present, it could not have been more than a small fraction of its present-day size'.[27] Their error was in assigning all of the increase in heavy oxygen to a reduction in water temperature.

Following its sojourn in the southwest Pacific, the *Glomar Challenger* ventured into the Southern Ocean, where it drilled 26 deep holes in the course of austral summers 1973 and 1974.[28] Results from the new sediment cores confirmed that the world had changed radically 33.5 million years ago and provided crucial evidence concerning the history of global ocean circulation and the glaciation of Antarctica. In a much-cited article of 1977,[29] James Kennett reviewed the results of the four Antarctic drilling legs, concluding that 'The evolution of the Southern Ocean itself has had a fundamental influence on the development of Antarctic glaciations, which in turn has affected the direction of global climatic evolution, including the north polar region'.

The Antarctic drilling confirmed that the Eocene–Oligocene boundary was a major threshold in climatic and glacial evolution, involving an abrupt drop in the temperature of bottom water accompanied by a major crisis in deep-sea faunas, the formation of Antarctic sea ice, the initiation of modern 'thermohaline'[30] circulation in the world's oceans, and a 2000 m deepening of the maximum depth at which calcium carbonate sediments can form (the 'calcium carbonate compensation

[26] Kennett and Shackleton (1976).

[27] Shackleton and Kennett (1975).

[28] DSDP Legs 28, 29, 35, and 36.

[29] Kennett (1977).

[30] 'Thermohaline circulation is generally taken to be that circulation driven by density differences imposed at the ocean surface by interaction with the atmosphere' (Hay et al., 1982).

depth'). In addition, previous drilling in the south-west Pacific[31] had revealed a gap in sedimentation throughout the region to the east of Australia, interpreted by Jim and colleagues as a consequence of erosion by strong bottom currents released following a major reorganization of ocean circulation in the early Oligocene. Sticking with his previous theory, Jim assumed that, while glaciers of limited extent may have existed from the beginning of the Oligocene, a major ice sheet did not exist on Antarctica before about 14 million years ago, when an ice cap like that of today formed quite rapidly, possibly as a result of the closure of the seaway through Indonesia. This assumption was challenged a few years later, when it became clear that the tropics had not participated in this cooling.[32] The drastic increase in the oxygen isotope ratio 33.5 million years ago therefore reflected rapid growth of an Antarctic ice sheet extending right to the coast, accompanied by a worldwide drop in sea level of 50–60 m.[33] Later work has shown that ice sheets did, indeed, form rapidly in two bursts of 40,000 years, separated by 200,000 years.[34] Rapid onset of extensive glaciation at this time has also been confirmed by the discovery of glacial sediments in cores drilled on later ODP legs and from drill rigs on floating ice in the Ross Sea.[35]

Amplifying his previous conclusions, Jim explained that the initiation of circumpolar circulation in the Southern Ocean required the opening of seaways between Australia and Antarctica (at the South Tasman Rise) and between South America and Antarctica (at Drake Passage). Conversely, the subsequent decline in equatorial circulation was a result of the closure of seaways between Africa and Eurasia, India and Eurasia, South East Asia and Australia and, finally, between North and South America. That is to say, the creation and destruction of seaways led to the establishment of the climate regime that we regard as 'normal', but which in reality is very different from that of earlier times. Plate reconstructions showed that, despite the fact that Australia first rifted from Antarctica during the Late Cretaceous, the South Tasman Rise would not have cleared the coast of Victoria Land until the Eocene–Oligocene boundary, about 34 million years ago (Figure 9.4). Likewise,

[31] Kennett et al. (1974).
[32] See Matthews and Poore (1980).
[33] See Miller et al. (2008).
[34] See Coxall et al. (2005).
[35] Barrett (1998).

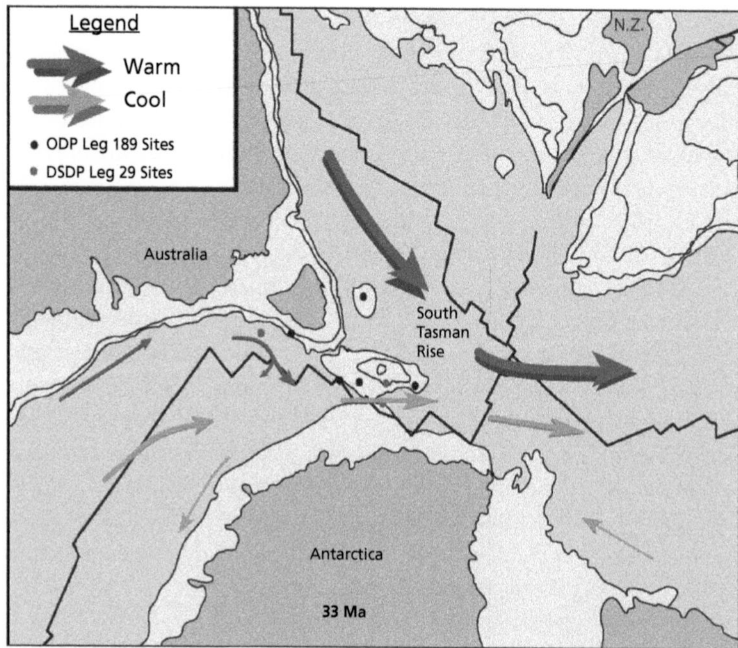

Figure 9.4 Reconstruction of Australia and Antarctica at 33 million years ago. At this time, the South Tasman Rise was just clearing the Antarctic margin, permitting the first circum-Antarctic flow of deep currents (shown by light grey arrows). [From Exon et al. (2004).]

although South America had been moving independently of Antarctica since the Jurassic (around 150 million years ago), separation in the region of Drake Passage only occurred around 30 million years ago.

Clearly, a Southern Ocean circulation like that of today could not have developed until these two 'gateways' had widened and deepened sufficiently to permit the growth of major currents around Antarctica. The rifting and northward movement of Australia and South America thus created a circum-Antarctic seaway within the latitude band of the 'furious fifties', which led eventually to the birth of 'Earth's mightiest ocean current',[36] driven by these winds. This is the Antarctic Circumpolar Current (referred to by oceanographers as the 'ACC'), formerly known to oceanographers and seamen as the 'West Wind Drift', which today

[36] Scher et al. (2015).

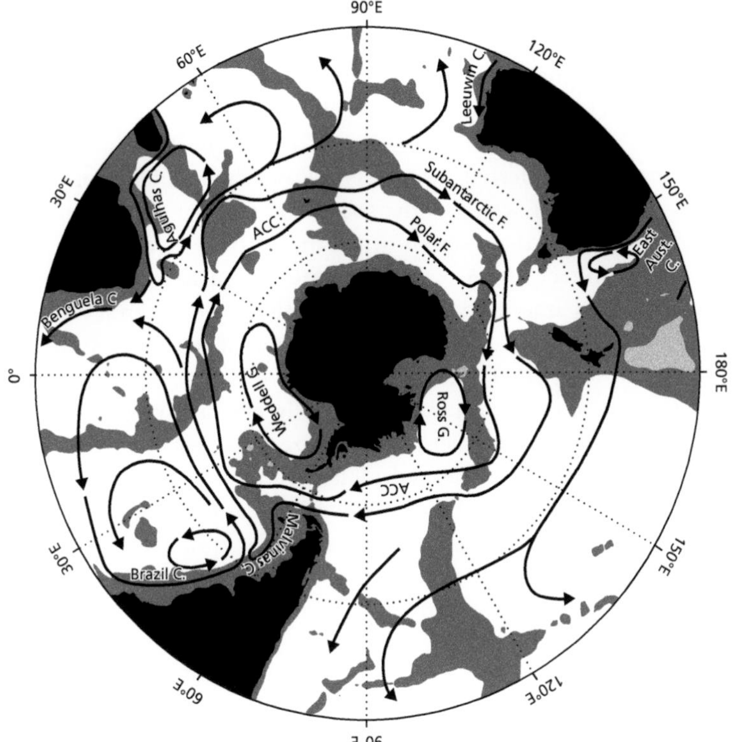

Figure 9.5 Ocean currents in the Southern Ocean. The Antarctic Circumpolar Current, labelled 'ACC', follows the polar front. Areas of shallow bathymetry are shown in grey. [From Rintoul et al. (2001).]

shifts over 130 million cubic metres of water every second[37] eastward through Drake Passage, the narrowest constriction in its endless path around the planet (Figure 9.5). As the gateways continued to widen and deepen, Jim proposed, the ACC increased in size and influence, eventually isolating Antarctica from warming by currents flowing from the north. This increasing isolation was, he proposed, reflected in its glacial history throughout the Oligocene.

The ACC is a major component of world ocean circulation today, having a profound influence on the exchange of gases between the surface and the deep ocean, as well as on global surface heat distribution.

[37] For comparison, this is more than four times that of the Gulf Stream through the Florida Straits.

Upwelling of nutrient-rich water deep water associated with the giant current system has a major influence on the strength of ocean currents in the Atlantic (the Atlantic Meridional Overturning Circulation), and thus on global ocean circulation. In each of the three major oceans, the ACC flows parallel to circulating currents further north, allowing heat and salt to be transferred. Its course follows the boundary or 'front' between cold polar water to the south and warmer water from the tropics, a boundary that slopes down to the north, rather like a subduction zone.[38] In fact, the ACC is not so much a current as a system of 'jets', travelling at around 1 knot around the planet. Each jet is composed of a series of eddies with typical diameters of less than 100 km, the whole resembling the flows observed on Jupiter. Since they extend all the way to the sea bed, the jets are 'steered' by bottom topography, such as that in Drake Passage, where they squeeze through a deep gap 800 km wide between the tips of Tierra del Fuego and the Antarctic Peninsula. The eddies that constitute the ACC jets are actually the principal means by which heat is transferred to the south polar region, and also bring heat energy from depth to the surface—hence the ACC is not such a barrier as Jim thought. Nevertheless, there is no doubt that the opening of the two Southern Ocean gateways marked a fundamental change in ocean and atmospheric circulation leading to the cool climate of the modern world.

Computer simulations by researchers at the universities of Leeds and Bristol,[39] using a version of the Hadley climate model developed by the UK Meteorological Office, have supported Jim Kennett's idea that a strong ACC did not develop immediately following the opening of Southern Ocean gateways. They concluded that further changes in geography, involving the evolution of topography within the gateways, were required, and thus a 'coherent' ACC was not possible during the Oligocene. That is, an ACC like that of today developed more recently than 25 million years ago, once a suitable circumpolar deep-water pathway had developed. Nevertheless, 'significant flows of Pacific water into the Atlantic were established as soon as there was a route through the Drake Passage during the late Eocene'. This suggests that gateway opening was more likely associated with the long-term decline in ocean temperatures from the Early Eocene 'Climate Optimum', that continued

[38] This is known as the 'Polar Front'.

[39] Hill et al. (2013).

through much of the Oligocene, than with the abrupt changes at 33–34 million years ago. This is supported by evidence from marine geophysics and proxies, including the study of neodymium isotopes in fossil fish teeth, pointing to opening of the Southern Ocean gateways to shallow circulation during the middle Eocene (around 50 million years ago). Jim had also suggested that a second threshold was crossed around 3.5 million years ago, when the northern continents entered the ice age that has persisted and deepened ever since. Could changes in ocean circulation related to ocean gateways be involved here too?

Global Coiling

Around 3.5 million years ago, planktonic foraminifera of the genus *Pulleniatina*, having quite happily coiled their calcite shells in a clockwise sense for hundreds of thousands of years, began winding them in the opposite direction. Half a million years later, the genus disappeared altogether from the Atlantic, while, in the Pacific, it didn't know which way to turn, flipping from left to right and back again. This may not sound like a momentous step in the evolution of life, or even a particularly interesting phenomenon, yet such changes of bias in coiling direction have been employed for many years by paleoceanographers as an indication of past sea temperatures. The basic assumption is that, for reasons unknown, cooler waters induce these tiny creatures to grow and coil their shells in an anticlockwise or sinistral sense, while warmer waters, at temperatures over about 8°C say, induce clockwise or dextral coiling. As it turns out, this assumption is, to say the least, simplistic, and a variety of other factors needs to be taken into account (including recent evidence from molecular biology that sinistral and dextral forms are actually distinct species). Nevertheless, the analysis by Lloyd Keigwin at Rhode Island appeared to show a significant difference in coiling bias between the Pacific and Caribbean sides of the Panama isthmus after about 3.1 million years ago, a difference that he interpreted as a consequence of the creation of a barrier to ocean currents between the two—that is, the isthmus itself—and the closure of the former Central American Seaway.[40] Later analyses of stable oxygen isotopes in planktonic foraminifera from DSDP sediment cores recovered in the eastern Pacific and North Atlantic confirmed a divergence in temperature

[40] Keigwin (1978).

and salinity between 4 and 3 million years ago, while carbon isotopes from deep-living foraminifera indicated divergence beginning about 6 million years ago, reaching present-day levels by 3 million years ago, supporting Lloyd's earlier conclusions.[41] The timing of this closure was tantalisingly close to the date of onset of glaciations on the northern continents, and hence the formation of a barrier to eastward flow at this time may have been as important a control on global circulation and climate as the opening of Drake Passage.

Close inspection of Teddy Bullard's reconstruction of the Atlantic-bordering continents[42] reveals an embarrassing overlap between North America and South America that became even more embarrassing in some later reconstructions. It was necessary to get rid of Central America somehow, but how? Fortunately, the history of relative motion between the North American and South American plates is well known, since both can be accurately reconstructed to the African plate by means of the excellent record of magnetic anomalies and fracture zones in the Atlantic. Following initial separation in the Jurassic, North America and South America have separated by more than 3000 km in the region of the present-day Caribbean, creating new oceanic crust in which a record of plate motions was preserved in sea-floor magnetic anomalies and fracture zones. Unfortunately, this ocean crust no longer exists, having been subducted as the Caribbean plate migrated eastward from the Pacific during the past 60 million years, interposing itself between the two major plates (see Figure 9.6). Eastward movement was accompanied by a slow-motion collision of the volcanic arc at its leading edge, known as the 'Great Arc of the Caribbean', with the South American continent, a collision that continues today in the eastern Caribbean, where the arc remains active as the Lesser Antilles island arc.[43] Around 14 million years ago, the Panama arc, on the trailing edge of the Caribbean plate, began to collide with what is now western Colombia, beginning the process of closure of the seaway and the creation of a land bridge between the two Americas, a land bridge that was to have profound implications for biological evolution in both continents.

The timing of final closure, much like that of initial opening of the Drake Passage seaway, is difficult to pin down from direct geological or

[41] Keigwin (1982).
[42] See Chapter 2.
[43] For details of this fascinating evolution, see Escalona and Mann (2011).

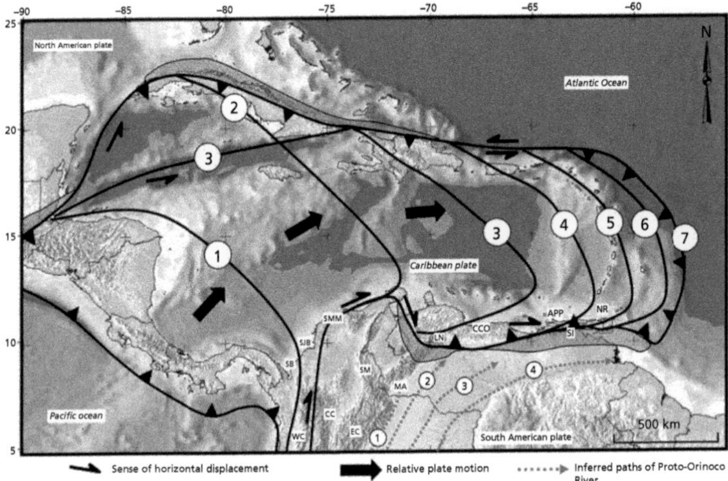

Figure 9.6 Eastward migration of the Caribbean plate relative to the North and South American plates since the Late Cretaceous (80 million years ago). Numbers refer to dates as follows: (1) 80 million years ago; (2) 60 million years ago; (3) 44 million years ago; (4) 30 million years ago; (5) 14 million years ago; (6) 5 million years ago; (7) present day. [From Escalona and Mann (2011).]

geophysical evidence because of inconvenient complexities, and it is by no means certain that we are talking about a single event: more likely, transient land connections were made and broken on several occasions before the final establishment of a sturdy land bridge. Likewise, biogeographic evidence is equivocal: despite the occurrence of the 'Great American Interchange' 3.5 million years ago,[44] complexities such as changes in sea level and the ability of some species to swim or island-hop mean that evidence for species migrations from North America to South America (or the reverse) exists at much earlier times. The situation is comparable to the recent evolution of South East Asia, where connections via various islands have been made and broken many times during the ongoing collision of Asia and Australia. Tony Coates, a geologist at the Smithsonian Tropical Research Institute in Panama, has demonstrated the surprising similarities between the recent evolution of the 'Indonesian gateway' and the likely developments in the Central American Seaway 3 million years ago. As he points out, shallow oceanic

[44] Marshall et al. (1982).

gaps between Borneo and Australia, mainly in the vicinity of Sulawesi (or Celebes, as it was formerly known), allow a significant flow of ocean currents between the Indian and Pacific oceans, while maintaining a significant barrier to the migration of mammals and other land animals. Indeed, closure of the Indonesian gateway has itself been implicated in the onset of northern hemisphere glaciations. Lamont scientist Mark Cane, together with Peter Molnar (now at MIT), speculated that northward motion of the Australian plate brought about global cooling by allowing the passage of much cooler water from the Pacific westward into the Indian Ocean.[45] Lower sea-surface temperatures in the western Indian Ocean also led to a change to cooler and drier conditions in East Africa, with spectacular consequences for primate evolution.

Marine sedimentary rocks now outcropping in northern Colombia[46] suggest that open ocean conditions existed prior to 16 million years ago, implying that the Central American Seaway was then fully open. Changes in sedimentation at that time indicated shoaling of the seaway, and then, 12–13 million years ago, major uplift occurred, reducing ocean depths to an estimated 1000 m and creating a partial barrier to currents flowing between the Pacific and Caribbean. Land animals of the two continents began to intermingle and distinct differences in deep-living foraminifera developed. After a brief increase in water depths between 6 and 7 million years ago, the Panama isthmus emerged completely by about 2.8 million years ago.[47]

With such remarkable agreement between the timing of sudden global cooling, the onset of glaciations, and the opening and closing of ocean gateways, it was inevitable that things couldn't last and that alternative explanations would soon have to be sought for these major climatic changes. Recent work on stable oxygen and carbon isotopes from ODP cores in the Atlantic[48] have led to the conclusion that 'the early Pliocene shoaling of the CAS [the Central American Seaway] had no profound impact on the evolution of climate', while computer simulations[49] suggest that, while modelled closure of the Central American Seaway led to more intense ocean circulation and greater snowfall in Greenland and North

[45] Cane and Molnar (2001).
[46] See Duque-Caro (1990).
[47] See Lessios (2008).
[48] Bell et al. (2015).
[49] Lunt et al. (2008).

America, the effect on the volume of ice on the northern continents was minor, so that closure was not a major forcing mechanism for glaciations.

It's a Gas

In 2003, an influential paper[50] was published in *Nature*, claiming that Antarctic glaciation was a consequence of a long-term decline in atmospheric carbon dioxide concentration, and minimizing the influence of gateway opening. The authors, Robert DeConto, working at the University of Massachusetts, and David Pollard, at Pennsylvania State University, came to their conclusions on the basis of the results of computer simulations using version 2 of the GENESIS general circulation model employed previously by Eric Barron. Despite the limitations in the computer simulation—for example, each element in the model ocean represented a generous surface area of $2.0° \times 2.0°$ (at the equator, this would be $220\ km \times 220\ km$), while the ocean was represented by a 50 m layer simulating the surface ocean without any deep circulation and no bottom topography—Bob and Dave felt confident enough to conclude that the opening of the Drake Passage (which they had not included explicitly in their model) was of secondary importance. They even went so far as to specify a CO_2 concentration below which extensive ice sheets would form on Antarctica—750 ppm. The idea of CO_2 being implicated in long-term climate changes was by no means new, but the publication of this article in such an authoritative journal did help to shift the balance of opinion on the causes of major ice ages.

In 2008, Bob and Dave modelled northern hemisphere glaciations in the same way, coming to the conclusion that here a much lower threshold CO_2 concentration of 280 ppm was required. Carbon dioxide is not, of course, a major constituent of the atmosphere, representing only around 400 ppm by volume (i.e. approximately 0.04 per cent), equivalent to a total of 750 gigatonnes of carbon (1 GtC = 1 billion tonnes of carbon). There is a lot more carbon in the oceans, in the form of carbonate and bicarbonate ions, accounting for another 40,000 GtC. But there is far more still stored in the Earth's crust, some sequestered by living organisms as oil, gas, and coal, but mostly in sedimentary rocks such as limestone, amounting to over 100,000,000 GtC (i.e. a hundred thousand million million tonnes).

[50] DeConto and Pollard (2003). By early 2015, this article had clocked up over 500 citations.

This provides a constant supply of CO_2 through the agency of volcanism at mid-ocean ridges, volcanic arcs, and hotspots. In fact, if CO_2 were not removed constantly from the air (by processes that we'll come to), volcanism would double the concentration of this gas in just 600,000 years, and the Earth's climate would accelerate towards a 'super-greenhouse' condition like Venus, where the atmosphere is composed of more than 96 per cent CO_2 and the surface temperature is around 460°C. So what are the processes that remove CO_2 from the atmosphere to maintain some sort of balance?

Back in 1859, the Irish physicist John Tyndall demonstrated experimentally that some atmospheric gases, such as water vapour and CO_2, were highly effective in storing heat reflected from the Earth's surface, while others, such as oxygen and nitrogen, were virtually transparent, allowing solar energy to be returned to space. Increases in the concentrations of the former group, now referred to as 'greenhouse gases', would thus lead to warming of the atmosphere, while their removal would result in cooling. John was doubtful about whether such variations could produce effects profound enough to plunge the world into ice ages, however. Forty years later, the Chicago University geologist Thomas Chamberlin[51] presented, in three instalments totalling nearly 100 pages, a 'working hypothesis', proposing just that.[52] Thomas suggested that reductions in the atmospheric concentration of CO_2 had led to global cooling in the geological past, sufficient to trigger glaciations on the scale observed in the polar regions today, while, at the other extreme, increases had resulted in warming, producing a totally ice-free world such as that of the Cretaceous. Many of these ideas had, in fact, been published a few years earlier by the mining engineer Jacques-Joseph Èbelmen. Unfortunately, being a Frenchman, Jacques made the error of publishing in French, with the result that his great insights were ignored.[53]

Thomas hypothesized that the primary mechanism for the removal of CO_2 from the atmosphere was the deposition of 'limestones and carbonaceous deposits' and predicted that, without replenishment, further removal by this mechanism would plunge the Earth into a frigid

[51] The same Thomas Chamberlin who had criticized William Thomson's flawed estimates of the Earth's age—see Chapter 5.

[52] See Chamberlin (1899).

[53] See Berner and Maasch (1996).

future. Limestones are produced by the precipitation of calcium carbonate from seawater by a wide range of animals and plants,[54] such as the foraminifera that proved so useful in working out changes in ocean temperature. The carbonate is supplied via a reaction between rocks containing silicate minerals, exposed at or near the surface, and rainwater acidified by the dissolution of atmospheric CO_2, and is transported by rivers draining the continents. Rock weathering is a complex process, but can be represented by a series of simplified equations, beginning with the breakdown of a calcium silicate mineral:[55]

$$CaSiO_3 + 2CO_2 + 3H_2O \rightarrow Ca^{2+} + 2HCO_3^- + H_4SiO_4$$

On exposure at or near the surface, such minerals react with water and carbon dioxide[56] and are reduced to clay minerals with the release of calcium, magnesium and bicarbonate ions, and also silicic acid (H_4SiO_4). Calcium and bicarbonate ions are then transported by rivers to the ocean where they are combined to form the calcite ($CaCO_3$) from which many marine organisms construct their homes:[57]

$$Ca^{2+} + 2HCO_3^- \rightarrow CaCO_3 + CO_2 + H_2O$$

The point to notice here is that, even though a molecule of CO_2 is liberated by the precipitation of calcite in the second equation, *two* molecules of CO_2 are removed by weathering in the first. Hence, the combined processes of weathering and calcite precipitation result (in theory) to the net removal of one molecule of CO_2 from the atmosphere, as expressed in Harold Urey's equation[58] summarizing the entire process:

[54] Some limestones are also produced by direct (inorganic) precipitation from seawater in warm, tropical regions.

[55] The calcium silicate here is, strictly, a mineral known as 'wollastonite', but is intended to represent a range of calcium silicate minerals found in rocks. Of course, rocks such as granite contain a variety of silicate minerals containing other elements, but the same principle applies to most of them. The process is greatly assisted by the activities of plants in fixing carbon from the air and then transferring it in the form of organic acids to the soil.

[56] In reality, with organic acids in the soil derived from CO_2 fixed from the atmosphere by plants.

[57] The silicic acid may be used by other organisms, principally diatoms, to produce their shells.

[58] See Urey (1952).

$$CaSiO_3 + CO_2 \rightarrow CaCO_3 + SiO_2$$

If this calcite is then buried in sea-floor sediments, the carbon is locked away from the atmosphere, at least until it reaches a subduction zone (which, of course, Thomas knew nothing about).

In considering the tectonic implications of his hypothesis, Thomas realized that the rate of silicate weathering would be affected by temperature and moisture, and also by tectonic uplift. Pointing out that 'The rate of chemical action of the atmosphere on the surface of the rocks is believed to have been intimately connected with the extent and height of the land area', he noted 'It is obvious that the greater the surface area of rock exposed to the effective action of the atmosphere, the more rapid will be the rate of disintegration, other things being equal, and the more rapid the consumption of carbon dioxide'. Despite complications, such as the influence of higher temperatures and thicker vegetation at low altitudes, and the greater effectiveness of physical weathering (e.g. frost action) at higher altitudes, Thomas still believed that 'Making all allowances that seem required for the offsetting factors, it would still appear that the elevated condition increases the activity of decomposition in a very notable degree'. Sadly, however, Thomas lost faith in his ideas and began to disown them, deciding that he had overemphasized the role of CO_2.[59] In the end, his working hypothesis suffered a fate similar to that of Jacques Èbelmen, and was forgotten, at least until the post-war twentieth century.

Rise and Fall

Seventy-one years after the publication of Thomas Chamberlin's working hypothesis, three USGS geologists, Zell Peterman, Carl Hedge, and Harry Tourtelot, wrote a paper[60] about strontium isotopes in seawater. This was not quite as esoteric as it might, at first, appear. The trio had exploited the fact that one of the four natural isotopes of strontium is produced by slow radioactive decay of unstable rubidium (^{87}Rb) with the usefully long half-life of 49 million years. This is strontium-87 (^{87}Sr), the concentration of which in a rock will thus be dependent on the amount of unstable ^{87}Rb originally present, and will

[59] See Summerhayes (2015).
[60] Peterman et al. (1970).

increase slowly with time relative to its non-radiogenic cousin, ^{86}Sr. Radiogenic ^{87}Sr constitutes around 7.0 per cent of naturally occurring strontium today, while ^{86}Sr accounts for 9.86 per cent.[61] Ratios of ^{87}Sr to ^{86}Sr measured in seawater in today's oceans are close to 0.7092. The important point is that mantle rocks and their products (i.e. oceanic basalts) have low Rb concentrations, and therefore low $^{87}Sr/^{86}Sr$ ratios (around 0.7025). Continental rocks have higher concentrations of Rb and are, in any case, much older on average, and so their $^{87}Sr/^{86}Sr$ ratios are higher—not much higher (around 0.7119), but enough to measure with 1960s technology. Weathering of continental rocks produces strontium ions that are carried, along with the calcium and bicarbonate ions, down the world's rivers to the sea, where they become evenly mixed by the global ocean circulation. Once again, we have tiny sea creatures to thank—in this case, molluscs—for storing away a record of past strontium isotope ratios. Because of their resemblance to calcium, strontium ions are incorporated into the calcite shells of these creatures and preserved in fossils after their death. Since both strontium isotopes behave in much the same way chemically, the $^{87}Sr/^{86}Sr$ ratios measured in ancient marine sedimentary rocks will reflect the ratio in seawater at the time of growth.

You should now be able to see where all this is leading. Variations in the $^{87}Sr/^{86}Sr$ ratio should reflect changes in continental weathering rates: when rates are high (for whatever reason), the $^{87}Sr/^{86}Sr$ ratio will be high in seawater; when weathering is subdued, the ratio will be low. Similarly, high rates of input from the mantle by hydrothermal reactions at mid-ocean ridges—say, when global rates of spreading were high—would lower the ratio in seawater, while low rates of spreading would increase the ratio. Zell, Carl, and Harry plotted ratios measured from mollusc shells worldwide for the past 500 million years, observing that they had dipped to a low during the Jurassic and Cretaceous, before increasing rapidly during the Cenozoic (the past 50–70 million years). They proposed that 'Lowering of $^{87}Sr/^{86}Sr$ in the oceans could be accomplished by an increased contribution of strontium from young volcanic rocks whereas times of increasing $^{87}Sr/^{86}Sr$ may be related to times of continental emergence resulting in a greater contribution of more radiogenic strontium derived from older crystalline rocks'. In other words, the low ratios of the Jurassic and Cretaceous were the

[61] The most abundant isotope is ^{88}Sr, which makes up around 83% of the total.

result of unusually high levels of volcanism, while the high ratios that followed reflected increased rates of rock weathering. During the 1980s, a number of groups confirmed both the general form of Zell and Co.'s curve, and their conclusions that the primary controls on $^{87}Sr/^{86}Sr$ ratio were sea-floor spreading and hydrothermal activity at mid-ocean ridges, recycling of carbonate from previously deposited sediments, and continental weathering[62].

In 1988, William Ruddiman and his research student at Lamont, Maureen Raymo, compared the strontium isotope record for the past 60 million years with estimates of the rate of uplift of Tibet, concluding that the flux of dissolved strontium and calcium salts in the major rivers draining the Tibetan Plateau and Himalayas did, indeed, reflect the rate of chemical weathering in these elevated regions. This, in turn, was 'strongly dependent on rates of uplift'. They suggested that increasing rates of uplift over the past 5 million years had resulted in a rapid decrease in atmospheric CO_2 concentration, sufficient to bring about glaciation of the northern continents. They concluded: 'Ultimately, the onset of Northern Hemisphere glaciation may have been related to mountain uplift through the impact of high plateaus and mountain ranges on atmospheric circulation patterns [...] and the impact of enhanced rates of chemical erosion on the oceanic C cycle and atmospheric CO_2'.

Summing up their 'uplift–weathering' hypothesis in a 1992 review article in *Nature*,[63] Raymo and Ruddiman suggested that 'over the past 40 Myr, uplift of the Tibetan plateau has resulted in stronger deflections of the atmospheric jet stream, more intense monsoonal circulation, increased rainfall on the front slopes of Himalayas, greater rates of chemical weathering and, ultimately, lower atmospheric CO_2 concentrations'. In the same year, Frank Richter, who had previously worked with Dan McKenzie on the problem of mantle convection, got together with University of Chicago colleague David Rowley and Berkeley geochemist Donald DePaolo[64] to perform calculations showing that reductions in the supply of low-^{87}Sr magma and hydrothermal fluids to the oceans could not explain the rapid increase in seawater $^{87}Sr/^{86}Sr$ since the Eocene. Instead, they presented further evidence confirming that the cause was an abrupt and significant increase in the supply of strontium

[62] See e.g. Burke et al. (1982) and Palmer and Elderfield (1985).

[63] Raymo and Ruddiman (1992).

[64] Richter et al. (1992).

dissolved in river water draining uplifted mountain regions and, following Raymo and Ruddiman, focused on one mountain region in particular: the Himalayas and Tibetan Plateau. The collision of India with Eurasia had caused rapid uplift of the Tibetan Plateau, beginning around 10 million years ago, resulting in a sudden increase in chemical weathering of silicate minerals and thus an increase in the input of strontium with high $^{87}Sr/^{86}Sr$ ratios to the oceans. The Brahmaputra, Indus, Ganges, Mekong, and Yangtze rivers draining the region supply around 20 per cent of the flux of Sr to the oceans today, and hence changes in supply from the Himalayan region could have global implications. Frank and colleagues concluded that the amount of strontium transported in rivers and its $^{87}Sr/^{86}Sr$ ratio are both consequences of 'the intensity of mountain building resulting from continent–continent collisions'.

The introduction of CO_2 as a primary modulator of climate via continental silicate weathering dramatically alters the expected effect of changing land and sea distribution resulting from plate motions. Since the rate of weathering and CO_2 extraction depends on temperature and rainfall, continents in low latitudes will experience higher rates of weathering and thus CO_2 will be removed more rapidly, leading to cooling. On the other hand, in higher latitudes, weathering will be inhibited by lower temperatures and reduced precipitation, slowing CO_2 removal and maintaining higher global temperatures. This, as you will have noticed, is the exact opposite of what Charles Lyell expected on the basis of simple differences in the rate of temperature rise between land and sea. In addition, the rate of CO_2 removal will be affected by the size of a continent: a large continent in tropical latitudes will experience low rainfall in its interior, limiting weathering and CO_2 removal, whereas a string of small continents will experience high rates of removal. In this way, plate tectonics serves to alter the 'weatherability' of rocks, rather than affecting the energy balance directly.

Ultimate Causes

In a sense, the uplift–weathering model was a bit too good. If the elevation of silicate rocks in the Himalayas and Tibet resulted in increased weathering, and if that led, ultimately, to removal of CO_2 from the atmosphere, then, since CO_2 is only a trace gas, it would not take long—perhaps 300,000 years—to strip it out of the atmosphere entirely, bringing life on Earth to a shivering halt. Likewise, the vast amounts of coal, oil, and, above all, limestone stashed beneath our feet testify to CO_2 removal on an industrial scale in the past, yet there is still enough CO_2 left in the atmosphere to keep things ticking over nicely.

Thomas Chamberlin had the answer to this conundrum more than a century ago. In order to maintain a reasonably constant balance of CO_2 in the atmosphere, 'sources of permanent supply' must exist 'of a competency approximately equal to the sources of loss'. A prime candidate was the igneous rock of the Earth's crust, which contained volumes of gases[65] equivalent to a 'multitude of atmospheres'. While much of this derived from previous burial in the form of limestones and organic deposits, part of it, Thomas speculated, originated in the deep mantle and therefore represented a 'real contribution to the earth's atmosphere and hydrosphere'. The means by which these gases were transferred to the atmosphere was volcanism, which, he pointed out, had varied in intensity over geological time, perhaps in a periodic manner. Thus, Thomas wrote, 'By the terms of the hypothesis the state of the atmosphere at any time is dependent upon the relative rates of loss and gain'. Supply by means of volcanism and removal by silicate weathering tended to neutralize one another, and 'it was only when one agency fell behind the other in its competency that its specific results became manifest, and then only by the difference in their respective effects'.

Thomas' working hypothesis was revived in 1983 and used as the foundation for the construction of a numerical model of changes in atmospheric CO_2 concentration through geological time. The architect was Yale University geochemist Robert Arbuckle Berner, who, together with colleagues Antonio Lasaga and Robert Garrels, produced a computer model of the carbon cycle for the past 100 million years.[66] This numerical representation, which became known as the 'BLAG' model,[67] included all the processes they thought were important for the cycling of carbon, enabling them to predict[68] past levels of CO_2 in the atmosphere, and also of calcium, magnesium, and bicarbonate ions in the oceans (Figure 9.7). The processes they considered included carbonate

[65] That is, substances like carbon and sulphur oxides, combined chemically in minerals rather than existing in a gaseous state.

[66] Berner, Lasaga, and Garrells (1983).

[67] Derived from the initials of the authors.

[68] For the sake of pedants, I acknowledge that predicting the past involves a degree of illogicality. However, I prefer to avoid ugly terms like 'hindcast' or 'postdict'.

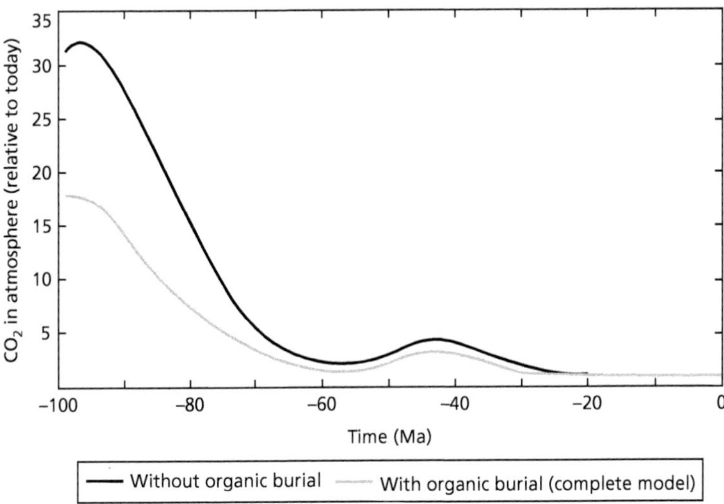

Figure 9.7 CO_2 concentration (as percentage) and temperature for the past 100 million years predicted by the BLAG model.

and silicate removal by weathering, together with precipitation and burial of carbonates from seawater by living organisms and inorganic reactions. CO_2 was supplied to the atmosphere by volcanic 'degassing' associated with plate subduction and sea-floor spreading. The model was adjusted to account for changes in continental land area as they affected weathering and CO_2 removal, and also changes in sea-floor spreading rates as they affected the supply of CO_2. Berner, Lasaga, and Garrells' results indicated that 'the CO_2 content of the atmosphere is highly sensitive to changes in seafloor spreading rate and continental land area'. Assuming a correlation between atmospheric CO_2 concentration and past temperature, their predictions were in fair agreement with 'hothouse' temperatures of the Cretaceous estimated from proxies. 'Consequently', they concluded, 'our results point to plate tectonics, as it affects both metamorphic–magmatic decarbonation[69] and changes in continental land area, as a major control of world climate'.

The project turned out to be rather like the painting of the Forth Bridge, a never-ending enterprise, with constant updates and improvements keeping Bob Berner in business until his death in January 2015.

[69] By 'decarbonation', they meant the liberation of CO_2 from rocks.

A sequel to BLAG, subsequently named 'GEOCARB', appeared in 1991, and was soon followed by GEOCARB II in 1994 and GEOCARB III in 1996. The final version (so far as Bob was concerned) was GEOCARBSULF in 2002, by which time competitors had begun setting up their own models. In his calculations, Bob made a distinction between the 'short-term' carbon cycle that involved transfers of carbon between the atmosphere, ocean, lakes, rivers, plants, animals, soils, and sediments; and the 'long-term' carbon cycle controlled by plate tectonics. In essence, the long-term cycle could be regarded as a balance of fluxes between two carbon 'reservoirs', one at the surface (incorporating the oceans, atmosphere and biosphere), and one in the solid Earth (incorporating crustal rocks and the deep mantle).

To cut a long story short, the GEOCARB models ended up predicting a very high level of CO_2 (more than 20 times the present concentration—see Figure 9.8) in the Cambrian atmosphere (541 million years ago), declining to low concentrations similar to today in the Carboniferous and Permian (350 to 250 million years ago), before increasing in the

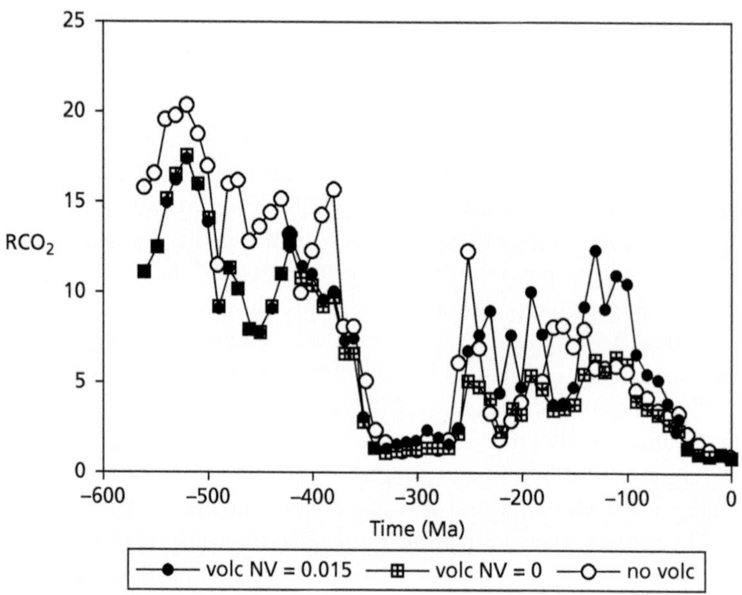

Figure 9.8 Concentration of atmospheric CO_2 relative to present day as predicted by GEOCARB models. NV is an arbitrary parameter used by Berner (see Berner, 2006).

Triassic and Jurassic (250 to 140 million years ago), reaching values of roughly five times present values during the Cretaceous hothouse (around 100 million years ago), before declining again, with a blip at around 50 million years ago, to the present (pre-industrial) value of 280 ppm.

Perhaps surprisingly, Bob found that one of his most vexing problems was the estimation of the rate of 'degassing' (i.e. release to the atmosphere) of CO_2 during igneous and metamorphic activity.[70] His solution was to assume that degassing occurred solely at volcanic arcs associated with subduction zones. Since the Earth does not appear to be growing or shrinking, the global rate of crustal loss by subduction must be approximately the same as the global rate of ocean crust creation by sea-floor spreading. Hence, by calculating the latter, he could derive the former. As it turned out, this was crude at best and downright inaccurate at worst. The calculations were based on the area of oceanic crust produced in a given interval. However, stepping back in time involved uncertainty resulting from the loss of ocean lithosphere by subduction, so that the pattern of mid-ocean ridges and subduction zones prior to 100 million years ago was poorly known, and, before 150 million years ago, virtually unknown. For example, in the shrinking Pacific Ocean, spreading centres have been overridden by continents and subducted, just like the Juan de Fuca Ridge today. In the past, a network of mid-ocean ridges existed, separating plates with strange names, like 'Izanagi', 'Kula', and 'Farallon'. Without taking into account the CO_2 output from these ridges, the estimates built into Bob's GEOCARB models were bound to underestimate atmospheric concentrations. An alternative for earlier periods was to use changes in global sea level as a proxy for sea-floor spreading rates on the assumption that higher rates led to warmer and therefore shallower ocean floors[71] and hence higher sea levels. A major problem with this assumption, as we have seen, is that sea levels are seriously affected by other phenomena, such as the growth and melting of continental ice sheets. To make matters worse, early estimates of sea level prior to the Cretaceous turned out to be serious overestimates: later estimates showed that sea level in the Cretaceous reached about 100 m higher than today, rather than the 230 m calculated previously. Nevertheless, Bob concluded that the primary control on the concentration of CO_2

[70] Berner and Lasaga (1989).

[71] Recall the 'age–depth' relationship from Chapter 2.

in the atmosphere, and therefore of global temperatures, was the rate of tectonic release of greenhouse gases.

Despite their important role in the long term, volcanoes emit only about 0.2 GtC per year at present (compared with our own generous contribution of 8.9 GtC per year[72]). In the main, their distribution reflects the boundaries between the plates, in particular mid-ocean ridges and subduction zones (volcanic arcs). In addition, volcanic hotspots make a significant, but variable, contribution. Mid-ocean ridges could be regarded as a single, 65,000 km-long, volcano, erupting sufficient lava to floor two-thirds of the planet. Since eruption is accompanied by the release of considerable amounts of CO_2 and other volatiles, it stands to reason that this must be the major source of carbon in the oceans. Estimates based on direct and indirect methods suggest that the global output of CO_2 from mid-ocean ridges to the oceans may be around 2×10^{12} moles of CO_2 per year.[73] Notwithstanding that this means nothing to most of us,[74] it turns out that the hydrothermal circulation found in young oceanic crust (producing the hot water vents observed by the *Alvin* divers in Chapter 1) also *removes* CO_2 from seawater. In fact, it removes around 3×10^{12} moles of CO_2 per year, more than wiping out the contribution from mid-ocean ridge volcanism. Hence, mid-ocean ridges are more likely a sink for CO_2 rather than a source.

What about volcanoes on land? The best current estimates for the global rate of degassing from subaerial volcanoes (i.e. all volcanic arcs and hotspots on land) is $2.0–2.5 \times 10^{12}$ moles of CO_2 per year,[75] which is still less than the 3×10^{12} moles of CO_2 per year removed by silicate weathering. However, as residents of Sumbawa in the Dutch East Indies discovered in 1815, and the rest of the northern hemisphere found the following year,[76] volcanic output is hardly constant. Some very large eruptions, such as that of Toba, 74,000 years ago, can exceed the annual output of all other volcanoes combined. In addition, eruptions of different magma types with different water contents produce wide variations in the ratio of CO_2 to magma volume, while a great deal of CO_2 is produced during non-eruptive (i.e. passive) phases of some volcanoes.

[72] IPCC (2013).
[73] Kerrick (2001).
[74] By my shaky maths, this is 88 million tonnes, but don't rely on it.
[75] Kerrick (2001).
[76] This was 'eighteen-hundred-and-froze-to-death', the 'year without a summer' following the eruption of Tambora, an arc volcano on Sambawa, the previous year.

Presently, more than half the global output of CO_2 from subaerial volcanoes is contributed by just two volcanoes—Etna and Popocatépetl—one of which (Etna) is atypical of subduction-zone volcanoes. In this situation, the addition or extinction of one or more volcanoes of similar size and output could make a huge difference. Estimates of past output based on present global averages are virtually useless therefore.

In a fascinating article published in 2015,[77] Cin-Ty Lee and colleagues highlight the contrast, insofar as the carbon cycle is concerned, between volcanic arcs at subduction zones along the edges of continents (i.e. 'continental arcs', such as the Andean arc), and those located within oceanic regions ('island arcs', such as the South Sandwich arc). Because continental crust is (almost) unsinkable, its vast store of carbonates (e.g. limestones) and organic sediments (e.g. coal and oil shales) remain intact for very long periods (unless exploited by humans). Magma generated beneath continental arcs will incorporate some of this stored carbon and transport it to the surface, where it is degassed to the atmosphere as CO_2. Island arc magmas, on the other hand, have low concentrations of carbon and contribute little to the atmosphere. Hence, changes in the pattern of subduction zones from, say, a time during which a great length of continental arcs existed to a time when subduction was dominated by island arcs could result in a major decrease in the amount of CO_2 degassed. Such a transition occurred around 50 million years ago, when back-arc basins such as the Lau Basin began to open in the western Pacific, replacing a long line of continental arcs marking the eastern margin of Eurasia. Cin-Ty and colleagues suggest that this reduction may have been sufficient to trigger the long period of global cooling that followed. Speculating further, they suggest that continental arcs may be typical of supercontinent break-up and dispersal, while island arcs reflect continent convergence and supercontinent assembly. Hence, global climate may be bound in to the supercontinent cycle.

Global output from hotspot volcanoes can be even more variable than that from arc eruptions. Massive outpourings, usually of basaltic lavas and widely associated with the arrival of hot mantle plumes at the Earth's surface, have occasionally wreaked havoc with both climate and biota, wiping out much of the life on the planet and providing geologists with convenient makers for their stratigraphical boundaries. These large igneous provinces, (or 'LIPs', as they are known) may be truly vast:

[77] Lee (2015).

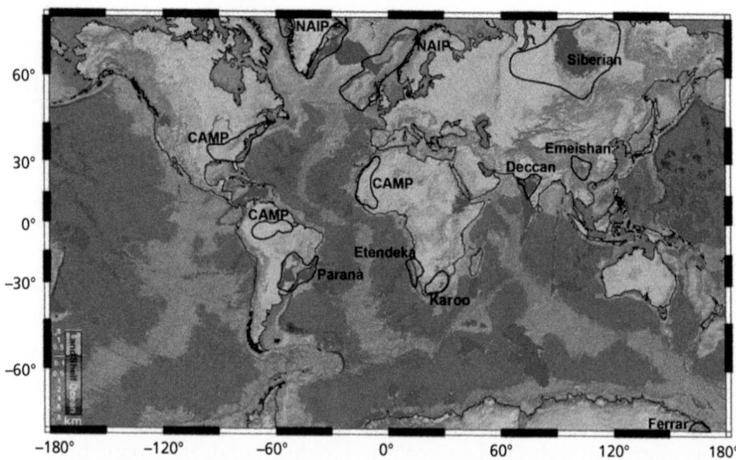

Figure 9.9 Large igneous provinces (LIPs) of the world: CAMP, Central Atlantic Magmatic Province; NAIP, North Atlantic Igneous Province. [After Jones et al. (2016).]

one estimate suggests that the largest may cover as much as 10 million km^2 (or 4 million square miles in old money), releasing enough gases during eruption to seriously disrupt the carbon and sulphur cycles of the planet. Examples are the Deccan Traps and Siberian Traps that we visited in Chapter 6 (Figure 9.9). Most of the CO_2 in lavas is released to the atmosphere during eruption, since it is not very soluble in magma and is incompatible with most rock-forming minerals. Recent work suggests that the mass of CO_2 released by the Siberian plume at the end of the Permian (251 million years ago) could have been greater than 170,000 GtC.[78] The Cretaceous is known to have been a time of enhanced LIP activity, with the production in the Pacific of the giant Ontong Java and Kerguelen plateaus. This has been used to argue that the Cretaceous 'hothouse' was a result of CO_2 output associated with plume activity.

More recently, an ambitious attempt has been made to quantify the history of volcanic CO_2 outgassing associated with plate tectonics by reconstructing ancient subduction zones using seismic wave-speed anomalies in the lower mantle, thought to represent subducted slabs of oceanic plate. The results of this work, carried out by a mainly Dutch group led by the geochemist, Douwe van der Meer, were

[78] Jones et al. (2016).

published in 2014.[79] Their estimates of degassing from volcanic arcs and mid-ocean ridges were based on developments in seismic tomography which are gradually improving the resolution with which the deep mantle can be imaged.[80] Sinking slabs at progressively greater depths should obviously represent older subduction events, providing a vertical history of subduction zones for the past 250 million years, even where no present-day trenches or volcanic arcs remain, such as within the giant ocean that encircled Pangea. Total subduction-zone length calculated in this way varied from about 42,400 km measured from the uppermost 200 km of the Earth (compared with an actual length of 48,800 km today) to a maximum of about 75,000 km at 2000 km depth (estimated to represent an age of about 170 million years) and then decreased to about 65,500 km at 2815 km depth (dated at about 235 million years). In other words, up to 75 per cent more ocean plate was being subducted in the past than today. Assuming a constant rate of sinking, Douwe *en vrienden* generated a curve showing changes in subduction rate with time, from which they estimated global ocean $^{87}Sr/^{86}Sr$ ratios using a model in which tectonic output (with low $^{87}Sr/^{86}Sr$ ratios) is mixed with river output (with high $^{87}Sr/^{86}Sr$ ratios), obtaining a reasonable fit for the past 250 million years. They then assumed that the rate of CO_2 release per kilometre of subducting slab has been constant through time, in order to generate a 'degassing parameter' which they used as input to Bob Berner's GEOCARBSULF model. The resulting global balance of CO_2 agreed far better with proxy data estimates than did Bob's standard model. It therefore seemed that total subduction rate was itself a reasonable first-order proxy for atmospheric CO_2 concentration, despite all the assumptions made.

Soon afterwards, another group, working at the University of Exeter, refined these estimates further.[81] Postdoc Benjamin Mills, together with Stuart Daines and Timothy Lenton, incorporated the degassing rates computed by van der Meer *en vrienden* into a development of GEOCARB constructed at the University of East Anglia and dubbed COPSE (Carbon–Oxygen–Phosphorus–Sulfur Evolution). Like Bob Berner, they separated basalt weathering from granite weathering, and also accounted separately for weathering on land and on the seafloor.

[79] Van der Meer et al. (2014).

[80] We will return to this topic in the next chapter.

[81] Mills et al. (2014).

By estimating the area of LIPs, such as the Siberian Traps, they were able to calculate the amount of CO_2 degassing produced by basalt weathering. This, together with CO_2 removal by weathering of oceanic basalts (hydrothermal alteration), reduced their net CO_2 concentration in line with proxy estimates. They found that sea-floor weathering of basalts contributed equally with continental silicate weathering prior to Pangea break-up, following which the latter became dominant.

Thus, it appeared that Thomas Chamberlin had been right in attributing fluctuations in atmospheric CO_2 concentration primarily to changes in the balance between supply by volcanism and removal by rock weathering. In 1991, nearly a century after Thomas' great insights, Maureen Raymo, who had just been appointed as Associate Professor (i.e. junior lecturer) in the Department of Geology and Geophysics at the University of California at Berkeley, acknowledged his neglected hypothesis in an article entitled 'Geochemical Evidence Supporting T. C. Chamberlin's Theory of Glaciation',[82] published in the journal *Geology*, Despite the failure of Thomas' contemporaries to appreciate its significance, his hypothesis was now destined to become a fundamental tenet of Earth science in the twenty-first century.

A major omission in DeConto and Pollard's work invoking CO_2 decline as the cause of Antarctic glaciation at the Eocene–Oligocene boundary, was any explanation of what might have caused the decline. Following publication of their article in 2003, there has been a series of articles arguing variously that the opening of Southern Ocean gateways and onset of the ACC was a dominant, secondary, or irrelevant factor in promoting the growth of large ice sheets on Antarctica. Recent modelling has suggested a way in which the two arguments may be reconciled.[83] In this simulation, the development of a deep Southern Ocean pathway for the ACC leads to strengthening of the Atlantic Meridional Overturning Circulation, thereby transporting more heat into the northern hemisphere and raising temperatures over landmasses there. This, in turn, leads to increased weathering of exposed silicate rocks, and hence an increased rate of removal of CO_2 from the atmosphere, sufficient to bring about global cooling and the onset of Antarctic glaciation. This explanation remains to be thoroughly tested, but does illustrate the potentially extreme climatic consequences of

[82] See Raymo (1991).

[83] Elsworth et al. (2017); Scher (2017).

fluctuations in what has been called the rock weathering thermostat operated by plate tectonics. As we will discover in Chapter 11 (if you will pardon the spoiler), the operation of Thomas Chamberlin's thermostat is about the only mechanism we know of that can maintain surface temperatures within the narrow range required by multicellular life in the presence of an evolving star.

Afterthought

In 2001, climatologist Jim Zachos and colleagues at the University of California and elsewhere published an important review article in *Science* entitled 'Trends, Rhythms, and Aberrations in Global Climate 65 Ma to Present', which, by early 2018, had collected more than 4500 citations.[84] As the title suggests, they classified climate changes under three headings. 'Trends' refers to gradual long-term changes like the global cooling that has occurred since the Eocene, as witnessed by oxygen isotopes. 'Rhythms' are the cycles produced by variations in the Earth's orbit. These have very regular periods and are now known to be responsible for the dozens of cycles of glacial advance and retreat since extensive ice sheets first formed on the northern continents, 2.6 million years ago. 'Aberrations' are abrupt changes such as the major short-term climatic jerks revealed by oxygen isotope studies around 33, 23, and 3.5 million years ago.

In summing up the causes of long-term climate change, I cannot improve upon the eloquence of Jim and his fellow researchers, and will therefore leave the last word to them: 'The orbitally related rhythms [...] oscillate about a climatic mean that is constantly drifting in response to gradual changes in Earth's major boundary conditions. These include continental geography and topography, oceanic gateway locations and bathymetry, and the concentrations of atmospheric greenhouse gases. These boundary conditions are controlled largely by plate tectonics, and thus tend to change gradually, and for the most part, unidirectionally, on million-year time scales. Some of the more consequential changes in boundary conditions over the last 65 My include: North Atlantic rift volcanism, opening and widening of the

[84] Zachos et al. (2001). For anyone unfamiliar with scientific publishing, the citation count for a paper is the number of times it has been referred to by later peer-reviewed articles.

two Antarctic gateways, Tasmanian and Drake Passages; collision of India with Asia and subsequent uplift of the Himalayas and Tibetan Plateau; uplift of Panama and closure of the Central American Seaway; and a sharp decline in pCO_2.[85] Each of these tectonically driven events triggered a major shift in the dynamics of the global climate system.'

[85] pCO_2 is shorthand for the concentration (or partial pressure) of carbon dioxide in the atmosphere.

10

Ups and Downs

> The hardest thing of all to see is what is really there
>
> J. A. Baker (1967)

Despite the dumbing-down of education in recent years, it would be unusual to find a ten-year-old who could not name the major continents on a map of the world. Yet how many adults have the faintest idea of the structures that exist *within* the Earth? Understandably, knowledge is limited by the fact that the Earth's interior is less accessible than the surface of Pluto, mapped in 2016 by the NASA *New Horizons* spacecraft. Indeed, Pluto, 7.5 billion kilometres from Earth, was discovered six years earlier than the similar-sized inner core of our planet. Fortunately, modern seismic techniques enable us to image the mantle right down to the core, while laboratory experiments simulating the pressures and temperatures at great depth, combined with computer modelling of mantle convection, help identify its mineral and chemical composition. The results are providing the most rapid advances in our understanding of how this planet works since the great revolution of the 1960s.

*

Inner Space

For most people, rocks are finite entities—that is, fragments, blocks, or outcrops—that have definite compositions and origins, for example limestone or granite. They also tend to be hard (rock hard) and cool (stone cold). Within the mantle, however, this concept of 'rock' can be misleading. If you could descend into the mantle, you would find yourself in a crystalline sea of silicate minerals. Should you extract a sample from anywhere in the uppermost 410 km, you would find, on your return to the surface, that you had in your possession a rather hot piece of peridotite, a dense rock containing greenish olivines and silvery pyroxenes. Samples collected from greater depths would look slightly different and feel a bit

The Tectonic Plates Are Moving! Roy Livermore, Oxford University Press (2018). © Roy Livermore.
DOI: 10.1093/oso/9780198717867.001.0001

heavier (and also somewhat hotter). Had you acquired your specimen below 520 km, for example, the olivines might have a bluish tinge rather than the greenish hues of the shallower sample. And if you took a lump sourced from deeper than 660 km to the nearest petrologist, he would be stumped for a name to apply to your mystery rock. Yet, if you analysed all these samples, you would find that their chemical compositions differed very little, with oxygen, silicon, magnesium, iron, aluminium, and calcium the most abundant elements in much the same proportions, along with a host of minor and trace elements. The point I am getting at is that the mantle has a fairly constant chemical composition, resulting in a small number of different minerals that vary in their structure and abundance according to pressure (i.e. depth) and temperature.

Back in 1940, the famous New Zealand seismologist Keith Bullen divided the Earth's interior into seven regions or shells (Figure 10.1), based on tables of the travel times of earthquake waves he had compiled with his former supervisor, our old friend Harold Jeffreys. The boundaries between regions were defined by changes in the rates at which the speed of body waves (P- and S-waves) changed with depth—that is, their velocity gradients. For example, rapid increases in both P- and S-wave velocity were observed in some depth intervals, while, in others, P-wave speeds actually declined, and S-waves sometimes disappeared altogether. Being essentially a mathematician, Keith classified these regions using the letters A to G, with A corresponding to the crust, B, C, and D representing layers of the mantle, and E, F, and G corresponding to the core. Like the numbering scheme for magnetic reversals, this strategy was a hostage to fortune and, in 1950, as new information accumulated, Keith was obliged to split layer D, representing the lower mantle (from the transition zone to the core), into two shells. Following his mathematical instincts, he naturally named these D′ and D″ (D-prime and D-double-prime), the latter representing the lowermost 200 km, within which there was evidence for a diminution in velocity gradients. There is no need to remember his scheme, however, since Keith's zones have long since fallen into disuse, having been superseded by standard Earth models based on much larger datasets—apart from D″, that is, which refused to die, and took on a new life in the next century. A new model of the variations in seismic velocities and densities with depth was published in 1981 by Adam Dziewonski (at Harvard University) and Don Anderson (at Caltech), which they dubbed the

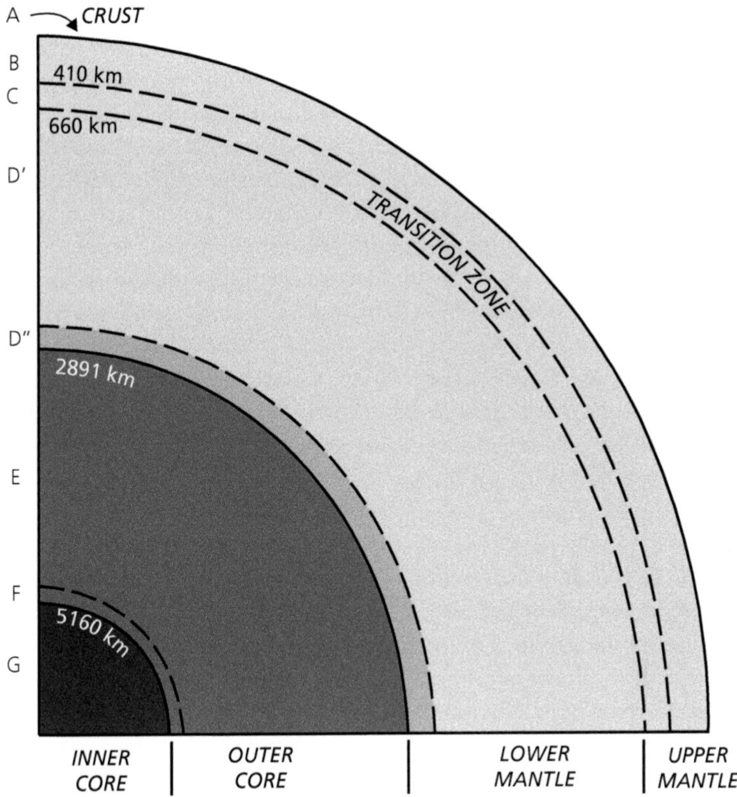

Figure 10.1 Keith Bullen's zones within the Earth. The lower mantle is divided into two zones, D′ and D″. Zone F does not exist in more recent models. Depths are current estimates.

'Preliminary Reference Earth Model', or 'PREM' for short. This has since become the standard reference model for much earthquake seismology and, in particular, for tomographic imaging[1] of mantle structure.

Middle Earth

A major focus of mantle research has been Keith Bullen's 'C region', now known as the 'transition zone', which lies between 410 and 660 km

[1] Explained later in this chapter.

depth. Indeed, one prominent mineralogist opined, as long ago as 1954, that 'the transitional layer is the key to what is going on in the mantle. When we understand its nature, we shall be well on the way to a grasp of the dynamics of the Earth's interior.'[2] Keith had previously detected[3] a rapid increase in the speed of both P- and S-waves between about 400 and 700 km depth, which Harold Jeffreys suggested might be caused, not by a change in composition, as at the Moho, but by a transformation in the crystal structure of olivine, under great pressure, to a denser configuration.[4] A second illustrious antipodean was responsible for confirming, during the 1960s and 1970s, that this was, indeed, the explanation, and that a whole series of such 'phase changes' occurred in the mantle. Probably the greatest Aussie geochemist of all time,[5] Ted Ringwood began his Earth Science career by carrying galena (lead ore) by packhorse from an abandoned silver mine in northeast Victoria and selling it in Melbourne for the production of lead shot.

During his years as Professor of Geophysics and, subsequently, Professor of Geochemistry at the Australian National University (ANU), Ted became the world's leading expert on mantle mineralogy, his success, like that of Joe Cann, most probably a consequence of learning some geophysics. One of his most enduring contributions was his estimate of the bulk chemical composition of the mantle. He noted that basalts form by partial melting of 'fertile' peridotite[6] in the upper mantle, leaving behind a composition depleted in low-melting-point and incompatible elements, such as calcium, sodium, potassium, uranium, and thorium. This depleted peridotite, known as 'harzburgite' (consisting mainly of the minerals olivine and orthopyroxene)[7] or dunite (consisting of more than 90 per cent olivine with minor pyroxene), thus forms the bulk of the plates beneath the basaltic oceanic crust. You could therefore estimate the composition of the primitive mantle prior to melting by adding together the compositions of basalt crust (the

[2] Birch (1954).

[3] Bullen (1938).

[4] An idea that Jeffreys had picked up from the famous Irish polymath John Bernal.

[5] I know what you're thinking, but Earth Science down under has become a force to be reckoned with.

[6] A mantle rock consisting of olivine, orthopyroxene, and clinopyroxene—named, you will remember, for the olivine it contains (see Chapter 1).

[7] By now, you have probably forgotten that harzburgite is named for the Harz Mountains in northern Germany.

melt) and the depleted mantle (the solid left behind). In 1962, Ted invented 'pyrolite' (an aggregation of 'pyro', from pyroxene, 'ol', from olivine, and the suffix 'ite', meaning rock or stone). Pyrolite was a theoretical composition intended to represent the chemistry of the upper mantle as a whole, produced by adding 4 parts of depleted mantle (dunite) to 1 part mid-ocean-ridge basalt. It should be emphasized here that the pyrolite model was concerned more with the concentrations of chemical elements than the minerals present, which would naturally vary with pressure and temperature. A number of later estimates of the bulk composition of the mantle using different assumptions gave results in agreement with Ted's, suggesting that both the upper and lower parts of the mantle had similar pyrolite compositions. This had major implications for mantle convection, and hence for plate tectonics.

But perhaps Ted's most important work was concerned with laboratory studies of the phase changes in mantle minerals brought about by increasing pressure and temperature. Ted and his colleagues at ANU were pioneers in establishing the mineralogical changes that might occur within the mantle transition zone. Their experiments focused primarily on olivine [$(Mg,Fe)_2SiO_4$],[8] as the most abundant mineral in the upper mantle, and garnet [$(Mg,Fe)_3Al_2(SiO_4)_3$], a magnesium–iron silicate containing significant amounts of aluminium. At a pressure of 14 GPa[9] (corresponding to a depth of 410 km), olivine, referred to as α-olivine, collapsed to a denser 'spinel' molecular structure known as β-olivine or 'wadsleyite'[10] (Figure 10.2), resulting in a sharp increase of about 5 per cent in P- and S-wave speeds. At a pressure of 17.5 GPa, corresponding to around 520 km depth, the wadsleyite collapsed to an even denser form known as—you guessed it—γ-olivine, later named in honour of Ted as 'ringwoodite', producing an increase in body wave speeds of around 1 per cent. Meanwhile, garnet, which increases in abundance with depth by the transformation of pyroxenes, and thus constitutes around 10 per cent of the uppermost transition zone, collapsed to a denser structure named after Ted's ANU colleague, Alan

[8] Chemists use the shorthand '(Mg,Fe)' to denote variable proportions of magnesium and iron in these minerals. The two elements form what is known as a 'solid solution', in which compositions may vary from 100 per cent iron to 100 per cent magnesium. Such variations affect the physical properties of the mineral.

[9] Gigapascals: 1 GPa is equivalent to 10,000 times atmospheric pressure at the surface of the Earth.

[10] Named after another antipodean geochemist, David Wadsley.

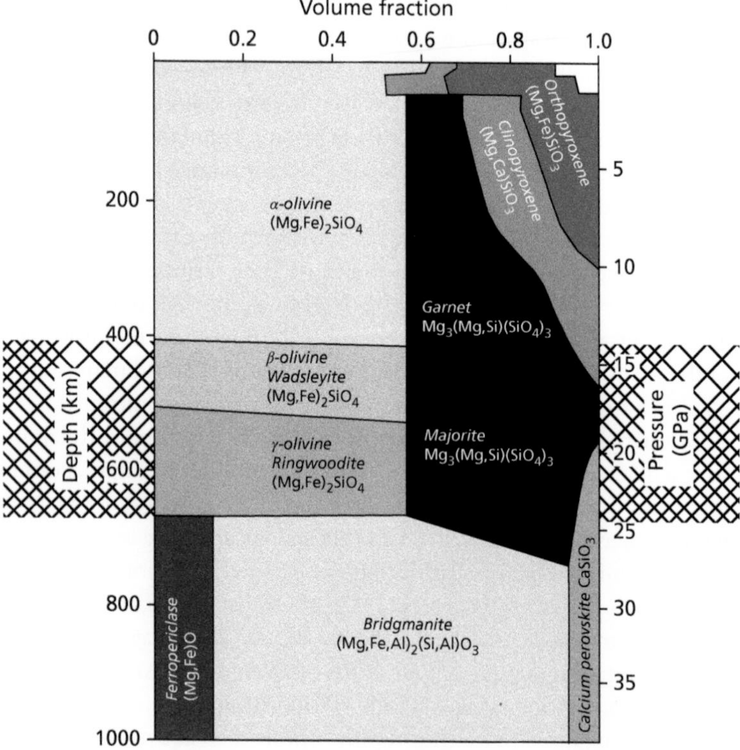

Figure 10.2 Mineral composition of the upper 1000 km of the mantle. The cross-hatched region is the mantle transition zone.

Major, namely 'majorite' [$Mg_3(Fe,Si,Al)_2(SiO_4)_3$], further increasing seismic wave speeds. Unlike the olivine transitions, the garnet phase change takes place gradually over a broad depth interval, resulting in a progressive increase in seismic wave speeds rather than a sharp discontinuity. Finally (at least as far as the transition zone is concerned), at 660 km and 24 GPa, ringwoodite breaks down abruptly to even denser mineral phases, producing a major jump in seismic wave speeds of around 10 per cent that marks the base of the transition zone.

Rock Bottom

In 2014, geochemists named a new mantle mineral. Or, rather, they gave a previously known mineral a new name. This was not, as you might think, some exotic crystal composed of rare elements, but what is very likely the most abundant mineral within the Earth. At the very high pressures of the 660 km boundary, ringwoodite and majorite undergo transformations to what, until 2014, had been known as magnesium silicate perovskite[11] or Mg-perovskite [$(Mg,Fe)SiO_3$], plus the oxide ferropericlase[12] [$(Mg,Fe)O$]. As we discovered in Chapter 5, mineralogists mix the names of specific minerals and their structures in cavalier fashion, and this was another example. Mantle perovskite is actually a magnesium–iron silicate with a structure that mimics that of true perovskite (which is calcium titanate [$CaTiO_3$]). Under the strict rules of the Commission on New Minerals and Mineral Names, no mineral can be named formally unless the structure and composition of a natural sample have been established. This presented a problem for mineral physicists hoping to name their lower-mantle perovskite mineral, since the nearest sample was at least 660 km beneath their feet. Fortunately, a stony meteorite blasted from an asteroid by some ancient impact provided them with the sample they needed, enabling them to finally name the mineral 'bridgmanite' (after Percy Bridgman, a Nobel Prize-winning Harvard physicist, and pioneer in the study of minerals at high pressure). The second most abundant mineral in the lower mantle, ferropericlase, has atoms arranged in a cubic structure, rather like rock salt, and forms rare inclusions in some diamonds. The precise proportions of bridgmanite and ferropericlase in the lower mantle are not well known, but recent mineralogical work[13] suggests that together they may constitute more than 93 per cent of the total by volume, with bridgmanite accounting for around 74 per cent and ferropericlase roughly 19 per cent. Making up the final 7 per cent or so of the deep mantle is believed to be calcium silicate perovskite (or Ca-perovskite [$CaSiO_3$]), a mineral with a structure similar to bridgmanite. Combined, these three minerals constitute more than half the volume of the Earth, yet, with the exception of ferropericlase, fewer people have ever seen them than have held rocks from Mars.[14]

[11] Named after Count Lev Aleksevich von Perovski, an early nineteenth-century Russian mineralogist. My apologies for all these terms, but just think how impressed your friends will be as you reel them off in the pub.

[12] Also known, incorrectly (since it contains more magnesium than iron), as magnesiowüstite.

[13] Murakami et al. (2012).

[14] Mars meteorites are rare (only around 100 have been found) and extremely valuable.

Ever since Keith Bullen was forced to revise his ill-fated scheme, there had been a suspicion that this was not the end of the story. That is, it seemed unlikely that bridgmanite (née Mg-perovskite), ferropericlase, and Ca-perovskite constituted the entire lower mantle all the way down to the core–mantle boundary at 2900 km depth. In 1983, Caltech seismologists[15] discovered a seismic discontinuity about 280 km above the core at which the S-wave speed increased by about 2.75 per cent. The discontinuity did not seem to be present everywhere and, where it existed, it appeared to undulate, varying in depth by around 40 km or so. Then, a couple of years later, seismologists at ANU discovered an abrupt jump in the speed of P-waves at the same depth.[16] The Caltech folk later suggested,[17] on the basis of their modelling, that the most likely cause of the discontinuity was a phase change comparable to those established for the transition zone, although precisely what this was they couldn't say.

In the winter of 2002, a dozen years prior to the naming of bridgmanite, Motohiko Murakami, a research student of Kei Hirose at the Tokyo Institute of Technology (TIT), dropped into his supervisor's office to give him some news. He had been using a synchrotron X-ray machine to determine the structure of crystals of Mg-perovskite (as it was then still known) at high pressures by observing their diffraction patterns. The patterns had changed abruptly at a pressure of 120 GPa, equivalent to a depth of 2600 km, suggesting a possible phase transformation. This was, to say the least, surprising, since it was widely believed that the perovskite structure was the densest possible for this chemical composition. Nevertheless, after checking the results at temperatures appropriate for this depth, Kei was convinced that they had discovered a new high-pressure silicate structure that might account for the presence of the Bullen D″ discontinuity at the bottom of the mantle. The following year, Kei's TIT colleague Katsuyuki Kawamura ran a computer simulation of the new high-pressure silicate in order to predict the diffraction pattern that would be expected. It matched that observed by Motohiko exactly. Their conclusions appeared in a *Science* article in May 2004, only just scooping a rival group, who published similar results in *Nature* two months later.[18] Since it had not been identified in nature, this new mineral

[15] Lay and Helmberger (1983).
[16] Wright et al. (1985).
[17] Sidorin et al. (1999).
[18] Oganov and Ono (2004).

could not be named officially, so Kei and colleagues gave it the rather mundane temporary label of 'post-perovskite', which should now be updated to 'post-bridgmanite'. This will have to do until someone finds a bit included within a diamond or meteorite.

A Riddle Wrapped in a Crystal Inside a Diamond

In 2014, a *Guardian* headline announced: 'Earth may have underground "ocean" three times that on surface', while the BBC asked 'Are there "oceans" hiding inside the Earth?' These, of course, are outrageous examples of the overselling that bedevils modern science: there is no 'ocean' of liquid water in the hot, high-pressure, interior, but there could be (somewhat less excitingly) an awful lot of hydrogen bound up in minerals. Just how much is of crucial importance not just for plate tectonics, but also for the existence of life.

These articles were referring to newly published research by a group of US scientists,[19] in which they combined high-pressure experiments on ringwoodite with recordings of P- and S-wave speeds in the transition zone[20] beneath North America, concluding that the mineral could harbour as much as 1 per cent H_2O (i.e. quite a lot) within its structure.[21] This, they believed, was evidence that the source of water filling the oceans was the Earth herself (rather than, say, incoming comets). In addition, the presence of water could induce partial melting, producing a small quantity of magma at the bottom of the transition zone at a depth of 660 km. Shortly before these results appeared in *Science*, Graham Pearson and colleagues at the University of Alberta published an equally eye-catching article in *Nature*,[22] describing their discovery of a tiny (about 40 μm) greenish crystal within a worthless diamond from Brazil. Using X-ray diffraction and other techniques, they showed that

[19] Schmandt et al. (2014).

[20] Not 'near the earth's core', as claimed by *Time* magazine. Ironically, the British headlines may be correct: it is believed that much more 'water'—perhaps more than all the surface oceans—resides in the *core* as a result of the solubility of hydrogen in iron, together with the huge volume of iron alloy present (the core is exactly twice the size of the Moon). Probably best not to mention this to the media, however. See Umemoto and Hirose (2015).

[21] In fact, this had been known for years (e.g. Richard et al., 2006).

[22] Pearson et al. (2014).

the crystal was ringwoodite—the first natural example ever to be observed, other than in meteorites. The diamond had been blasted up[23] from the transition zone, 550 km beneath, so rapidly that the ringwoodite had not had a chance to revert to olivine. Such eruptions produce the kimberlite[24] pipes that host most of the world's diamonds,[25] including the eponymous example at Kimberley, South Africa.[26] No-one has ever witnessed one of these violent eruptions, although our ancestors may have been around when the last one occurred about 100,000 years ago in what is now Tanzania.[27] As well as unveiling the first terrestrial ringwoodite crystal to an admiring world, Graham and his colleagues observed that their sample contained around 1.4 per cent water, reinforcing earlier suspicions that the transition zone was 'wet'.

It seems that what is really involved here is the substitution of two hydrogen ions (H^+) for magnesium (Mg^{2+}) or iron (Fe^{2+}), or four hydrogen ions for silicon (Si^{4+}) in the cubic molecular structure of ringwoodite, forming hydroxide ions (OH^-) as impurities, but, since geochemists are happy to call this water, we'll do the same. And, since both ringwoodite and wadsleyite are able to incorporate OH^- ions into their crystal structures to become 'hydrated', it could be that the transition zone contains a substantial quantity of 'water', forming a sort of invisible subterranean reservoir. On the other hand, these results may simply point to the *local* presence of a hydrous fluid that induced mantle melting (in a similar way to that which occurs beneath volcanic arcs), leading to the eruption of kimberlites. Another consequence of hydration is that the phase transitions that normally occur at 410 km

[23] Kimberlites rise through the upper mantle at up to 70 km per hour, but may emerge at ten times that speed.

[24] An ultramafic (enriched in magnesium and iron) igneous rock.

[25] Contrary to popular belief, diamonds are not formed by the compression of coal or graphite within the Earth. They form slowly from C-rich fluids circulating deep within continental plates. Pressures of at least 30 kbar and temperatures of at least 400°C are required, limiting their origin to depths greater than 150 km. Most diamonds are of Precambrian age (1–3 billion years) but transported in much younger (50–100 million years) kimberlite and related magmas. Kimberlites are found in all ancient cratons, including North America, South America, Africa, India, Australia, Greenland, and Antarctica (the latter only being discovered in 2013).

[26] Where mining of the pipe spawned the De Beers Consolidated Mines company and led to the excavation of a big hole, 200 m deep, known as 'the Big Hole'.

[27] Prior to that, the next youngest kimberlite occurs in Congo and is dated at 32 million years.

are accelerated, making the upper transition-zone boundary shallower, while those at 660 km are delayed, making the lower boundary deeper. The transition zone thus becomes thicker where minerals are hydrated, and this can be detected by earthquake seismology. A recent attempt to map the 410 and 660 km discontinuities[28] suggested that only isolated pockets of hydrated minerals exist within the transition zone, which therefore may contain rather less water than the surface oceans.

For many years, it was widely believed that, while the transition zone might (or might not) act as a sponge, soaking up oceans of water, the *lower* mantle was dry, the solubility of hydrogen in bridgmanite, ferropericlase, and calcium perovskite being very low. Yet basalts erupted at ocean islands, thought to result from melting of hot plumes rising from deep in the mantle, tended to have higher H_2O contents than mid-ocean-ridge basalts. So, where was the reservoir of hydrated mantle minerals that provided the source for these ocean island basalts? Back in 1967, Ted Ringwood and Alan Major reported some unsuccessful experiments designed to establish whether certain dense magnesium silicate compounds could be stable under mantle conditions.[29] They concluded that their equipment was not up to the job, but published their preliminary results anyway. The compounds included three new mineral phases that, adopting Keith Bullen's logic, they named 'phase A' [$Mg_7Si_2O_8(OH)_6$], 'phase B' [$Mg_{24}Si_8O_{38}(OH)_4$], and 'phase C' [$Mg_{10}SiO_3O_{14}(OH)_4$]. Other phases were soon discovered and given equally memorable names, such as 'phase D' [$Mg_2SiO_4(OH)_2$] and so on, leading to their description as 'alphabet phases'. The important thing about these phases is that they contain a lot of hydrogen in their molecular structure and so, if they exist, could conceivably act as camel humps, transporting 'water' into the parched lower-mantle desert, the camel being oceanic crust forming the upper part of subducting plates. Further research showed that phases C, D, and H [$MgSiO_2(OH)_2$] all appeared likely candidates for this job. Unfortunately, they all tended to break down at pressures and temperatures far short of those of the lower mantle.

Other dense hydrous silicates have more imaginative names. One aluminous phase in particular, known as 'phase Egg',[30] with the chemical

[28] Houser (2016).

[29] Ringwood and Major (1967).

[30] I kid you not: it was named for its discoverer, Tony Eggleton, another ANU mineralogist.

formula [$AlSiO_3(OH)$] was found to be sufficiently stable to survive at least down to the 660 km discontinuity, and might therefore, it was speculated, accompany plates sinking even deeper—but this was all theory. Theory, that is, until 2007, when a German group, led by Richard Wirth of the GeoForschungsZentrum (GFZ), in Potsdam, discovered nanometre-sized crystals of phase Egg included within another 'superdeep' Brazilian diamond, suggesting that it may have formed in the hydrous transition zone.[31] It appears that the presence of aluminium has the effect of stabilizing these hydrous phases, allowing them to survive in the lower mantle. Hence, the aluminium-rich version of phase D [$AlSiO_4(OH)_2$] may be the primary water carrier to depths of over 1000 km,[32] while, below 1250 km, phase D transforms to an aluminium-rich version of phase H, which could thus be the dominant hydrous phase in the lowermost mantle.

The effect of small quantities of hydrogen/water (and also of carbon dioxide) on the physical properties of the mantle is very noticeable. Incorporation of OH^- ions into the crystal lattice of mantle minerals such as ringwoodite increases volume, decreases density, reduces viscosity, and lowers the elastic moduli (a fancy way of saying that seismic waves travel more slowly). Hydrogen ions (protons) diffuse through silicate minerals creating 'point defects', which accommodate diffusion creep, while the movement of dislocations (dislocation creep) also becomes easier in 'wet' minerals, resulting in lower viscosity and more vigorous convection. Higher pressures limit the capacity of mantle minerals to incorporate hydrogen in the lowermost mantle, with the result that viscosity increases gradually towards the core–mantle boundary.

Putting a Spin on It

Iron, as 1980s Austin Metro owners will confirm, has an affinity for oxygen. On a planet like the Earth, which is nearly 30 per cent oxygen,[33] it therefore tends to occur as an oxide. Its chemical properties, like those of all elements, are largely dictated by the arrangement of its outermost

[31] Wirth et al. (2007).

[32] Nishi et al. (2014).

[33] Iron is the most abundant element, accounting for nearly a third of Earth's total mass, although (interesting fact) there are more than three times as many oxygen atoms as iron atoms.

electrons. Depending on the availability of oxygen, iron forms two different ions: ferric iron,[34] with five electrons in its outer shell (usually written as Fe^{3+}), and ferrous iron,[35] with six (usually written as Fe^{2+}). Ferric oxide [Fe_2O_3] is the compound that put a hole in resale values of the poor old Metro. It occurs naturally as the mineral haematite, while ferrous oxide [FeO] is found as the mineral known as wüstite.[36] Magnetite, you may remember, is the iron oxide mineral that stores the magnetic record of plate tectonics beneath the oceans. Its formula, Fe_3O_4, shows that the iron within is neither pure ferric nor pure ferrous, but a bit of each.[37] In the deep mantle, iron exists within both bridgmanite and ferropericlase, mainly in the ferrous form.

As long ago as 1960, yet another great antipodean, the New Zealand geochemist William Fyfe, then at the University of California at Berkeley, made a remarkable prediction concerning the behaviour of iron at the pressures and temperatures encountered in the deep mantle. He proposed that, under these conditions, ferrous iron might undergo a transition in which the atomic orbitals collapse from a 'high-spin' to a 'low-spin' state.[38] Leaving aside the question of what, exactly, the spin of an electron might be (even physicists don't seem able to tell us), the important point is that two electrons can share the same orbital,[39] but only if they have opposite spin. In such a case, the electrons are said to be 'paired' and their total energy is reduced. In ferrous iron, the six outermost electrons can occupy all of the five 3*d* orbitals, four of which contain just a single electron, while the two in the lowest orbital are paired. This is the 'high-spin' state. Alternatively, given the right conditions, they may all be paired within the three lowest-energy orbitals, leaving the outer orbitals empty, resulting in the 'low-spin' state.[40] The 'right conditions' may occur in the deep mantle, causing

[34] Known to politically correct types as 'iron (III)'.

[35] Or 'iron(II)' in modern textbooks.

[36] Named for Fritz Wüst, founding director of the Kaiser-Wilhelm-Institut für Eisenforschung, in Aachen.

[37] Indeed, as Wikipedia will tell you, its common name is 'ferric-ferrous oxide'.

[38] Fyfe (1960).

[39] For non-physicists, an orbital is defined in the Oxford Dictionary as 'Each of the actual or potential patterns of electron density which may be formed in an atom or molecule by one or more electrons, and can be represented as a wave function', which I doubt will help very much.

[40] If this sounds like Double Dutch, I can only recommend that you consult a good physics textbook.

the iron within bridgmanite and ferropericlase to switch from the high-spin to the low-spin state. As a result of this transition, Fyfe believed, the physical and chemical properties of the iron would change, and the volume of the ferrous ion could be reduced (and hence density increased) by as much as 45 per cent.

In the early twenty-first century, this idea was resurrected by French mineral physicists and studied using a diamond anvil placed in a synchrotron to measure X-ray scattering in ferropericlase subjected to mantle pressures.[41] The results confirmed the existence of an iron-spin crossover from high- to low-spin at pressures corresponding to a depth of around 2000 km. This demonstration excited great interest among mineral physicists, who rushed to perform similar experiments on pyrolite compositions. The upshot of all this activity was that ferropericlase, forming roughly 20 per cent of the lower mantle, was found to be affected more than bridgmanite by the transition, which was predicted to occur over a broad range of depths, starting at about 1500 km, rather than as a sharp boundary like the post-bridgmanite transition. The net effect of the crossover is a gradual increase of around 4 per cent in density, resulting in an increase in the buoyancy of plumes rising through the lower mantle, and a reduction in the ratio of P-wave speed to S-wave speed. It also appears to affect the affinity of iron for lower mantle minerals, resulting in its migration from bridgmanite to ferropericlase. This was therefore another factor to be taken into account in mantle convection models.

One Lump or Two?

The magnitude 8 earthquake that occurred on 30 May 2015 beneath Chichi-jima in the Japanese Bonin Island group (the 'father' island we visited in Chapter 6) caused significant disruption in Tokyo. About 200 people who had gone to see a Star Wars exhibit on the 52nd floor of the Roppongi Hills shopping complex were forced to climb down stairs as lifts stopped working, and, in the suburb of Saitama, a woman in her 70s sustained a minor head injury when a dinner plate fell from a cupboard. At the epicentre, on Chichi-jima itself, buildings shook, but no-one was hurt and, despite the magnitude of the quake, no tsunami warning was issued.

[41] Badro et al. (2003).

The Chichi-jima quake was very unusual. Of course, any magnitude 8 quake is unusual, there being on average just one per year. But this quake was exceptional in another way. When the USGS routinely recalculated the location of the quake (finding that its magnitude was, in fact, 7.9, the same as the Great Kanto Earthquake experienced by Kiyoo Wadati in Chapter 4), they discovered that the source had been at a depth of 664 km, making it the deepest quake ever recorded. This was the reason why the release of so much energy had resulted in so little harm (other than to the lady with the dinner plate). The hypocentre was associated with the subducting Pacific plate, although the quake was not, as you might expect, a 'megathrust' type, caused by the locking of the subducting Pacific slab against the Philippine plate to the west, but rather a consequence of slow distortion *within* the slab itself, as if its descent had met with some resistance. During the past century, just 66 large earthquakes with a magnitude of 7 or more have occurred at depths greater than 500 km, all of them associated with subducting slabs. The largest, the 2013 quake beneath the Kuril–Kamchatka arc to the north of Japan, had a magnitude of 8.3, with a hypocentre depth of 598 km, again involving the sinking edge of the Pacific plate. Although felt as far away as Moscow, Delhi, and Seattle, it likewise caused little serious damage. Thus, it appears that there is a maximum depth, somewhere around 660 km, beneath which earthquakes do not occur, suggesting a fundamental change in the properties of the mantle. It won't have escaped your notice that this depth corresponds to the bottom of the transition zone, now known to reflect abrupt changes in the structure of mantle minerals as a result of increasing pressure and temperature.

The existence of the 660 km boundary dividing the upper and lower mantle led to a pedantic controversy over whether convection occurs throughout the mantle ('whole-mantle convection') or in two distinct layers, above and below the boundary ('layered convection'). Actually, the argument was very far from pedantic, since it had a strong bearing on the question of 'what makes the plates go'.[42] It soon turned into a confrontation between geochemists, who were adamant that the transition marked an impervious wall, and geophysicists, who pointed to the evidence from theoretical studies and the developing field of seismic tomography,[43] suggesting that it was possible for convecting mantle to

[42] Davies (1977).
[43] See below.

pass through the 660 km boundary. At the root of all this was the question of whether the phase changes observed in the transition zone were purely a result of increasing pressure and temperature on a pyrolite composition, or whether they marked a change in chemical composition. It had previously been shown that, while it was possible for sinking slabs and rising plumes to pass through a purely thermal boundary, a change in composition could represent an impenetrable barrier, perhaps confining convection to the upper mantle. By 1977, most geophysicists were convinced that convective heat loss involved the entire mantle,[44] although one prominent seismologist, Don Anderson at Caltech, stood apart. Accepting that the 660 km boundary would not work as the limit of upper-mantle convection, he argued that another boundary existed at a depth of around 1000 km, beneath which the mantle was stagnant.[45]

To their credit, the geochemists have still not given in, despite the mounting evidence for mantle-wide convection. They point to the differences in the composition of mid-ocean ridge basalt, which they refer to as 'MORB', and basalt erupted on oceanic islands like Hawaii, which they refer to as 'OIB'. While MORB contains low concentrations of incompatible elements (i.e. elements that do not fit easily into common mantle minerals and therefore go into the melt) such as rubidium, strontium, and the rare earths, as well as the heat-producing elements uranium, thorium, and potassium, OIB is enriched in these same elements. Geochemists explain this by noting that MORB results from mantle melting at shallow depths beneath mid-ocean ridges, while OIB is produced by melting of plumes from the deep mantle. Hence, they believe, a 'reservoir' of 'fertile' mantle rock exists at great depth, isolated from the upper mantle by the 660 km boundary, and providing the source for mantle plumes. If such a reservoir has existed since the Earth's formation, it would also provide an explanation for the observed enrichment in the isotope ^{3}He observed within OIB minerals, since all ^{3}He is believed to have been incorporated during Earth accretion.[46] One way to test these ideas is to image the mantle at higher resolution using

[44] Davies (1977).

[45] Wen and Anderson (1997).

[46] The concentration of ^{3}He in MORB is reduced because this isotope was lost to space during previous eruptions, showing that the mantle source for MORB has been recycled.

seismic tomography to see if there is evidence of slabs or plumes passing right through the transition zone, thereby linking circulation in the upper and lower mantle. In order to appreciate this technique, which dominates much of mantle research today, we need to pause and look more closely at just what is involved.

Thinking Inside the Box

In 1967, while rambling in the London countryside, Godfrey Hounsfield was pondering the problem of how to determine the contents of a sealed box. This may sound like an odd thing to be thinking about when surrounded by the wonders of nature, but it's a good thing Godfrey had nothing better to contemplate. His idea was that, by interrogating the box with numerous crossing X-rays, you could build up a picture of what was inside. At that time, first-generation plate tectonics was reaching its zenith and the Beatles were already, in their own words, 'more popular than Jesus'—and that too, was a good thing. For Godfrey was an electrical engineer working in the 'E' division of the famous British 'Electric and Musical Industries' company (EMI), and, thanks to the enormous revenues achieved by the Liverpool pop group for the 'M' part of the empire on its Parlophone record label, Godfrey was allowed to develop his ideas in the basement of a small Wimbledon hospital. The result was the world's first 'computed tomography' (CT) scanner, which he used to distinguish grey matter from white matter in a preserved human brain. The ability of the new device to discriminate between different kinds of soft tissue inside the sealed box of the skull led to dramatic advances in radiology and a Nobel Prize and knighthood for Godfrey.[47]

Geophysicists soon adopted a similar approach to imaging the interior of the Earth. X-rays would, of course, be somewhat impractical, but there were other 'rays' that would do as well, rays that were generated by the Earth herself—seismic rays. You may not have thought of seismic waves as rays, but then you might not have considered X-rays as waves. Rays are fictitious concepts defined as points on a wavefront that move with a wave, and their function is to simplify the mathematics of

[47] Sadly, despite the head start provided by this success, EMI failed to maintain its advantage and, even before Hounsfield received his Prize in 1979, was brought to its knees by competitors, to whom it finally sold its interest in medical scanners.

travelling waves. Seismic waves are elastic waves that pass though solid rock and, in the case of P-waves,[48] through liquids, too. Like X-rays, they suffer attenuation, that is, they gradually lose energy and die out as they travel (otherwise they would carry on forever), but, while CT scans depend on differences in the *absorption* of X-rays, seismic tomography commonly depends on differences in seismic *wave speed*, which is determined by the properties of the rocks through which the waves pass. The computation of images uses much the same approach in both, however. So, just as X-rays allow us to see inside people, seismic waves allow us to see inside planets. There were no Nobel Prizes for the seismologists involved because, as we discovered in Chapter 1, Nobel Prizes are not given to Earth scientists. Nevertheless, like Vine and Matthews, they had the satisfaction of opening up a whole new branch of science.

Seismic tomography began in the 1970s, during the first generation of plate tectonics, but only arrived at the forefront of geophysics research in the 1990s, driven by improvements in data quality and coverage, in tandem with rapid increases in the speed and memory of computers. Techniques continue to evolve in the twenty-first century, and we are now beginning to see consistent pictures of structures in the lower mantle, particularly in the region just above the core. Nevertheless, it is important to remember the limitations of the technique: what we would really like to image are variations in physical properties such as density, chemical composition, temperature and viscosity. What we are able to image are variations in seismic wave speed and attenuation. Consequently, there is a need for additional information from mineral physics and numerical modelling of mantle convection in order to interpret tomographic imagery.

Ultimately, image resolution is limited by the number of earthquake sources and detectors—the closer the spacing, the sharper the image. Wide gaps in the arrangement of surface sensors can result in small features being missed altogether. In this respect, seismic tomography has one big drawback compared with medical applications of CT scanners. For, whereas the radiologist can determine where X-ray sources and receivers should be placed around his patient to focus on the area of interest, the seismologist has no control over the distribution of the

[48] P-waves, as you will remember from Chapter 1, are compressional waves, like sound waves in air.

sources of earthquake waves and only limited influence on the siting of receivers. Earthquake locations are, of course, dictated by Mother Earth herself, being concentrated along plate boundaries.[49] The majority of events therefore tend to occur in zones affected by previous earthquakes, rather than in the large unsampled regions that seismologists would really like. And, for obvious reasons, permanent receivers tend to be on land, clustering around populated coastal regions near subduction zones. The basis of global studies is the IRIS Global Seismographic Network[50] of 150+ digital broadband seismic recording stations. In the USA, this has been supplemented by the USArray, a set of 400 transportable seismic stations deployed across North America with the aim of imaging the mantle at higher resolution.

In essence, seismic tomography involves the calculation of a 'model' of the physical properties of the mantle that could give rise to the observed data, typically the time taken for earthquake waves to arrive at a set of receivers, or the actual shapes and sizes of the waves recorded. I say 'could give rise' because, owing to the limited distribution of data together with errors of observation, it is possible that more than one model could fit the data equally well. Geophysicists refer to this as 'inverse theory', since the data—resulting from the convolution of earthquake waves with the Earth—are used to establish the properties of the rocks through which the waves have travelled. The resulting tomographic images are typically shown as velocity anomalies a few per cent higher or lower than normal—'normal' being a reference model (such as PREM) of averaged seismic velocities within the Earth. Body waves—that is, P-waves and S-waves—are the bread and butter of seismologists studying the deep mantle. While both contribute valuable information on the Earth's interior, S-waves (shear or 'shake' waves) have proven particularly useful in revealing major and minor heterogeneities in the lower mantle. The propagation of a shear wave can be visualized by watching the locomotion of a snake,[51] which advances by passing a series of waves down its body: the movement is sideways, but the snake moves forward. Likewise, in crystalline rocks, shear waves propagate by tiny sideways elastic movements within mineral crystals perpendicular to the direction of wave propagation. At the higher pressures of the lower mantle, particles are

[49] Although, on the plus side, Mother Earth also provides the energy source for free.

[50] IRIS stands for 'Incorporated Research Institutions for Seismology'.

[51] Or, if a snake is not handy, try shaking a Slinky.

forced closer together and sideways motion is restricted, resulting in wave energy being propagated faster.

One of the first attempts to apply tomography on a global scale was by Adam Dziewonski and colleagues in 1977.[52] Their 'feasibility study' used the limited global data set then available to compute seismic velocity anomalies within five concentric mantle shells down to the core–mantle boundary, revealing a pattern of S-wave variations in the lowermost mantle that was rather like that of a pool ball: that is, two antipodal low-velocity regions separated by a globe-encircling band of higher velocity.[53] As the quality and volume of earthquake data increased, several other groups constructed tomographic models of the global mantle, hoping to image features such as sinking plates and rising plumes, and to explore structures in the lower mantle.

Slab Graveyards

Even before the advent of seismic tomography, Thomas Jordan, at Princeton University, detected what he interpreted as 'descending material in the convecting mantle', producing anomalously high P- and S-wave speeds beneath the Caribbean.[54] This material may have penetrated to 1400 km or even deeper—more than twice the depth of the transition zone. His startling interpretation implied that the subducted slab of the lost Farallon plate[55] was still intact within the deep mantle.

Two decades later, S-wave tomography presented by Stephen Grand, at the University of Texas, imaged the Farallon slab further north.[56] Incredibly, the slab sloped down to the east right under North America, passing about 1500 km beneath New York. Even more incredibly, the old plate seemed to reach almost to the core–mantle boundary, nearly 2900 km below the surface. Soon, other obsolete plates began to be imaged in the deep mantle, including a slab that had formerly formed the floor of the ancient Tethys Ocean, now discovered lurking beneath Eurasia, as well as subterranean curtains of former Pacific plates beneath the present-day Aleutian and Tongan subduction zones. Recent

[52] Dziewonski et al. (1977).

[53] For anyone unfamiliar with American billiards, pool balls carry two opposing white circles, each containing a number, contrasting with the colour of the rest of the ball.

[54] Jordan (1974).

[55] See Chapter 1.

[56] Grand (1994).

improvements in seismic data coverage and in tomographic techniques have revealed other ancient slabs, such as the one extending down to the core–mantle boundary beneath the Kerguelen Plateau in the Indian Ocean.[57] All these old plates have passed right through the transition zone, demonstrating that the cool, downward limb of convection involves the lower mantle too. In some cases, such as the Middle America Trench (Figure 10.3) and the southern IBM arc, the slab seems to penetrate the transition zone without being deflected, while in many others, for example, the well-studied Japan subduction zone, the descending plate appears to baulk at the 660 km boundary, where it stagnates and thickens, before resuming its journey to the centre of the Earth.[58] Such variations may be due to heterogeneities in the 660 km boundary, or to differences in the sinking slab from place to place. On the basis of these results, a hybrid model of mantle convection, involving both whole-mantle and layered overturning, has been suggested.[59] In this model, slabs that fail to pass beneath the 660 km boundary are cycled in an upper mantle cell, while those that reach the lower mantle form part of a mantle-wide circulation.

The latest laboratory methods allow mineral physicists to investigate not just the composition, but also the creep behaviour of minerals under lower mantle pressures and temperatures.[60] Bridgmanite, it appears, is both stronger and more viscous than ferropericlase, and thus provides the majority of the resistance to slabs sinking through the 660 km boundary. Most recent studies suggest that the rate of sinking through the lower mantle is around 15 km per million years—somewhat slower than in the less-viscous upper mantle—so that a plate, having made it through the transition zone, will take around 150 million years to reach the core–mantle boundary. It is thus a very remarkable fact that the hulks of plates that entered subduction zones in the Cretaceous, when they were already tens or even hundreds of millions of years old, are still more or less intact today and can be imaged almost 3000 km beneath the surface. These discoveries were, however, very far from the end of the story regarding deep-mantle structure.

Geophysicists had suspected for years that the deepest (D″) layer just above the core was much less homogeneous than implied by Keith

[57] Simmons et al. (2015).
[58] Fukao (2015).
[59] See Chen (2016).
[60] Girard et al. (2016).

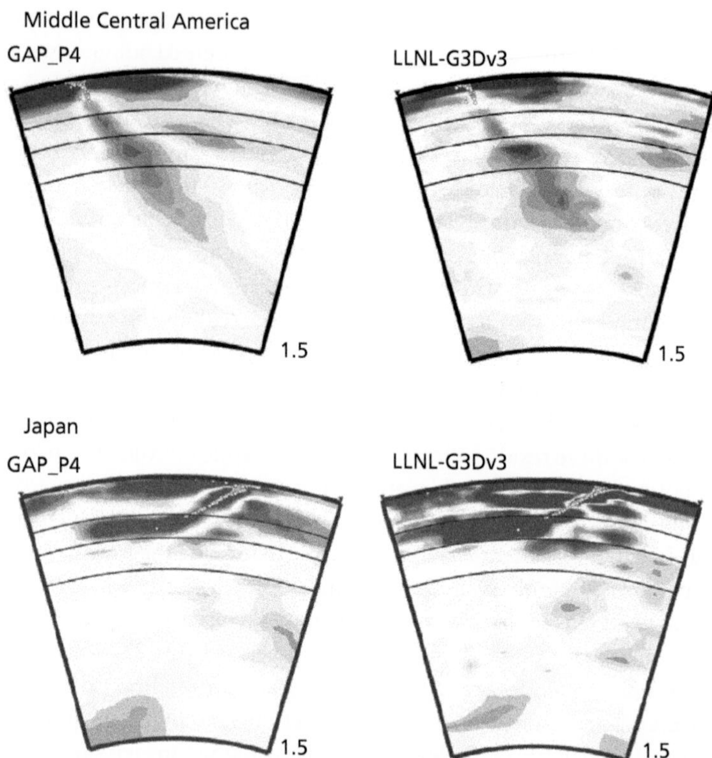

Figure 10.3 P-wave tomographic images across the Central America Trench showing 'fast' regions (dark) apparently penetrating directly into the lower mantle (top two sections), and across the Japan Trench, where slabs flatten out in the transition zone (bottom two sections). The concentric curves on each section represent the top and bottom of the transition zone at 410 and 660 km, with a third curve at a depth of 1000 km. Smearing of the high-velocity slab in the lower mantle is partly a result of a reduction in resolution at depth. [After Fukao and Obayashi (2015).]

Bullen's scheme. As far back as 1979, Clement Chase (whom we met previously) made some remarkable observations based on an estimate of the geoid, the sea-level surface of the Earth.[61] Plotting the deviation of the geoid from the shape expected of a homogeneous rotating planet, Clem noticed two things. Firstly, geoid anomalies clearly reflected the

[61] Chase (1979).

presence of subduction zones, which were associated with geoid highs of up to 100 m, but did not reflect other plate boundaries. Secondly, after removing the effects of the subduction zones, his 'residual' geoid looked rather like Adam Dziewonski's pool ball, with two large positive anomalies, each covering more than 10 per cent of the Earth's surface, in which the geoid was up to 80 m higher than expected. The two highs were almost antipodal: one centred over the western Pacific, the other covering much of Africa. Noting a tendency for surface hotspots such as Hawaii to lie within these highs, and assuming that they resulted from plumes rising from deep, unusually hot, regions, he concluded that the only credible explanation was that the geoid highs resulted from density variations in the lower mantle, caused by hotter-than-normal rock involved in deep convection. Residual geoid lows, on the other hand, presumably reflected lower temperatures in the deep mantle. In other words, the broad undulations of the Earth's surface resulted from temperature contrasts associated with the slow circulation of the mantle thousands of kilometres beneath.

A couple of years later, in 1981, Clem published another important article, this time looking at the geochemical difference between mid-ocean ridge basalt (MORB) and ocean island basalt (OIB).[62] Noting that many OIBs contain lead isotopes apparently indicating ages of formation millions of years in the *future* (pointing to enrichment of the mantle source in uranium, the parent isotope of lead), he suggested that a likely explanation for the enrichment was partial melting of the mantle, just like that occurring beneath mid-ocean ridges. In fact, Clem proposed, the melting did indeed occur beneath ancient mid-ocean ridges. He argued that the observed isotope ratios could be explained by the subduction of ancient oceanic crust deep into the mantle, where it subsequently (after a billion years or so) formed the source for Jason Morgan's mantle plumes that then rose to the surface, forming oceanic islands and plateaux. By one of those curious coincidences, Albrecht Hofmann and William White, at the Carnegie Institution of Washington, arrived simultaneously at more or less the same conclusion, but by a slightly different route.[63] They proposed that the oceanic crust forming the upper parts of subducted plates converted to denser eclogite as a result of pressure-induced phase changes, and thus became negatively

[62] Chase (1981).

[63] Hofmann and White (1982).

buoyant throughout the mantle, sinking all the way to the core–mantle boundary, where it formed an 'irregular layer of "degenerate crust" '. This basaltic layer would gradually thicken and its temperature would be raised by radioactive decay as well as by heat escaping from the core immediately below. Eventually, the degenerate crustal material would become buoyant and begin to rise towards the surface.

Soon after, Caltech seismologist Don Anderson published his famous paper invoking continental insulation as a cause of supercontinent break-up (see Chapter 7). In the same article, entitled 'Hotspots, Polar Wander, Mesozoic Convection, and the Geoid',[64] Don followed Clem in noting the strong correspondence between hotspots and the large geoid highs over the African and Pacific plates. In addition, he found that the extensive Cretaceous volcanism on the Pacific plate (also visited in Chapter 7), when reconstructed in the 'fixed-hotspots' frame of Jason Morgan, relocated to the region of present-day hotspots in the Pacific and hence also appeared to have erupted at the Pacific geoid high. Pushing back even further in time, Don postulated that basalts as old as 300 million years in North America and Eurasia might, likewise, have formed as these regions passed over the Pacific high, suggesting that whatever it was in the deep mantle that caused hotspots had been there for a very long time. Don also showed that Pangea, when reconstructed in the hot-spots frame, fitted neatly over the African geoid high about 250 million years ago. This, of course, supported his notion of 'thermal blanketing' of the mantle leading to the break-up of Pangea.

At about the same time, Clem Chase made a global plate reconstruction for 125 million years ago, assuming fixed hotspots, again demonstrating the good agreement between the location of Pangea (at that time beginning to fragment) and the African geoid high.[65] His reconstruction also showed good agreement between the estimated locations of ancient subduction zones and the major geoid low that separated the Pacific and African highs. This suggested to Clem that structures in the lower mantle reflected Cretaceous rather than present-day plate tectonics, implying a delay of 100 million years or more between changes in surface plate tectonics and the deep mantle. This was not surprising, since plates are able to change their configuration and motion rapidly (on a geological timescale), while mantle circulation

[64] Anderson (1982).
[65] Chase and Sprowl (1982).

takes tens of millions of years to catch up. Evidence thus pointed to a strong, but delayed, link between the surface pattern of plate tectonics and the major geoid anomalies, which most probably reflected deep-mantle structures.

In 1985, Brad Hager and colleagues at Caltech showed that the major geoid anomalies could be reproduced using a computer model of mantle properties derived from Adam Dziewonski's seismic tomography, coupled with simple assumptions about the rheology of the mantle.[66] Their convection model demonstrated that, while hotter bodies in the lower mantle might be expected to be less dense than their surroundings, mantle convection induced by the temperature differences caused the surface of the Earth to bulge,[67] resulting in positive rather than negative geoid and gravity anomalies. Thus, geoid lows reflected cooler regions in the lower mantle, while highs resulted from warmer regions that could be the sources of rising mantle plumes. These contrasts, they suggested, resulted from thermal convection driven by subduction and involving the entire mantle. That is, sinking plates in the lower mantle were related to cool (relatively speaking) regions in the deep mantle, while the intervening hotter regions were the sources of rising plumes.

Piles on the Bottom

So far, everyone had assumed that these effects were consequences of temperature variations in a mantle of uniform composition, but, the following year, Kenneth Creager and Thomas Jordan at the Massachusetts Institute of Technology (MIT), reported results[68] from seismic tomography revealing variations in P-wave speed at the core–mantle boundary that they believed were too large to be explained solely by temperature variations within a homogeneous Bullen D″ layer. They hypothesized that these variations resulted from 'large-scale accumulations on the surface of the core' that 'may in some ways be analogous to continents at the earth's surface'. Soon after, Geoff Davies and Mike Gurnis, at ANU, ran computer models looking at the effect of a thin, dense, layer of chemically distinct material at the core–mantle boundary—what they referred to as 'dregs'—on deep-mantle

66 Hager et al. (1985).
67 Known as 'dynamic topography'.
68 Creager and Jordan (1986).

convection.[69] They found that convection would sweep the dense layer into discrete 'piles', as observed by the seismologists. These piles would then modulate heat flow out of the core, affecting the geodynamo and hence the frequency of magnetic reversals. Ulrich Hansen (University of Cologne) and David Yuen (University of Minnesota) also incorporated 'thermal–chemical heterogeneities' in their own computer models of mantle convection.[70] Over time, these heterogeneities would tend to stagnate, and could remain in existence for billions of years as long as the material in them was replenished occasionally. This might occur, as Clem Chase suggested, by the arrival of subducted slabs, or, alternatively, by reaction with the hot outer core, both of which would increase their iron content, thereby raising their density[71] and causing them to remain at the bottom of the mantle.

Perhaps the most widely cited tomographic imagery today is derived from the Caltech global mantle model, known as S20RTS, and its successor, S40RTS.[72] These models were constructed by inverting[73] global S-wave travel times, together with Rayleigh waves and 'normal modes' (vibrations of the whole Earth excited by large earthquakes), to obtain variations in S-wave speed throughout the mantle. Regions in which S-wave velocity was slightly lower or higher than predicted by PREM were then mapped in red or blue, respectively (suggesting higher or lower than 'normal' temperatures). Sensing the growing importance of the subject, a number of groups have produced their own tomographic models of the mantle during the past decade using a variety of datasets, approximations, and assumptions. Despite the differences in approach, they all show two giant regions or 'piles' of anomalously low shear-wave speed (2–3 per cent slower than average) in the lowermost mantle beneath Africa and the Pacific Ocean (Figure 10.4), confirming the results of Creager and Jordan. These two piles, together occupying roughly 30 per cent of the core–mantle boundary and extending vertically to heights of 1000 km or so, were given the clumsy and unpronounceable

[69] Davies and Gurnis (1986).

[70] Hansen and Yuen (1988).

[71] Iron is by far the densest major element in the mantle, having a density 4.5 times that of magnesium.

[72] Ritsema et al. (1999); Ritsema et al. (2011). The 'S' stands for 'S-waves'. For those into spherical harmonics, '20', '40', and '12' refer to the degree of the model.

[73] That is, finding a best-fit model.

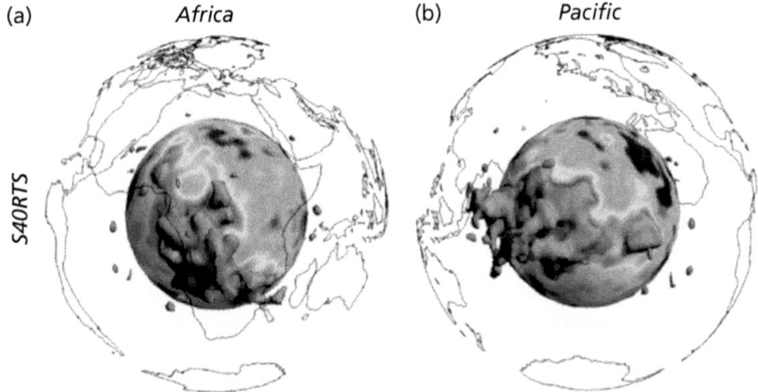

Figure 10.4 Three-dimensional perspective view of LLSVPs as defined by S-wave tomography (model S40RTS: Ritsema et al., 2011).

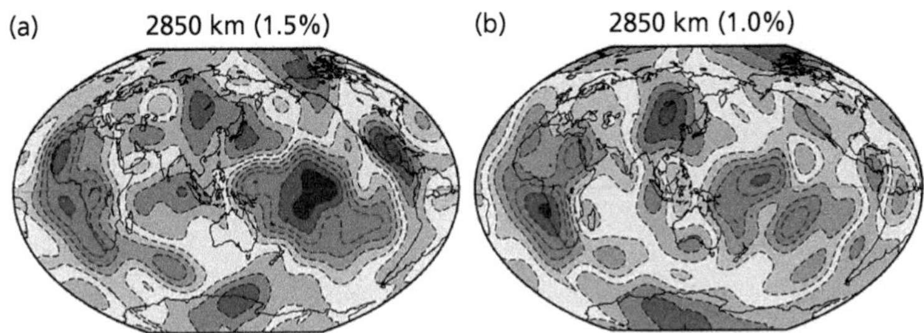

Figure 10.5 Lower-mantle slices through the SP12RTS tomographic model of Koelemeijer et al. (2016). (a) S-wave velocity anomalies. (b) P-wave velocity anomalies. Dark areas beneath the Pacific and Africa are low-velocity anomalies interpreted as piles of anomalous material at the core–mantle boundary (LLSVPs).

acronym 'LLSVPs' (i.e. 'large low-shear-velocity provinces') by Thorne Lay in 2007.[74]

A recent development of the S20RTS and S40RTS models, dubbed SP12RTS[75] (Figure 10.5), confirms the presence and locations of the two LLSVPs and demonstrates that they are also characterized by anomalously low P-wave speeds (although not quite as low as the S-wave anomalies).

[74] Lay (2015). They have also been referred to, predictably, as 'superplumes'.
[75] Koelemeijer et al. (2016).

The acronym could therefore be shortened to the only slightly less clumsy 'LLVPs' ('large low-velocity provinces'). However, to avoid any more confusion, I suggest retaining LLSVP for the time being, since it has now become entrenched in the academic literature, and simply altering the definition to 'large low seismic velocity provinces'.

From a careful analysis of S-waves traversing the African low-velocity zone, Mike Gurnis (now at Caltech) and his group showed that the eastern side of the pile was defined by a sharp and steep boundary[76] less than 50 km in width, at which S-wave speed dropped by 3 per cent.[77] If the pile were simply hotter than normal, its boundaries would be gradational: sharp boundaries suggested that it was, like the crust, composed of somewhat different material to the surrounding mantle. This was confirmed by numerical modelling of the recorded S-waves, incorporating a rising, buoyant, plume with chemically distinct composition and slightly increased density.[78] Similarly sharp boundaries were later demonstrated for the Pacific pile, too, reinforcing evidence from P- and S-wave anomalies (speed and amplitudes) in the two piles, and from studies of normal modes,[79] all pointing to a higher density in LLSVPs. Several groups developing similar models noticed a strong correlation between these low S-wave velocity regions and the locations of surface hotspots.[80] Moreover, it appeared that the sources of the plumes feeding the hotspots (assuming they originated in the lower mantle and rose vertically) tended to lie not within the lowest-velocity regions, as you might expect, but over the sharp boundaries of the LLSVPs.[81]

This was confirmed in 2004, when geophysicists in Arizona and Texas[82] observed that active hotspots were twice as likely to overlie the exceptionally high lateral gradients in S-wave speeds that bound the African and Pacific piles. But what about ancient hotspot eruptions, such as those that produced large igneous provinces (LIPs) like the Deccan Traps? The problem with this, of course, is that LIPs have been shunted around by plate tectonics. So, while active hotspots such as

[76] Ni et al. (2002).

[77] They also noted that both the eastern and western boundaries seemed to slope or lean to the east with height.

[78] The effects of temperature overcame those of density to produce the buoyancy.

[79] Ishii and Tromp (1999).

[80] The earliest of which may be Masters et al. (2000).

[81] Castle et al. (2000).

[82] Thorne et al. (2004).

Iceland and Hawaii may still lie above the sources of their heat, older, extinct LIPs need to be returned to their original locations before we can begin to search the deep mantle for their origins. Eruption of a LIP is frequently followed by tens of millions of years of reduced activity as the plume tail continues to supply magma to the surface, leaving a hotspot trail that, like footprints in the snow, connects the crime to the criminal, which in many cases is a still-active volcano fed by the same plume. The obvious example is, of course, Hawaii, but there are several others, such as the Chagos Ridge that links the Deccan Traps with the young volcanic island of Réunion in the Indian Ocean, or the Walvis Ridge, linked to the Tristan hotspot in the Atlantic. But even where there is no trail, perhaps because the hotspot track has been subducted, or because the flood basalts erupted on continental crust, LIPs younger than 180 million years old can be restored using ocean-floor data (magnetic anomalies and fracture zones). Thus, if long-lived plumes originate from fixed regions in the mantle, the reconstructed positions of LIPs should give a good idea of their whereabouts.

Our old friend Kevin Burke, who previously (Chapter 6) had a hand in converting recalcitrant geologists to plate tectonics, did just this, plotting LIPs on plate reconstructions corresponding to the time of their eruption. The LIPs not only correlated with LLSVPs, but, like present-day hotspots, plotted directly over the *edges* of piles.[83] Together with co-author Trond Torsvik, Kevin subsequently published a series of papers in which he developed this idea, sprinkling acronyms with abandon. The margins of the LLSVPs became 'plume generation zones' (PGZs), from which pipes of hot, buoyant, rock rose towards the surface *à la* Jason Morgan, while LIPs were classified as OPs (oceanic plateaus) and CLIPs (continental large igneous provinces). Kevin and Trond also interpreted Late Carboniferous volcanic rocks in northwest Europe as a LIP,[84] which they reconstructed to lie directly above the margins of the African LLSVP, 300 million years ago, suggesting that this pile had hardly moved since the formation of Pangea. They also showed that 80 per cent of kimberlites, when reconstructed like the LIPs, overlie the margins of the African pile.[85] By 2014, Kevin and Trond had reconstructed LIPs and kimberlites as old as early Paleozoic (540 million

[83] Burke and Torsvik (2004).

[84] Christened 'SCLIP' (i.e. Skagerrak-centred LIP) by Kevin.

[85] Torsvik et al. (2010).

years), finding that these likewise tended to cluster above the PGZs around LLSVPs. Finding the label LLSVP cumbersome, Kevin then came up with the bright idea of naming each of the piles after one of the plume pioneers. Thus, the pile beneath Africa was christened 'TUZO', while that beneath the Pacific became 'JASON'. He even contrived definitions to fit (more or less) these new acronyms: TUZO, **T**he **U**nmoved **Z**one **O**f Earth's deep mantle, and JASON, **J**ust **A**s **S**table **ON** the opposite meridian.[86] If only other geologists were as imaginative as Kevin.

One slight problem with their hypothesis is the assumption that plumes rise vertically from the plume-generation zones to the surface, whereas there is good evidence from tomography that this is not always so. For example, S-wave tomography beneath the North Atlantic shows the plume beneath Iceland sloping down to the east.[87] If correct, this interpretation means that a hot mantle plume exists 1000 km beneath London (Figure 10.6). Another problem is that the largest of all LIPs, the Siberian Traps, lies far from the two large piles on any reconstruction. Recently, however, a third, miniature, pile, around 1000 km across and rising 500 km into the lower mantle, has been identified beneath Russia and given the name 'PERM' (which is not an acronym, referring simply to its location 2500 km below the city of Perm). Despite its small size,

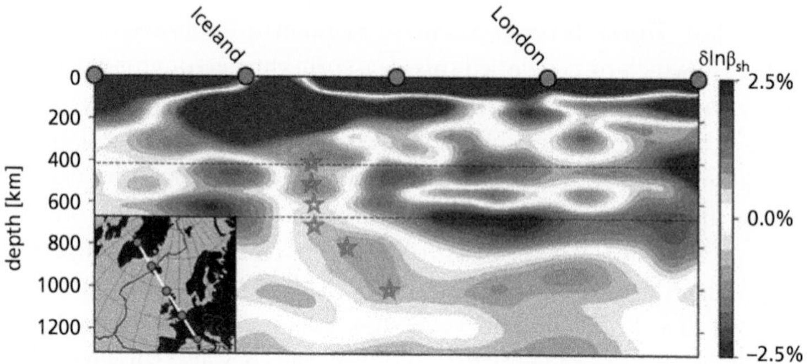

Figure 10.6 Slice through the S-wave tomographic model of Rickers et al. (2015) beneath the North Atlantic. Note the low-velocity region at around 1000 km beneath London that appears to rise underneath Iceland.

86 These acronyms may not, however, be appropriate if it turns out that the two piles have not been around for as long as Kevin and Trond suppose.

87 Rickers et al. (2013).

PERM appears to share the seismic characteristics of its big brothers,[88] and so is classed as an LSVP (low seismic velocity province). The one remaining LIP that does not fit the hypothesis is the Columbia River Basalt Group, the youngest LIP by far, which appears to overlie a region of deep mantle exhibiting faster than average seismic velocities. Meanwhile, Kevin and Trond continue to fend off attacks from other scientists who argue that there is no statistical support for their claim[89] and that plumes are just as likely to rise from anywhere on the surfaces of LLSVPs, or that their reconstructions contain unsupported assumptions.

Computer models of mantle convection incorporating observed plate motions and starting with a dense layer of different composition a few hundred kilometres thick at the base of the mantle, are able to reproduce the general form of the piles quite nicely, although there is uncertainty about some of the parameters included in the models.[90] Plate motions determine the locations of downward flow (i.e. subduction zones) and result in the deep dense layer being 'swept' into two large piles beneath the Pacific and Africa, as observed in the tomography. It was found that, in the case where the bottom layer is simply hot mantle (i.e. with the same chemical composition, but at a higher temperature), a different pattern resulted, which did not match the structures observed. A group at ANU have disputed the conclusion that LLSVPs must be chemically distinct from the surrounding mantle on the grounds that their own geodynamic modelling suggests that the seismic observations can be explained by convection driven purely by temperature variations and that the effects of any differences in chemical composition are minor.[91] A majority of geophysicists, however, currently favour an interpretation involving differences both in temperature and composition.

Caltech scientists Eh Tan and Mike Gurnis constructed a series of 3D geodynamic models of mantle convection based on plausible estimates of mantle properties, incorporating dense piles in the lower mantle.[92] They managed to reproduce the dome-like structure of LLSVPs, together with their steep sides and long-lived nature. Plumes were again generated preferentially from the edges of the piles, and these were

[88] Lekic et al. (2012).

[89] For example, Austermann et al. (2014), Julian et al. (2015), and Doubrovine et al. (2016).

[90] McNamara and Zhong (2005).

[91] Davies et al. (2012).

[92] Tan and Gurnis (2005); Tan et al. (2011).

hotter, and therefore more buoyant, than plumes generated from the tops of domes. Tan and Gurnis also found that the domes remained stationary as long as subduction was steady, but migrated rapidly when new subduction zones were initiated. Hence, at present, it looks as though Kevin's hypothesis is supported for the past 130 million years, but the permanence of TUZO and JASON at earlier times remains unproven. Fixity or otherwise of the piles is important since their location controls the orientation of the rotation axis via changes in the Earth's moment of inertia. As Clem Chase pointed out,[93] if TUZO were to become much larger than JASON, Greenland would shift to the equator.

Lowest of the Low

Not content with two giant 'continents' sitting on the bottom of the mantle, Caltech seismologists uncovered other, smaller, zones just above the core–mantle boundary, in which P- and S-wave speeds were even more anomalous. To overcome the limited resolution of tomographic models, Edward Garnero and Thomas Helmberger (at the University of California, Santa Cruz and Caltech, respectively) used waveform modelling (i.e. calculating the shapes of earthquake waves that would be detected at the surface after travelling through different mantle structures) to discover a drop in P-wave speeds of around 10 per cent right at the base of the Pacific pile (JASON)—a major decrease they believed might be explained by the presence of melt in a localised region less than 100 km wide and about 40 km high.[94] Other researchers found similar deep-mantle zones in which S-wave speed was reduced by 20 per cent or more, many close to the steep edges of JASON, TUZO, and PERM. As well as serious reductions in seismic wave speeds, results suggested somewhat higher than normal density. These ultra-low-velocity zones (yes, 'ULVZs'[95]) are still being discovered. In 2013, a particularly large ULVZ was detected by Edward Garnero and colleagues within the JASON pile beneath the Pacific.[96] Around 800 km long by 250

[93] Chase (1979).

[94] Garnero and Helmberger (1995).

[95] Since much of the work on these zones, including their discovery, was done by Edward Garnero, a more pronounceable name for them than 'ULVZs' might be 'Garnero zones' although, unlike Kevin Burke, I am unable to turn it into an acronym.

[96] Thorne et al. (2013).

km wide, but just 10–15 km thick, it was characterized by P-wave speed reductions of around 15 per cent, S-wave speed reductions of around 45 per cent, and density increases of around 10 per cent, relative to the PREM. This large zone of low seismic wave speeds could, they suggested, be the consequence of a merger between two smaller ULVZs on opposite margins of coalescing LLSVPs. Overdoing the superlatives, Edward and colleagues referred to this as a 'mega ultra low velocity zone'. Another ULVZ, around 180 km across and 10 km high, was detected beneath the South Sandwich Islands in 2016.[97] Here, P-wave and S-wave speeds were depressed by about 10 per cent and 30 per cent, respectively, while the density was 10 per cent higher.

Many ingenious and exciting explanations have been suggested for the causes of ULVZs, boiling down to two fundamental ideas: either they contain a significant proportion (up to 30 per cent) of melt or they have a high iron content, one or both of which could reduce seismic wave speeds and increase density as observed.[98] You might expect to find partial melt everywhere on the core–mantle boundary at which the melting temperature (the solidus) is exceeded. Moreover, since the liquid outer core beneath these zones must be very good at ironing out (excuse the pun) any lateral temperature variations, this would result in either a continuous melt layer covering the entire core–mantle boundary or else no melt at all. The solidus (melting) temperature for the composition of the lower mantle has been estimated at around 4400°C, whereas recent estimates of the temperature of the core–mantle boundary are somewhat lower (around 3800°C) and, in addition, there are numerous places on the core–mantle boundary without ULVZs. This all points to the likelihood that these zones consist of iron-enriched material (although, of course, it may be that not all ULVZs are the same), but does not exclude the possibility of melt being present in small quantities. Given all this confusing complexity, it would, perhaps, be wise to pause again at this point, to illustrate current ideas with a diagram (Figure 10.7).

[97] Vanacore et al. (2016).

[98] Under lower-mantle conditions, partial melts are likely to be denser than the surrounding solid rock, and will therefore remain on the bottom. Similarly, during the Hadean, molten iron, being denser than solid silicate rock, made its way to the bottom, forming the core.

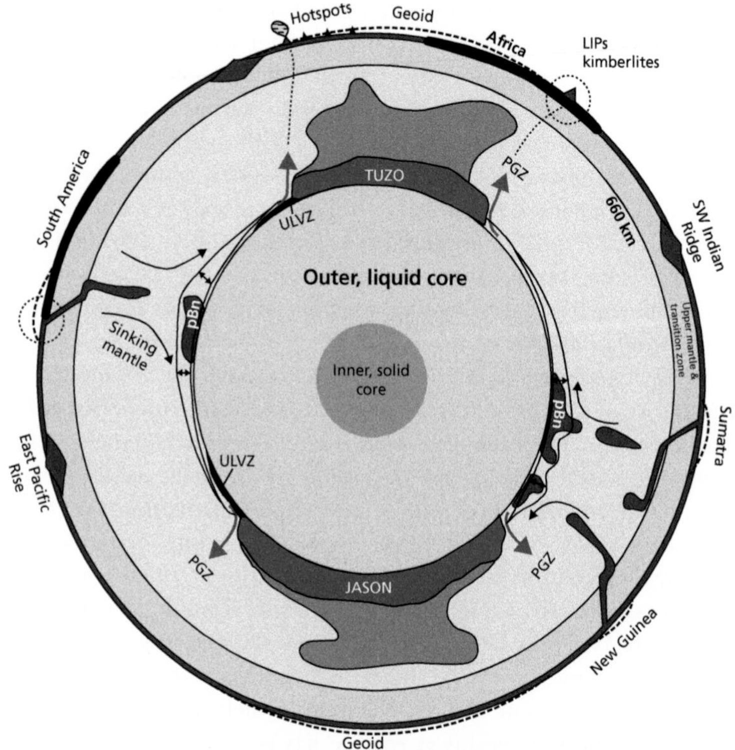

Figure 10.7 Interpretation of tomographic imagery showing LLSVPs in the lower mantle (named 'TUZO' and 'JASON' by Kevin Burke), in some cases bounded by smaller ULVZs. The margins of the LLSVPs are believed to be the generation zones for mantle plumes (PGZs), producing LIPs and kimberlites. Regions marked 'pBm' are interpreted as areas of post-bridgmanite. This interpretation has been referred to as the 'Burkian' Earth by Trond Torsvik (Torsvik et al., 2016).

The Anti-Crust

The story so far: we have two steep-sided, continent-sized (thousands of km across and up to 1000 m high), piles of hot rock at the core–mantle boundary, with lower seismic wave speeds than the surrounding mantle. They have slightly higher density than 'normal' mantle, are most probably of a different chemical composition, and may have been lurking in more or less their present locations for hundreds of millions of years. Then we have several small, thin (typically 100 km across and 10 km high) zones of greatly reduced seismic wave speed, located preferentially near the margins of the LLSVPs, that probably contain iron-enriched minerals and, possibly, some melt. The next questions are 'What the heck are they?' and 'How did they get there?'

Perhaps the most obvious explanation is that the bottom of the mantle, sitting right above the hot, liquid iron, core, became contaminated with metal alloy percolating upwards into the silicate mantle. Elise Knittle and Raymond Jeanloz performed diamond-cell experiments at Berkeley in the 1990s, which suggested that reactions between iron in the outer core and silicate minerals in the lowermost mantle could create a dense chemical boundary layer that might explain the heterogeneity observed in seismic studies.[99] Laboratory experiments suggested that core material might be able to penetrate several tens of kilometres into the lower mantle by diffusion along the boundaries between mineral grains.[100] More recent high-pressure laboratory experiments[101] suggest that blobs of molten iron alloy from the core may react with ferropericlase and penetrate the lowermost mantle to a height of 50 km or so, resulting in a reduction in seismic wave speeds and an increase in density. However, this height is much less than that of the LLSVPs, while the drop in seismic speed is greater; hence this may be a better explanation for ULVZs. Alternatively, dense 'upside-down' silicate 'sediments' may be deposited beneath the core–mantle boundary[102] by light elements precipitated from the liquid outer core as the solid inner core (which we are coming to) grows.

An explanation popular with French scientists is that the LLSVPs could be remnants of deep oceans[103] of seething magma formed following the collision around 4.5 billion years ago between two young planets, one the size of Venus and one the size of Mars, that resulted in the Earth–Moon system. Such a collision may well have vaporized part of the young Earth and caused the remainder to melt completely, creating a 'primordial magma ocean'.[104] It was at this time that iron and its

[99] Knittle and Jeanloz (1991).

[100] Hayden and Watson (2007).

[101] Otsuka and Karato (2012).

[102] Buffett et al. (2000).

[103] Plural, because the surface may have been melted many times by impacting asteroids and comets.

[104] See e.g. Labrosse et al. (2007).

'siderophile' shipmates (i.e. those elements that tend to form alloys with iron, such as nickel, cobalt, manganese, platinum, and gold) made their way to the bottom, forming the core, a process that released enough heat to raise the Earth's temperature by 2000°C.[105] It was suggested recently[106] that some of the dense droplets, incorporating nickel and sulphur, may have become trapped within bridgmanite and ferropericlase crystals forming at the bottom of the magma ocean, increasing their density and reducing the S-wave velocity as observed in the LLSVPs.

While it seems natural that magma formed by partial melting of mantle peridotite should be less dense than the residuum and therefore rise to the surface, forming a crust, crystallization of a deep magma ocean, on the other hand, may have produced a residual magma enriched in iron and thus denser, so that it remained at the bottom of the mantle. This would eventually form a layer containing the elements that did not fit easily into the molecular structures of the crystallizing minerals (i.e. the incompatible trace elements). Being denser than the mantle above, this basal layer was able to resist entrainment by convection currents, and was swept into piles such as JASON and TUZO by the arrival of subducted plates.[107] Alternative models have been suggested, in which crystallization begins at the bottom (i.e. the solid is denser than the melt) and a magma ocean exists at shallower depths. However, recent experiments on magnesium silicates indicate that, at deep-mantle pressures, melts are indeed denser than bridgmanite and ferropericlase, and hence the basal magma ocean concept appears to be viable. If so, then the LLSVPs could be composed of minerals precipitated from the increasingly iron-enriched magma, while the ULVZs could be remnants of the dense ocean itself.[108] This, of course, is good news for the geochemists, providing as it does a ready 'primordial reservoir' for their missing elements, and also for Kevin Burke, since the piles would have existed for more than 4 billion years, consistent with his hypothesis.

The third class of explanations involves the accumulation of former oceanic crust delivered by subducting plates to the lowermost mantle,

[105] The Earth was also heated by decay of radioactive elements (including short-lived isotopes such as ^{26}Al and ^{60}Fe), as well as core formation and tidal forces.

[106] Zhang et al. (2016).

[107] Labrosse et al. (2007).

[108] Petitgirard et al. (2015).

as Clem Chase and Albrecht Hofmann suggested, back in the early days of seismic tomography. Rather vague estimates have put the volume of oceanic crust formed through Earth history at between 10 per cent and 53 per cent of total mantle volume. If, as is generally believed, most of this found its way into subduction zones, then a significant proportion of the mantle is composed of ancient oceanic crust. It was once thought that this would be rapidly 'resorbed' into the mantle, or dispersed to produce a 'marble cake' structure. More recent work has shown that subducting plates remain intact for long enough to reach the bottom of the mantle. Basaltic crust forming the upper 6–7 km of a slab is dehydrated and converted to eclogite within the first 100 km of its descent, the presence of iron-rich garnet increasing the density of the eclogite above that of the surrounding mantle, thereby providing much of the 'slab-pull' force that drives the plates. The eclogite former crust is carried with the subducting slab down through the transition zone, where it undergoes further phase changes, producing perovskite-structure minerals that are likewise denser than ambient mantle. Eventually, therefore, old plates reach the lowermost mantle, where they enter the 'slab graveyard'. But what happens then?

Death Warmed Up

Fascinating laboratory experiments performed by Peter Olson (at Johns Hopkins University) and Christopher Kincaid (at the Carnegie Institution) gave some insight into what might occur.[109] By using a tank of concentrated sugar solution, they were able to view the processes involved in the millions of years following arrival of a slab at the base of the mantle on a more manageable timescale of about an hour. The viscosities and dimensions of the experiment were, of course, chosen to scale as closely as possible to conditions within the mantle, and the slab was constructed from frozen sugar solution, with an upper layer slightly denser than the surrounding fluid, and a lower layer slightly less dense (simulating eclogite and harzburgite, respectively). They observed that the slab remained intact until it reached the hot boundary layer at the bottom, where it buckled and collapsed into a pile. After heating up, the pile spread out with the denser component on top. The lighter layer (cf.

[109] Olson and Kincaid (1991).

harzburgite) then separated from the pile and rose as *thermochemical* plumes (i.e. their buoyancy was partly the result of their lighter chemical composition), while the denser layer (cf. eclogite) settled on the bottom, eventually becoming entrained in rising *thermal* plumes (i.e. their buoyancy was mainly a result of higher temperature).

A few years later, Ulrich Christensen and Albrecht Hofmann modelled this process on their computer at the Max-Planck-Institut für Chemie at Mainz. They found that almost 20 per cent of subducted oceanic crust could accumulate at the core–mantle boundary over geological time, and that 'segregation and re-entrainment of subducted crust is of fundamental importance for the dynamics and chemistry of mantle plumes'. More recently, Paul Tackley, at ETH Zürich, modelled 'living dead' slabs entering the graveyard at the bottom of the mantle, producing slightly unsettling 3D images of possible scenarios.[110] Results varied, depending on the angle at which the slab arrived, but he observed a tendency for slabs to settle 'butter-side-down', that is, with the denser (former crustal) layer at the bottom of the pile (Figure 10.8). Where the slab encountered a dense layer above the core–mantle boundary, Paul found that as much as 50 per cent of the former basaltic crust could be segregated from the slab to form piles. And, as with Peter and Christopher's sugar experiments, plume heads composed of lighter harzburgite then emerged from the edges of the piles, entraining a substantial fraction of former crust. Further modelling[111] suggests that folding of slabs as they settle produces more complexity, with phase changes becoming reversed. Overall, it took around 2 billion years for the convecting mantle to settle down to a state in which large, dense, piles were maintained above the core–mantle boundary, with the supply of basaltic crust by subduction balanced by entrainment in ascending plumes.

So, which of the three classes of explanation is most likely? All three imply remarkable phenomena in the Earth's interior, and all go some way to explaining seismological results obtained to date. Personally, I find the idea of relics of the Earth's early magma ocean now forming 'primordial reservoirs' at the bottom of the mantle highly intriguing, but, on balance, would favour the 'slab graveyard' explanation. As things stand, however, all three remain on the table—you pays your money and you takes your choice.

[110] Tackley (2011).

[111] Nakagawa and Tackley (2015).

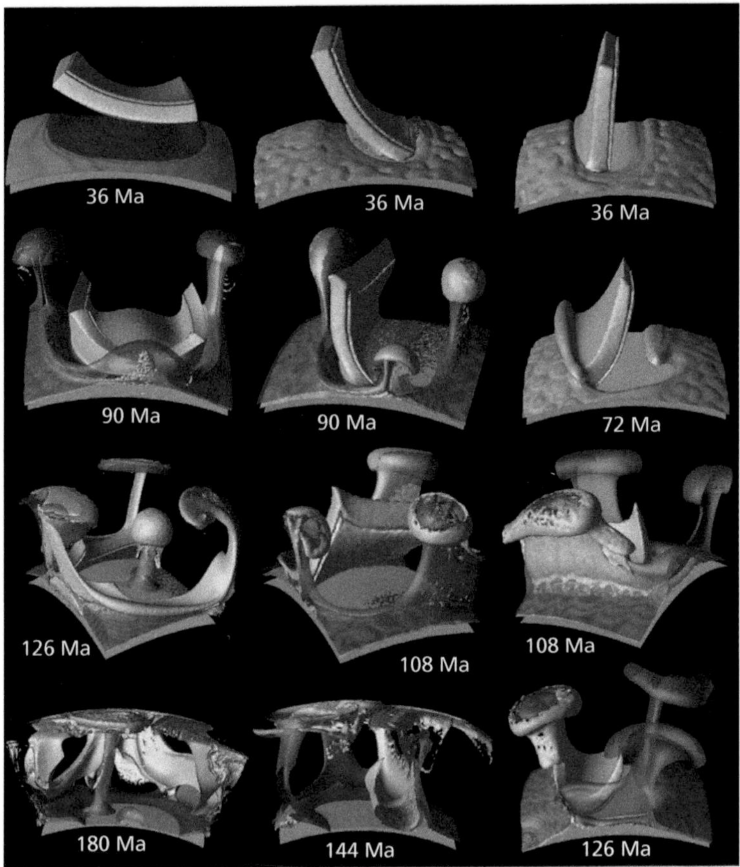

Figure 10.8 Computer simulation of the interaction of a compositionally stratified slab with a 150 km-thick layer above the core–mantle boundary in three dimensions, for three different initial slab dips. [After Tackley (2011).]

Core Issues

If it seems surprising that plate tectonics can exert such a strong influence on events right at the bottom of the mantle, then you are in for an even bigger surprise. You may remember that Walter Elsasser and Teddy Bullard showed many years ago that the geomagnetic field is generated by convection in the outer core, operating as a self-sustaining dynamo (see Chapter 3). The fuel for the geodynamo, as it is called, is ultimately the escape of heat trapped or generated within the core, but establishing just how it works is an unsolved problem.

An article[112] by Peter Olson and colleagues, published in 2015, began with this startling sentence: 'The geodynamo owes its existence to convection in the mantle.' It was not an error, and they had not muddled the mantle and core. They went on to explain that the power required to maintain the geodynamo (somewhere between 10 and 16 terawatts[113]) is so large that 'it would likely have ceased to operate long ago were it not for the heat extracted from the core by the circulation of the mantle'. What they were saying was that the core dynamo envisaged by Walter and Teddy could be maintained through geological time only if heat was removed by mantle circulation sufficiently fast to sustain a temperature gradient in the outer core that was steep enough to drive convection. The idea was not new, however: the notion that mantle convection could influence the geodynamo by altering the temperature distribution at the core–mantle boundary had been suggested as early as 1977 by Glynn Jones of the University of California at Berkeley.[114]

The mantle is effectively a gigantic engine, fuelled by the heat from radioactive decay of uranium, thorium, and potassium isotopes (providing around 60 per cent of the heat flow here at the surface), and also by heat supplied from below (contributing the other 40 per cent).[115] But there is a second engine, within the molten iron outer core, fuelled by latent heat and light elements released during the growth of the solid inner core. The place at which the two heat engines are coupled is, of course, the core–mantle boundary, probably the most chemically reactive place on or in the planet, where silicate and oxide minerals are in contact with molten iron alloy at a temperature of around 4000°C.[116] Poorly understood variations in the transfer of matter and energy across the core–mantle boundary give rise to a range of short-term fluctuations in the geomagnetic field known as the secular variation, most of which average out in less than a million years, leaving the axial dipole field that allows paleomagicians to draw their salaries. The existence of a surface

[112] Olson et al. (2015).

[113] For comparison, the total rate of heat loss through the Earth's surface is 46 terawatts.

[114] Jones (1977).

[115] Olson (2016).

[116] Buffett (2015).

geomagnetic field also allows paleomagicians, along with all other eukaryotes, to exist, by excluding the high-energy particles of the solar wind that would otherwise bring Mr Attenborough's seemingly endless series of documentaries to a swift and unhappy conclusion. By way of contrast, Mars, with its piffling magnetic field, has been ravaged by the solar wind, losing two-thirds of its atmosphere and all of its oceans, to the dismay of any hapless Martians that may have evolved during its brief 'habitable' period.[117] The means by which the geomagnetic field is generated and sustained is therefore of some interest to those of us living on the Earth's surface.

Within the core, the slowly cooling molten iron–nickel alloy (containing 5–10 per cent lighter elements—most likely silicon, oxygen, and sulphur[118]) is gradually freezing onto the solid inner core,[119] which consequently is growing at around 1 mm per year. While this stretches the everyday notion of freezing to its limits (the temperature is around 5400°C, close to that of the surface of the Sun), it does release a significant amount of latent heat that raises the temperature in the outer core still further, while crystallization of pure iron onto the surface of the inner core leaves the lighter elements in the melt, reducing its overall density. The combination of the two effects—heating by release of latent heat, and enrichment in lighter elements—creates buoyancy in the liquid iron alloy near the base of the outer core, with the result that material rises towards the core–mantle boundary, where it cools and precipitates some of the lighter elements, forming a convection cell driven both by differences in temperature and chemical composition. As the convecting molten alloy (a strong conductor of electricity) rises and falls, it is influenced by the Earth's rotation, producing a pattern of spiralling currents by the Coriolis effect, the same phenomenon that

[117] NASA are seriously proposing to launch an artificial magnetosphere to shield Mars and recover an atmosphere for the Red Planet, thereby giving our descendants somewhere to retreat when they have finished ruining the Earth.

[118] These elements are dissolved in the molten iron in the same way that the surface ocean contains many dissolved species.

[119] You may be wondering why, if temperature increases with depth, the core should begin crystallizing at the centre and not at the core–mantle boundary. If so, I congratulate you on your insight, and should explain that this is because the effect of increasing pressure predominates over increasing temperature. That is, the 'freezing point', or 'liquidus temperature', of the iron alloy increases with pressure (depth) faster than temperature increases within the core.

causes atmospheric depressions (cyclones) to rotate anticlockwise in the northern hemisphere and clockwise in the southern hemisphere. If it is not immediately obvious that such motions would inevitably give rise to a dipole field at the Earth's surface, aligned with the rotation axis, then fear not. Back in 1995, a computer model of the geodynamo, using realistic properties for the outer and inner cores, within a rotating planet, managed to simulate the dipole field and even produced a polarity reversal.[120] However, it required more than 2000 hours of CPU time on a Cray supercomputer to achieve this result, so it is hardly intuitive.

Like clouds, the core–mantle boundary can be viewed from both sides. On the mantle side, everything happens in slow motion because of the high viscosity of post-bridgmanite and ferropericlase, and it takes millions of years to see very much change. Below stairs, so to speak, down in the outer core, where the viscosity of liquid iron alloy is orders of magnitude less, things are much more lively. Flow here can iron out (there we go again!) temperature anomalies in a few centuries, such motions being easily detected in the secular variation. Changes in lower-mantle temperature will therefore have an almost instantaneous (geologically speaking) effect on the core, while changes in outer-core temperature will have little effect on the lower mantle. In other words, the mantle sees the core as an ocean of constant temperature, while the core sees the mantle as a heat sink.[121] As a result, and perhaps counter-intuitively, it is the mantle that controls the core on the time-scale of plate tectonics, by sucking out heat energy at a rate that depends on the temperature gradient in the D″ layer at the bottom of the mantle—a strong gradient giving rise to faster heat flow.[122] A (relatively) cool lowermost mantle will create such a gradient, which, in turn, will drive core convection and generate a magnetic field, while a (relatively) hot lower mantle will produce a low gradient, causing core motions to become sluggish, and the geodynamo to cease. Based on current understanding, temperatures at the core–mantle boundary beneath slab graveyards will be lower than beneath the LLSVPs, since subducted plates arriving at the core–mantle boundary will be cooler than the ambient mantle by several hundred degrees, while the LLSVP bottom

[120] Glatzmaier and Roberts (1995).
[121] Olson (2016).
[122] Just as your cup of tea cools faster on a cold winter's day.

piles may be hotter by a similar amount. As a result, temperature gradients beneath slab graveyards will be higher, extracting heat rapidly from the core. Beneath LLSVPs, by contrast, temperature gradients will be lower and less heat will be extracted. This will produce lateral differences in temperature in the outer core that can help drive convection and thus affect the geodynamo.

Of the approaches currently being taken to the problem of just how mantle convection influences the geodynamo, perhaps the most revealing is the development of increasingly complex computer models constructed by geodynamicists. These are not unlike the general circulation models (GCMs) used in climate studies. Indeed, the strong similarity between theories of mantle convection and atmospheric circulation was recognized at the very dawn of plate tectonics by the eccentric British physicist, D. C. Tozer, who wrote, in 1967, 'We therefore have considerable a priori reasons for thinking that a complete theory of motions in the Earth's interior will have a great complexity, comparable, perhaps with that for the atmospheric motions. As in meteorology one tries to break the total problem up into manageable pieces and what is normally described as a "Theory of Mantle Convection" has its parallel in the problem of the general atmospheric circulation.'[123] The parallel is now even more marked since, just as GCMs of the atmospheric have been coupled to models of ocean circulation in order to predict climate, so convection within the mantle has been coupled to flow within the core to create global circulation models (also GCMs) of the Earth's interior in the past.

Noting that mantle convection is the primary control on heat flow at the Earth's surface, and also at the core–mantle boundary, Nan Zhang and Shijie Zhong, at the University of Colorado, used a GCM to simulate changes in the patterns of heat flow through these boundaries during the past 450 million years, the period of formation and break-up of Pangea.[124] For the most recent 120 million years, a previously published plate motion model was used, while, for earlier times, less accurate estimates based on continental reconstructions were employed. The GCM reproduced present-day surface heat flow quite well, and, at the core–mantle boundary, predicted low heat flow today beneath Africa and the Pacific Ocean (corresponding to the locations of TUZO and JASON).

[123] Tozer (1967).
[124] Zhang and Zhong (2011).

Core–mantle boundary heat flow varied not only with location but also through time, reaching minima 270 and 100 million years ago. This was highly suggestive, since the rate of geomagnetic reversals is known to have declined to zero at both of these times, producing periods of 40 million years or so during which a single polarity prevailed (the 'Kiaman' reversed superchron, 316 to 262 million years ago, and the Cretaceous normal superchron, 118 to 83 million years ago). This work was extended and developed by Peter Olson and colleagues, who calculated the changes in the core required by Nan and Shijie's heat-flow results for the past 450 million years, including its structure, temperature, and composition, and also the size of the inner core.[125] They then used these properties as input to a computer model of the geodynamo (known as a 'numerical dynamo') to investigate their effect on the geomagnetic field and its polarity reversal history. In parallel with this, they ran their numerical dynamo on a second model in which core–mantle heat flow remained constant throughout, in a pattern determined by the presence of TUZO and JASON, in order to test the suggestion of Kevin Burke and Trond Torsvik that these piles have remained fixed. As you might expect, the latter model did not result in major fluctuations in geodynamo behaviour, but the model based on Zhang and Zhong's variable heat flow produced wide fluctuations in reversal frequency, with stable polarity around 475 million years ago (the age of a proposed 20-million-year reversed-polarity interval known as the Moyero superchron) and around 275 million years ago (the Kiaman reversed-polarity interval mentioned above). No constant-polarity interval matching the Cretaceous normal superchron at 100 million years was found in either model, however, which Peter and colleagues suspected was related to overestimates of Pacific plate motions at that time. By reducing heat flow further, they found they could induce a long normal-polarity interval as observed.

These results suggested that the piles on the bottom are involved in mantle convection and are therefore not static as proposed by Kevin and Trond. According to Nan and Shijie, the present pattern, with two antipodal dense piles, emerged only after the creation of subduction zones surrounding Pangea, prior to which a single warm pile existed beneath the Pacific, while the lower mantle beneath Africa was relatively cold. Trond and friends did not take this lying down and produced their

[125] Olson et al. (2013).

own mantle convection model,[126] which, to nobody's surprise, suggested that both TUZO and JASON had indeed remained where they were for the past 410 million years at least. The Colorado folk responded by repeating their modelling using the same plate reconstructions as Trond's group, to see if differences in poorly resolved plate motions prior to 200 million years could explain the discrepancy.[127] The results confirmed their earlier conclusions that the African pile (TUZO) had appeared only after about 200 million years ago. At the time of writing (late 2017), the discrepancy remains to be resolved, hopefully without bloodshed.

Peter Olson has recently repeated the exercise,[128] using a more recent CGM[129] incorporating plate motions for the past 200 million years from a comprehensive model published in 2012,[130] thereby avoiding the uncertainties of earlier motions based only on paleomagnetically based continental reconstructions. Once again, the estimated heat flux at the core–mantle boundary was coupled to a numerical dynamo and, once again, subduction was the major influence on the pattern of convection, producing two large holes at the core–mantle boundary where cool slabs swept aside the denser material at the base of the mantle, forming two major piles similar to the LLSVPs observed in seismic tomography. This produced extreme variations in heat flux through the core–mantle boundary: beneath the slabs, heat flux was as high as 130 mW m^{-2}, while, beneath the piles, it was reduced to around 30 mW m^{-2}. Hence, while the mean heat flux from the core was around 85 mW m^{-2}, variations of up to 100 mW m^{-2} existed. The outer core would tend to be unstable (and hence convect) beneath the subducted slabs, where heat flow exceeded that which could be transmitted by conduction alone, but stable (and hence stratified) beneath the dense piles, where heat flow was much less. The resulting time-averaged geomagnetic field contained significant non-dipole components, although, for the past 100 million years, it was not far from the assumed axial dipole field. At 200 million years, however, the altered pattern of plate motions reflecting the existence of Pangea resulted in a dipole field tilted by 20° from the rotation axis,

[126] Bull et al. (2014).
[127] Zhong and Liu (2016).
[128] Olson (2016).
[129] Zhong and Rudolph (2015).
[130] Seton et al. (2012).

another blow for Trond's group, since, if Peter's simulations reflected the genuine effects of pangeas on core convection and the geomagnetic field, then reconstructions of past supercontinents based on paleomagnetism could be subject to large errors.

Regardless of these uncertainties, the extraction of heat by plate tectonics appears to be crucial for the operation of the geodynamo. Plates cooled by heat loss at the surface descend into the mantle, where they induce convection and the removal of deep-seated heat, eventually reaching the core–mantle boundary, where they create regions of high heat flux, extracting heat efficiently from the outer core, aided by the transition to post-perovskite that increases its thermal conductivity by 50 per cent or so. This leads to cooling and downwelling within the molten core, eventually forming a solid inner core. Hence, as Peter puts it, 'the geodynamo is actually maintained by plate tectonics, telemetered deep into the core by the global mantle circulation'.[131]

If the geodynamo is powered primarily by heat released from inner-core growth, it is important to know when this growth began, and hence for how long the Earth's atmosphere and inhabitants have been shielded from the solar wind by a strong geomagnetic field. More prosaically, this would also give paleomagicians an upper age limit to their studies. If we know the rate of heat loss through the core–mantle boundary, we can calculate the rate of inner-core growth and so estimate the time at which growth began. Unfortunately, because of uncertainties, including the amount of heat flowing through the core–mantle boundary, the heat generated by radioactive decay in the core, and the identity of light elements, estimates of the age of the inner core remain imprecise. In 2012, researchers in the UK put the cat amongst the pigeons by calculating a much higher value for the thermal conductivity of the outer core, based on laboratory experiments using diamond anvils, implying more rapid cooling and hence a younger age for the inner core of around 600 million years[132]—surprisingly young considering its size (about the same as Pluto). This set in train another major controversy rather like that concerning the age of the Earth a century before, only in reverse—a controversy that has yet to be resolved.[133] Such a young age for the inner core begged the question of how the geomagnetic field was generated at earlier times, and even

[131] Olson (2016),
[132] Pozzo et al. (2012).
[133] Dobson (2016).

whether such a field existed, which would have important consequences for biological evolution. Studies of the intensity of magnetisation of Archean rock samples suggest that a dipole field may have been in existence since 3.5 billion years ago,[134] with an increase in intensity between 1.5 and 1 billion years ago perhaps reflecting core nucleation. To confuse the issue further, more recent laboratory studies using diamond anvils have produced much lower estimates of thermal conductivity,[135] in direct conflict with the UK group, providing support for the existence of a geomagnetic field driven by inner-core nucleation since the Archean (4.2 billion years ago).

To summarize this somewhat complicated story, the geomagnetic field is generated by vigorous convection in the outer core (the geodynamo). Variations in heat flow at the core–mantle boundary are produced by mantle convection, driven by subduction, bringing cool plates to the base of the mantle and sweeping denser material into piles. Beneath the cool slabs, the temperature gradient is high and convection in the outer core is vigorous. Beneath warm, dense, piles, the gradient is low and convection sluggish. This pattern can generate a geomagnetic field like that observed here at the surface of the planet, but the pattern will change as mantle convection responds to changes in plate motions, in particular, the formation and break-up of Pangea. Hence, the configuration of slabs and piles at the bottom of the mantle will change very slowly on a timescale of tens of millions of years. This, in turn, will affect the evolution of the core and the operation of the geodynamo, altering the long-term shape and tilt of the geomagnetic field and changing the frequency of polarity reversals. This amounts to what geophysicists call the 'top-down' model of the Earth. You need to bear in mind that this model is currently at the frontier of research and these are not yet firm conclusions. In any case, this cannot be the full story, because it does not include the contribution from mantle plumes, whether or not they originate from Kevin Burke's PGZs.

Plumatics and Crackpots

So far, we have been entertained by disputes over continental drift, the age of the Earth, and the nature of the mantle transition zone. Another

[134] Biggin et al. (2015).
[135] Konôpková et al. (2016).

dispute, as bitter as any of these, currently rages over the role, and even the existence, of mantle plumes. By now, you will have formed the impression that such plumes constitute an important component of mantle convection, providing an upward return flow that transports heat from the core to the surface. Plumes, as we have seen, provide a satisfying explanation for hotspots, volcanic island chains, LIPs, and kimberlites, as originally suggested by Tuzo Wilson and Jason Morgan. The identification of new hotspots, or 'hotspotting' as it has become known, was a popular sport during the 1970s. So popular, in fact, that the well-known American geophysicists John Holden and Peter Vogt observed:[136] 'In 1971 there were 20 plumes [Morgan, 1971,1972]; in a mere half decade the population has risen to no less than 122 [Burke and Wilson, 1976]. Our extrapolations from these data show that there will be 1,000,000 hot spots by the year 2000. We hope someone proves that hot spots do not exist, before it is too late.' Their appeal did not fall on deaf ears, and, if hotspots have not been proven not to exist, it is not for the want of trying.

Most geophysicists would agree that plumes have played an important role in the evolution of the Earth, and would therefore be classed, in John and Peter's terminology, as 'plumatics'.[137] Yet there remains a small band of sceptics that refuses to accept the evidence from seismic tomography and convection modelling, and persists in opposing the entire concept of mantle plumes, making all manner of objections and even setting up their own 'anti-plume' website.[138] These heretics believe that plumes are a myth, just as Carey believed that subduction was a myth, and that hotspots and LIPs can be explained by shallow processes directly beneath plates (Figure 10.9). One of the most prominent ring-leaders of this subversive movement was Don Anderson,[139] a leading seismologist of the twentieth century, who made many major contributions to the study of plate tectonics from the earliest days (as we have already seen), but who was unconvinced by Jason Morgan's notion of deep-mantle plumes.

136 Holden and Vogt (1977).

137 Holden and Vogt (2007).

138 With the rather misleading URL, www.mantleplumes.org—rather as if the National Front were to set up a website with the URL www.liberalsocialism.com.

139 Don Anderson, who passed away in 2014 at the age of 81, was something of a maverick, who argued for two-layer convection.

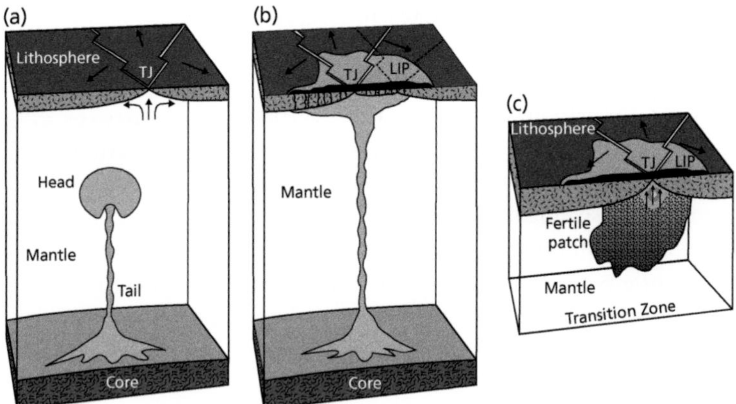

Figure 10.9 Mantle plume (a, b) and plate (c) models for oceanic plateaux. [From Sager et al. (2016).]

Warren Hamilton, an equally prominent American geologist with a history of opposition to generally accepted theories,[140] is even more adamant that plumes do not exist, and have never existed.[141] In numerous articles published over the past two decades, Warren has marshalled all the evidence he can muster against the detested plume hypothesis. His determined and sustained stance thus raises the question of whether he (like that other Warren, Sam Carey,[142] who occupied his final years promoting his expanding Earth theory), is spending his dotage fighting a lost cause, or whether he is a latter-day Tuzo Wilson, possessing insights that others are simply slow to appreciate. Given the current state of the science, you have to fear that the former may be the case. In 2008, Don and Warren combined forces to write a bitter diatribe for the 'mantle-plumes' website, describing plume theory as 'zombie science', defined as 'obsolete conjectures—propped up as conventional wisdom'. Following Don's death in 2014, his mantle (forgive yet another weak pun) has been assumed by Gillian Foulger, a geologist at Durham University.

Given their grim opposition to the concept of mantle plumes, you might imagine that the deniers had a better theory to put in its place,

140 Warren became a 'mobilist' in the early 1960s, while most American geologists remained fixists.

141 Hamilton (2011).

142 In case you have forgotten, Warren Carey was known to friends and colleagues as 'Sam'.

one that is supported by more evidence and provides greater illumination of planetary processes. What you find, however, is a ragbag of recycled ideas that amount to the melting of 'fusible' mantle beneath cracked plates.[143] Nevertheless, if the veracity of their arguments were to be judged on the quantity of publications, then you would have to say that the 'crackpots', as we shall call them, have a strong case. Numerous articles, books, and Web pages have appeared since the turn of the century, placing all manner of (largely ineffective) obstacles in the path of the mantle plume juggernaut. The beleaguered crackpots have therefore become increasingly frustrated. For example, in an online debate in 2003, Gillian accused 'most Earth scientists' of an 'unquestioning belief' in mantle plumes, and even charged seismologists with fiddling their data.[144] Journal editors and reviewers of research manuscripts did not avoid the lady's wrath either, standing accused of selective bias against the crackpots.

The great Stanford geophysicist Norman Sleep regards narrow plume conduits in the mid-mantle as the 'guts of the dynamic hypothesis' and resolving them with improved seismic data and techniques as a primary goal.[145] Imaging of narrow, low-velocity, pipes, passing up from the core–mantle boundary through the mantle beneath surface hotspots would, for all but the most die-hard sceptics, constitute unequivocal evidence (we resist the temptation to apply that tired cliché, 'smoking gun') for the existence of mantle plumes. Caltech seismologist Brandon Schmandt and colleagues believed they had done just this in 2012, when they imaged a plume beneath the Yellowstone volcano using data from the USArray seismic records.[146] Gillian, however, refused to accept their evidence, suggesting a non-plume explanation of her own. More recently, Scott French and Barbara Romanowicz, at the University of California at Berkeley, constructed a global tomographic model that managed to resolve low-seismic-velocity columns rising from the

[143] Referred to, confusingly, as 'the plate model'. It is well known that thousands of small volcanoes on the ocean floor probably do result from melting beneath cracks caused by plate deformation, but these are not classified as hotspots.

[144] For example, Gillian claimed that 'Carefully truncated cross sections, with colour scales cranked up, give noisy images the illusion of strong anomalies traversing the mantle. Proposed depths to plume heads have been increased in order to preclude measurable heat to flow at the surface.'

[145] Sleep (2006).

[146] Schmandt et al. (2012).

core–mantle boundary beneath the major hotspots,[147] some of which were rooted in ULVZs, providing evidence that these features might actually represent plume generation zones. The only problem was that the 'plumes' were rather too fat (more than 500 km in diameter), rather than the thin (less than 200 km wide) pipes suggested by theoretical models. Unfortunately, the resolution of seismic tomography in the lower mantle is limited, both by the characteristics of seismic waves and by the inevitable paucity of both earthquakes and recording stations in the mid-plate oceanic regions in which many hotspots occur. Presently, resolution of rising columns 200 km in diameter or less is beyond the capabilities of seismic tomography, and may remain at the limits of the technique for some years. Seismologists are a very bright lot, however, and it is only a matter of time before they find a way to detect plume conduits in the lower mantle—if they exist.

Despite these difficulties, it currently seems safe to say that the plumatics have the advantage, with around a dozen hotspots generally accepted as being fed by plumes from the lower mantle, including Iceland, Tahiti, East Africa, Réunion, Samoa, Kerguelen, the Canaries, the Azores, Hawaii, Cape Verde, and Easter Island. Should it turn out, however, that the crackpots were right all along, then a significant chunk of current plate tectonics theory (and this book) would have to be rewritten. Not only that, but our approach to the workings of other planets (Chapter 11) would also need to be completely rethought.

Afterthought

The rapid advances in solid Earth geophysics made since the turn of the century amount to a second revolution in our understanding of our home planet. Exploration of the Earth's deep interior has revealed a new world to rival the discoveries made at the surface during the fifteenth and sixteenth centuries by the likes of Cristóbal Colón, Fernão de Magalhães, and Francis Drake, including previously unknown 'continents' at the bottom of the mantle. Current research is leading us towards the conclusion that plate motions control mantle convection, which, in turn, controls core convection and the geodynamo. By mobilizing the entire planet, plate tectonics thus induces corresponding changes

[147] French and Romanowicz (2015).

in the behaviour of the geodynamo, one consequence being reversals in the polarity of the geomagnetic field—the very same reversals that created the bar-code that led to the discovery of plate tectonics in the first place. Thus, very remarkably, plate tectonics provides both the music and the tape recorder to create its own greatest hits.

As with so much of convection theory, the idea that the death of plates is linked with the birth of mantle plumes can be traced back to Arthur Holmes, who came very close to this concept in the first edition of his great textbook.[148] Having described how basaltic oceanic crust might suffer compression at great deeps, such as Tonga and Kermadec, where it might 'turn down' into the mantle, becoming transformed under pressure into dense eclogite, Arthur noted that 'a heavy root of eclogite would continue to develop downwards until it merged into and became part of the descending current'. He went on: 'The eclogite that founders into the depths will gradually be heated up as it shares in the convective circulation. By the time it reaches the bottom of the substratum it will have begun to fuse, so forming pockets of magma which, being of low density, must sooner or later rise to the top. Thus an adequate source is provided for the unprecedented flows of plateau basalt [what we now call LIPs] that broke through the continents during Jurassic and Tertiary times.' He may have got the part about melting wrong (although it is just possible that this could explain the ULVZs), since plumes rise mainly in the solid state, but everything else is brilliantly perceptive.

[148] Holmes (1944).

11

The Final Frontier

> An Earth-sized silicate planet is likely to experience several tectonic styles over its lifetime, as it cools and its lithosphere thickens, strengthens, and becomes denser.
>
> Robert Stern (2016)

After fifty years, you'd have thought that Harry Hess' gloomy prognosis would have come to pass and plate tectonics research would by now be reduced to a mopping-up exercise. Nothing, however, could be further from reality. Surprisingly, many of the outstanding problems at the frontier of current research are also the most fundamental—still unresolved after a generation of effort. For example, when did plate tectonics begin (and when might it cease)? What came before plate tectonics? How are plates formed? Does plate tectonics occur elsewhere in the solar system? Progress is being made on all these questions, and answers to some could well be found within our lifetime.

*

The LIPs of Venus

Huge expectation accompanied the *Magellan* spacecraft launch from the shuttle *Atlantis* in May 1989. The surface of Venus, completely obscured by opaque yellow clouds in visible light, was about to be revealed by radar mapping. Would Earth's twin resemble her sister, or would her appearance be strange and unfamiliar? One would certainly expect that two planets born together with similar size and weight, and orbiting the same star for four-and-a-half billion years, would at least show a family likeness.

The *Magellan* orbiter, like *GEOS-3* (see Chapter 8), was built using spare parts left over from earlier missions by a cash-strapped NASA, and, like *GEOS-3*, it carried a microwave radar altimeter with which it was planned to map the entire surface of the planet. It also sported an imaging system known as 'synthetic aperture radar' or 'SAR' (comparable

The Tectonic Plates Are Moving! Roy Livermore, Oxford University Press (2018). © Roy Livermore.
DOI: 10.1093/oso/9780198717867.001.0001

to sidescan sonar in the oceans), capable of providing 20 km-wide strips of surface imagery with a resolution of just 75 m on each pass. Furthermore, by measuring perturbations in its orbit, *Magellan* was able to derive Venus's gravity field, allowing geophysicists to investigate the planet's internal structure. Previous Soviet radar missions had mapped small areas of the surface at low resolution, suggesting the existence of mountains and volcanoes, but the big question was whether Venus, a planet of similar size and composition to the Earth, was likewise cooled by plate tectonics. Since it had no obscuring oceans, the solid surface of our sister planet was about to be revealed in a way that had only recently become possible for the Earth (thanks to GEOSAT and similar missions). If plate tectonics operated on Venus today, the scientists expected to see tennis-ball seams running around the planet and lines of deep trenches backed by volcanic arcs. If plate tectonics operated previously but had now ceased, they might still see large fracture zones recording the movement of the plates, even if trenches and mid-ocean ridges had disappeared.[1]

During its three years of mapping prior to being sacrificed in the planet's dense, hot, and acidic atmosphere in August 1994, *Magellan* successfully mapped 92 per cent of the Venusian surface, making it one of NASA's most successful missions ever. At first, it seemed that familiar features associated with plate tectonics might exist on Venus. The small number of impact craters pointed to a young age (relatively speaking) for the surface, indicating that some form of volcanic resurfacing had taken place, and, while no mid-ocean ridges could be resolved and transform faults were rare, it appeared that trenches were common, and what looked like abyssal hills were identified in one region.[2] A curved trough adjacent to a feature named, poetically, Latona Corona, was, it was claimed,[3] a dead ringer for the South Sandwich Trench: topographic profiles over the two features exhibiting uncanny similarities, while the corona itself was probably a back-arc basin comparable to the East Scotia Sea. Other troughs adjacent to the Artemis and Eithinoa coronae were also curved in plan view, and had dimensions suggesting that they too represented subduction zones. Three large plateaux, one near each pole and one in low latitudes, named, like the planet itself,

[1] These features are described by geophysicists as 'dynamic topography', being maintained by active processes.

[2] McKenzie et al. (1992).

[3] Sandwell and Schubert (1992).

after ancient goddesses (Ishtar, Lada, and Aphrodite), appeared to have some of the characteristics of continents, reinforcing the view that Venus had once been Earth-like.

Despite the initial excitement, it was soon apparent that whatever was occurring on Venus was nothing like Earth's plate tectonics. For one thing, impact craters were distributed uniformly over the surface, suggesting that it was much the same age everywhere. It appeared that Venus had been resurfaced within a short interval somewhere between 300 and 700 million years ago,[4] completely erasing the record of earlier crustal evolution. Volcanic features, including the giant volcano Maxwell Montes,[5] which rose to an altitude of 11 km, together with thousands of smaller cones, covered much of the planet's surface, although none appeared to be active.[6] On Earth, volcanic features would soon be eroded or else destroyed by subduction, but, on waterless Venus, ancient volcanoes appeared pristine in SAR images. Tectonic deformation, though common, was also distributed fairly evenly over the surface (Figure 11.1),

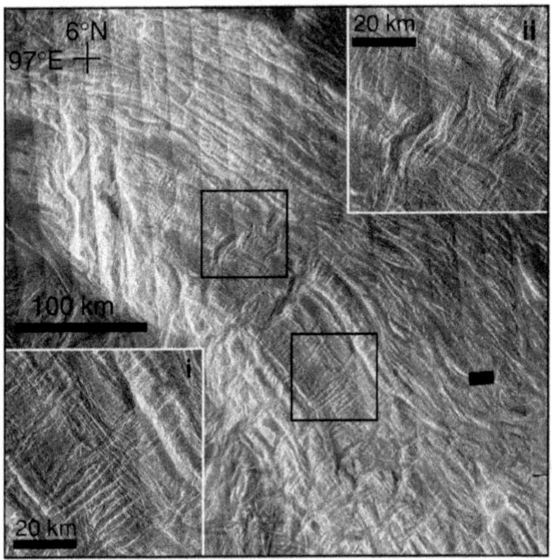

Figure 11.1 Ovda Regio, in the western region of Aphrodite Terra, showing numerous folds and faults. [From Phillips and Hansen (1998).]

[4] Turcotte (1993).

[5] Named after the great Scottish physicist.

[6] Recent evidence collected by the ESA's *Venus Express* mission suggests that some volcanoes may, after all, be active today.

rather than being concentrated in narrow plate boundaries as on Earth. But perhaps the clearest evidence of the non-existence of plate tectonics was the absence of a tennis-ball seam encircling the planet.

Lacking a Venusian network of seismic recorders,[7] researchers have only surface features and gravity measurements from which to deduce what the internal structure of the planet might be, akin to judging a book by its cover.[8] Nevertheless, a surprisingly extensive literature has been built on these slender foundations. The gravity field measurements made by *Magellan* have been particularly useful.[9] Recall that, where topography—for example highlands—is compensated isostatically by thicker crust, the observed gravity anomaly is small, but, where the surface is maintained dynamically at its present level by mantle convection, as at a subduction zone, a much larger anomaly is observed. Hence, the relationship between gravity and topography[10] can be used to decide whether a surface feature such as a corona is compensated by thicker crust or is being pushed up by forces within the planet. Studies on Venus have shown that only the Aphrodite and Ishtar terrae are compensated by thicker crust, most likely produced by volcanism generated by mantle plumes. Other surface features are therefore likely to reflect convection within the mantle, but not plate tectonics.

If not Venus, then what about Mars? Several missions have revealed a sharp contrast in surface features between the northern and southern hemispheres of Mars, a contrast that has become known as the 'Martian dichotomy' (Figure 11.2). The southern hemisphere consists of cratered highlands formed on crust around 60 km thick, while the northern hemisphere consists mainly of volcanic plains reminiscent of lunar maria. In 1994, Norman Sleep at Stanford University published an article boldly (or perhaps rashly) entitled, 'Martian Plate Tectonics',[11] in which he made confident identifications of mid-ocean ridges, transform faults, island arcs, and subduction zones. Audaciously (some might say

[7] Apart from the expense and difficulty of installing seismic recorders on the surface, the equipment would need to tolerate temperatures of 460°C and pressures of 90 bars (9 MPa) for a year or more.

[8] The simile is not mine: it formed part of the title of a session on this subject at the American Geophysical Union's Fall Meeting in 2015.

[9] McKenzie (1994).

[10] The ratio of the topography to the gravity anomaly is known as the 'admittance'.

[11] Sleep (1994).

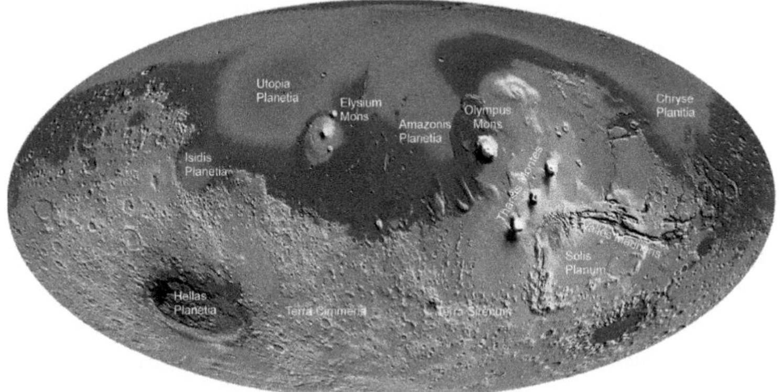

Figure 11.2 Surface of Mars. [After Yin (2012).]

recklessly), he even calculated Euler rotations and spreading rates for his putative plates, and presented reconstructions showing how the various features had evolved. Others were more cautious and, despite evidence for the possible existence of long transform faults and offset magnetic anomalies,[12] it is pretty clear that Mars does not exhibit plate tectonics today. Whether it did in the past is an open question, since it certainly had liquid water on its surface at one time.

Nevertheless, while there may be no life on its surface, Mars is very far from a dead planet. Remanence in ancient surface rocks suggests Mars had a magnetic field, and therefore a convecting core, very early in its evolution, but this was shut down, perhaps by massive impacts, about 4 billion years ago. Since then, its core has been stratified and no surface field generated. Evidence from radio tracking of the Mars Global Surveyor shows that it still has a large liquid iron core around half the planet's radius, believed to contain substantial concentrations of sulphur (perhaps 16 per cent). Recent interpretations attribute its spectacular volcanism (including the giant Olympus Mons volcano), like that of Venus, to rising plumes within the mantle of the planet. The vast size of such edifices reflects the fact that, unlike that of Earth, the crust of Mars is static, so that magma is delivered continually to the same spot at the surface. Thus, in a remarkable reversal of the continental drift saga, the assumption of horizontal motion has been replaced by vertical movements, and mobilism by fixism.

[12] Connerney et al. (2005).

Apart from Venus, only one other planet in the solar system has a mean density comparable to that of the Earth:[13] Mercury. In recent years, Mercury has been imaged at high resolution by NASA's Mercury *MESSENGER* spacecraft. The results show a surface much like the Moon, with heavy cratering reflecting events during the early stages of planetary evolution, around 4 billion years ago. Like the Earth, Mercury has a molten iron-rich core (much larger in proportion to the planet's size[14]) and also, possibly, a solid inner core. Convection in the molten part generates a weak dipole magnetic field, with strength around 1 per cent of the Earth's at the surface. Being something of a planetary minnow (its diameter is just 38 per cent of the Earth's), Mercury finds itself a victim of geometry: having a large ratio of surface area to volume, it cools effectively by conduction and perhaps weak convection. Surface mapping during the *MESSENGER* mission showed that 27 per cent of the planet's surface is covered by extensive lava flows (a single volcanic plain in the northern hemisphere covers 6 per cent of the planet), comparable to the flood basalts erupted on Earth in places like Siberia and India. Giant thrust faults, hundreds of kilometres in length and over a kilometre high in places, together with a host of smaller faults, show that Mercury is still tectonically active,[15] but the latest interpretations discount plate tectonics and instead invoke, ironically, the very mechanism assumed by Victorian geologists for the Earth—contraction of the planet.[16]

Hence, within our sample of eight planets plus sundry odds and sods of moons and dwarf planets, we now know for sure that only one body in the solar system exhibits plate tectonics today. The other rocky planets—Venus, Mars, and Mercury (Figure 11.3)—all have hot, convecting interiors, but their surfaces are not involved in this convection as on Earth. It seems clear that all four inner planets started off with enough heat generated by impacts and core formation to melt their surfaces, creating global magma oceans that quickly lost heat to space,

[13] That is, about 5430 kg m^{-3}, compared with 5514 kg m^{-3} for the Earth and 3933 kg m^{-3} for Mars.

[14] Results from *MESSENGER* suggest it occupies around 85% of the planet's radius. The molten core may be enclosed in an iron sulphide shell.

[15] Watters et al. (2016).

[16] Contraction of 5–7 km in radius over the life of Mercury has been suggested both by theoretical modelling and by *MESSENGER* observations. This may not sound a lot, but is sufficient to generate a wide range of geological features.

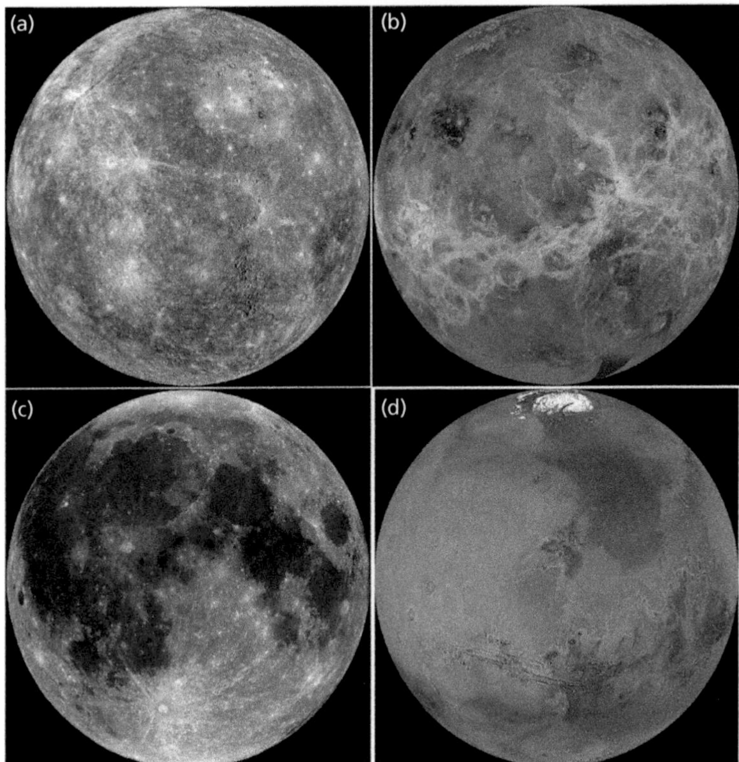

Figure 11.3 The rocky inner planets: (a) Mercury; (b) Venus; (c) the Earth's Moon; (d) Mars [After Platz et al. (2015).]

forming solid lids within a few tens of millions of years. Something unusual must then have occurred on Earth to break its lid into fragments, some of which began to sink into the mantle, putting the planet on the road to plate tectonics. Exactly what this was is a matter of current debate.

Keeping a Lid on It

Given all their similarities (similar diameters, similar densities, and similar ages), Earth and Venus ought to be like two rather large peas in a pod. Having formed at the same time, and assuming similar compositions, the interiors of both planets must be heated by comparable concentrations of radioactive elements, yet they dispose of their heat in very different ways. Why should this be?

A series of calculations at Caltech by Slava Solomatov emphasized the importance of changes in the viscosity of silicate rock with temperature.[17] As temperature decreases towards the surface of a cooling planet, so viscosity increases exponentially, leading to a cold, rigid, upper boundary layer.[18] Employing a 'viscoplastic rheology', in which failure of the lithospheric lid occurs once its 'yield stress' is exceeded, Slava showed that increasing the viscosity contrast between the surface layer and the hotter deep mantle could produce three different convection regimes. In the first, the mantle viscosity was low enough to allow fluid motion right to the surface, rather like Rayleigh–Bénard convection. As the viscosity contrast increased, a transitional 'mobile' regime was reached in which a rigid upper layer moved slowly over the surface, before sinking back into the mantle. Finally, as the upper layer cooled further, it became too viscous to move and no longer took part in mantle convection, forming a 'stagnant lid'.[19]

Soon after, two-dimensional computer modelling by Slava and Caltech colleague Louis Moresi suggested that Venus may have developed a rigid lithosphere as thick as 550 km, resulting in a stagnant-lid mode of convection.[20] Likewise, Mars and Mercury were also 'single-plate' planets, while the Earth appeared to be in something like the transitional regime, although the question of how the mobile lid fractured into plates remained unexplained. Later work by Louis and Slava[21] confirmed that the strength of the lithosphere played a crucial part in determining the style of tectonics: a strong lithosphere favoured the stagnant-lid mode, while a weaker lithosphere became incorporated into the convection regime as a mobile lid. Lithosphere with intermediate strength might cycle between the stagnant-lid and mobile-lid states, periodically collapsing into the mantle and then reforming once again, rather like the crust on the Kilauea lava lake described by Mark Twain (Chapter 5).

[17] Solomatov (1995).

[18] For a temperature difference of 100°C, the viscosity may change by a factor of 10 or more. Refer back to Chapter 5 if you have forgotten about boundary layers.

[19] Morris and Canright (1984); Fowler (1985).

[20] Solomatov and Moresi (1996).

[21] Moresi and Solomatov (1998).

Since the strength of the lid depended mainly on temperature, this raised the possibility that, as a planet cools, it may pass through more than one regime of convection. It is generally believed that all the rocky planets started out with magma oceans in which full-scale convection occurred. As they cooled, they may have passed into the mobile-lid regime, which, in the case of Mars and Mercury, was probably very brief, before their transition to the stagnant-lid mode seen today. Venus may have survived in the mobile-lid regime until around 500 million years ago, when the 'plates' became too thick and strong to break and it also entered the stagnant-lid mode, or perhaps it exists within the intermediate mode with periodic overturn every few hundred million years. Earth was an exception, remaining in the mobile-lid mode, enjoying the benefits of plate tectonics for billions of years. Louis and Slava's models were, however, rather low-resolution and highly simplified. Their mobile-lid regime resembled the Earth's plate tectonics in some respects, but failed to capture several crucial aspects. For example, plate boundaries in their models were much broader than on the Earth and tended to 'heal' over time, rather than persisting as on Earth. Moreover, 'downwellings' were most unlike subduction zones, being symmetrical and vertical,[22] rather than the asymmetrical sloping zones characteristic of plate tectonics. Furthermore, their model had no mechanism for the creation and maintenance of transform faults.

By the turn of the century, computer models of mantle convection were becoming increasingly sophisticated, but more than a decade of further effort would be needed before they were able to develop realistic plate tectonics. Some modellers, as we found in Chapter 10, forced plates on their models as initial conditions, but 'self-consistent' plate tectonics, in which rigid plates appeared naturally from the convecting system, remained, in the words of geodynamicist, Paul Tackley (who you may recall modelling slab graveyards in Chapter 10), the 'Holy Grail' of mantle convection modellers.[23] Since then, Paul (now at ETH Zürich) and his colleague Takashi Nakagawa (at the Japan Agency for Marine-Earth Science and Technology, Yokohama) have gradually developed a sophisticated three-dimensional spherical computer model of mantle convection that is able to reproduce many of the characteristics of plate tectonics. In their recent simulations, they investigated the three

[22] In many ways like that proposed long ago by Felix Vening-Meinesz (see Chapter 4).

[23] Tackley (2000a).

modes of mantle convection: stagnant-lid, like that suggested for Mars and Mercury, episodic-lid, like that suggested for Venus,[24] and mobile-lid, like that on Earth, incorporating a dense 'primordial' layer just above the core–mantle boundary.[25] Beginning with a stagnant lid, the mode of convection was established by adjusting the strength of the lithosphere, weaker values causing subduction to appear in the mobile and episodic cases within 60 million years of the start of the simulation. In the stagnant-lid mode, the model unsurprisingly retains a thick basaltic crust, which nonetheless is thinned by mantle convection through geological time. The episodic mode resulted in a very thick layer of eclogite being built up in the lower mantle from periodic 'overturn' of the basalt crust that occurred after subduction was initiated, resulting in the removal of large areas of the basalt crust (typically around half the surface area) within about 150 million years, after which the crust began to form once again. The accumulation of a thick eclogite layer reduced heat flow out of the core, leading to a failed geodynamo. In the mobile-lid (plate tectonics) mode, subduction continued throughout geological history, the subducted slabs sweeping eclogite former crust into one or two large piles, much like those discovered by seismic tomography.

Recent computer modelling by a group of Australian geophysicists led by Craig O'Neill at Macquarie University[26] gives further support to the ideas of Slava and Louis, suggesting that an early 'hot stagnant-lid' regime, in which vigorous convection takes place in the low-viscosity mantle beneath the stagnant lid, may have existed on Earth following solidification of the magma ocean at 4.5 billion years. As cooling proceeded, this then gave way to the 'episodic' plate tectonics regime, in which long periods of stability were interspersed with abrupt and catastrophic overturn of the lithosphere, followed, around 3 billion years ago, by modern-style plate tectonics. Eventually, their modelling predicts, plate tectonics is superseded by a 'cold stagnant-lid' regime in which convection becomes too weak to break the strong (highly viscous) lithosphere. If correct, their models imply that the Earth occupies a window in planetary evolution, and is doomed to enter the cold stagnant-lid mode at some point in the future, a prospect of more than

[24] Turcotte (1993).
[25] Nakagawa and Tackley (2015).
[26] Weller et al. (2015); O'Neill et al. (2016).

passing interest, since it is likely to make the planet uninhabitable. Encouragingly, this transition is not predicted to occur for another billion years or so. The Aussie group concluded that 'The ultimate tectonic state of a planet is a result of a balance between the coupling of the plates and the mantle beneath, and also the buoyancy forces driving convective motion'. The group warned, however, that the course of tectonic evolution was highly sensitive to the initial conditions assumed on the early Earth, in particular, the thermal state of the mantle following the magma ocean phase: slightly different assumptions could give a completely different evolutionary sequence. Recent simulations by Slava Solomatov and others have demonstrated that, even where the starting values are the same, the time of lid failure can vary greatly, owing to the chaotic nature of the convective system.[27]

It Cracks Me Up

It turns out that generating plate tectonics is by no means as straightforward as you might imagine, since a basic tenet is that the lithosphere is strong, as proposed all those years ago by the unfortunate Joseph Barrell. Models of mantle convection need to generate stresses greater than the yield stress of the lithosphere in order to produce plates, but there's the rub. For, stresses produced by such models are very much smaller (less than 10 MPa) than the yield stress measured on rock samples in the laboratory (around 800 MPa), implying that the 'stagnant-lid' mode exhibited by Venus and Mars should be universal on rocky planets. On Earth, therefore, something else must have occurred to reduce the yield stress of (i.e. weaken) the previously stagnant lid in order to create the first plate boundaries and, in particular, the first subduction zones. Numerous suggestions have been made as to what this might be (which only goes to show that, currently, nobody really knows), and we will look at just a few.

Like sex, plate tectonics depends on rigidity. Klaus Regeneuer-Lieb, a geophysicist at the Johannes Gutenberg-Universität in Mainz,[28] observed that 'The rigidity of plates, which is a fundamental prerequisite of plate tectonics, is directly related to the dehydration of the oceanic

[27] Wong and Solomatov (2016); Weller and Lenardic (2017).

[28] Now at the University of New South Wales, Sydney.

lithosphere'.[29] Hydrogen[30] is incompatible in mantle minerals and so, like many other trace elements, it gets partitioned into the melt when minerals are heated above their solidus temperature. Being highly soluble in magma, it is then transported to the surface, where, on eruption, it completes the water cycle by being released to the oceans and atmosphere. This leaves the mantle lithosphere dehydrated[31] and consequently more viscous, increasing its strength and making it difficult to break. To reduce its strength and create weak plate boundaries, therefore, some mechanism for locally rehydrating the stagnant lid is required.

In 1990, Z. U. Mian and D. C. Tozer (both at Newcastle University) wrote an article entitled 'No Water, No Plate Tectonics: Convective Heat Transfer and the Planetary Surfaces of Venus and Earth'.[32] In it, they made the case that the primary reason for the difference in the tectonics of Earth and Venus was the presence of a surface ocean on the former and its absence on the latter.[33] Their argument was that, during the early history of the solar system, rocks constituting the lid of the Earth were weakened by the presence of water, while, on hotter Venus, surface water was quickly lost and consequently its lid remained too strong to fracture into plates. Surface water promoted failure of the Earth's lid by reducing the amount of stress that needed to be applied by convective motions beneath the lithosphere in order to break the lid. Experimental work has tended to support this idea. Water (or hydrogen, if you prefer) is sequestered within mantle peridotites at concentrations of a few tens or hundreds of parts per million, the majority within the mineral lattices of pyroxene and olivine. While pyroxenes host more water, olivine is most abundant and therefore controls the overall strength of the mantle. At the atomic level, deformation such as the formation of plate boundaries involves creep, the rate of which depends on the speed at which silicon and oxygen atoms can diffuse through the crystal lattice of olivine, which, in turn, is dependent on

[29] Regeneuer-Lieb (2006).

[30] Once again, we conflate 'hydrogen' with 'water', remembering that, when geochemists refer to 'water', they really mean hydrogen (or protons), structurally bound within the mineral as hydroxyl (OH).

[31] H concentration is reduced from about 800 ppm/Si in the asthenosphere to about 80 ppm/Si in mantle lithosphere.

[32] Mian and Tozer (1990).

[33] A suggestion made by Dan McKenzie as long ago as 1977.

temperature and pressure. In addition, laboratory experiments showed that these rates are greatly accelerated by the addition of hydrogen ions, which, as mentioned in Chapter 10, create 'point defects' into which silicon and oxygen atoms can migrate. Water concentrations as low as 45 ppm have been found to be able to increase diffusion rates of silicon by a factor of 1000 compared with rates in dry olivine.[34]

These conclusions have been questioned by more recent laboratory studies conducted by Hongzhan Fei (at Universität Bayreuth) and colleagues, which suggest a much smaller dependence, resulting in only a minor change in viscosity.[35] If correct, this would mean that the presence of bound water within mantle minerals was unlikely to have weakened the Earth's stagnant lid sufficiently to permit plate tectonics in the first place. Hongzhan and colleagues attributed the large difference between their results and those of previous researchers to faulty methodology in the earlier studies, which did not go down well. Unsurprisingly, their conclusions have been challenged, for example, on grounds that they investigated a single olivine crystal containing only magnesium (i.e. forsterite), rather than a collection of grains containing both iron and magnesium, as found in the mantle. Hongzhan and colleagues have subsequently responded by publishing further results supporting their earlier findings, but the matter is far from settled.

Yet, even if water bound into mantle minerals is able to weaken the lid substantially, there are problems. For, although computer models starting with a weakened lid are able to cause failure of the stagnant lid by exceeding its reduced yield stress, they do not generate realistic plate tectonics. For a start, they cannot produce the very narrow and long-lived boundaries that define the plates today. And, secondly, the style of subduction generated in these models, involving symmetrical and vertical sinking of both plates, is completely unlike that seen on Earth (Figure 11.4). By contrast, genuine subduction is asymmetrical, with a single plate sinking at an angle that may vary from 20° to 90°. In order to reproduce realistic asymmetrical subduction, computer models need to have strong (but not strong enough to induce a stagnant lid) plates and a weak interface between the subducting and overriding plates.[36] The plate interface is weakened by the release of water from

[34] Costa and Chakraborty (2008).

[35] Fei et al. (2013).

[36] Gerya et al. (2008).

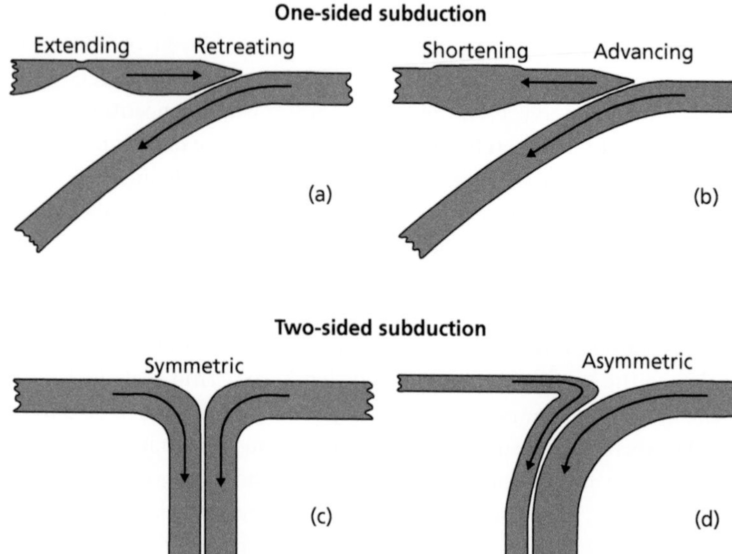

Figure 11.4 One-sided (a, b) and two-sided (c, d) geometries. [From Gerya et al. (2008).]

hydrated crust as it enters the subduction zone, resulting in serpentinization of peridotite (in which OH is incorporated into olivine to become the soft mineral serpentine). Serpentinite acts as a lubricant, reducing friction and permitting long-term subduction (Figure 11.5). In addition, the generation of one-sided subduction is assisted by allowing the upper surface of models to deform freely, allowing the formation of an outer high and deep trench. Thus, by assuming strong plates, surmounted by weak hydrated crust, together with a free upper surface, models can now produce a style of asymmetrical subduction that closely resembles that observed on Earth.[37]

There are other ways in which the presence of water can reduce the viscosity of mantle rocks. Continental crust, as we discovered in Chapter 6, owes its existence to mantle melting induced by the release of watery fluids from subducting slabs beneath volcanic arcs. The presence of water lowers the melting point of the hot rock dramatically, resulting in the production of magma, even tiny amounts of which, in

[37] Crameri and Tackley (2016).

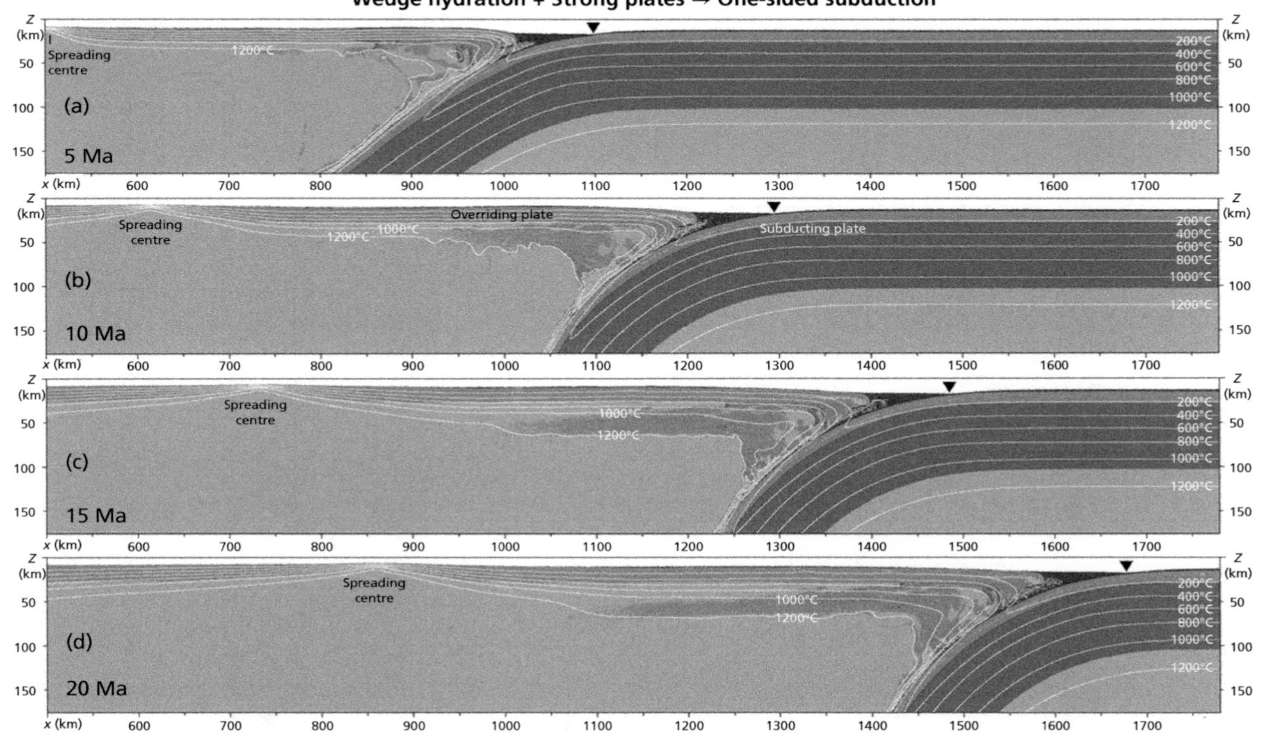

Figure 11.5 Results of numerical simulations achieving sustained, one-sided, subduction. [From Gerya et al. (2008).]

the spaces between olivine and pyroxene crystals, can drastically reduce the strength of mantle peridotite. Localized weakening caused by melt generation could thus lead to a major reduction in yield stress and failure of a stagnant lid. Such weakening is most likely to occur at convergent and divergent boundaries, both sites of mantle partial melting.

David Bercovici (at Yale University) and colleagues have pointed out a few difficulties in trying to extend the idea of water as a weakening mechanism to the localization of new plate boundaries,[38] noting that, while increasing pore pressure certainly assists brittle failure (by jerks), this only occurs at shallow depths (less than 10 km), where the lithosphere is already weakest. At mid-lithosphere depths, where the lid is strongest, deformation takes place by creep, and so friction reduction is unimportant. Since the lithosphere is dry, water can only be a factor if it enters from the surface to depths of 100 km or so, where the pressure is so great that cracks and pores are closed and permeability is virtually zero. Nevertheless, David and colleagues concede that weakening by water-filled cracks at shallow depths may nucleate deformation by other means at greater depths.

A second possibility, borrowed from a theory in engineering known as 'damage mechanics', has been promoted by David Bercovici and Yanick Ricard (Université de Lyon). In essence, their idea is that a pristine stagnant lid can be broken by causing 'damage', involving recrystallization and reduction in grain size, to the minerals from which it is constituted, weakening regions that may then evolve into plate boundaries. In an attempt to generate realistic plate boundaries in their convection model, David and Yanick incorporated two mineral phases, representing olivine and pyroxene (in the ratio 60:40), the minerals that together constitute over 75 per cent of mantle rocks. The sizes of the grains of the two minerals and the level of stress determine the type of flow induced in the mantle—coarser grains (larger than 100 μm) favour dislocation creep (non-linear) at larger stresses, while smaller grains (smaller than 20 μm) and low stress favour diffusion creep (Newtonian). Diffusion creep is particularly sensitive to grain size, declining as grain size cubed; hence, if the grain size can be reduced sufficiently to cross from the usual dislocation-creep regime to the diffusion-creep regime, the rate of deformation can be accelerated. Grain-size reduction will thus create weak zones that will act as plate boundaries, remaining as

[38] Bercovici et al. (2015).

zones of weakness that may be reactivated later, so long as they do not 'heal' by an increase in grain size. Placed under stress, mantle rocks will experience two opposed tendencies: grain growth or healing will tend to strengthen them, while grain-size reduction by damage will tend to weaken them.

A mantle with two phases (olivine and pyroxene) will behave differently from a mantle composed of a single phase when put under stress. In particular, grain growth is impeded by the blocking of migrating grain boundaries by the interfaces between the two different minerals (known as Zener pinning). At the same time, grain-size reduction by damage may push the mantle locally into the diffusion-creep regime, resulting in a zone of weakening at mid-lithosphere depths where plates are strongest. If this weakness persists over geological time, stable boundaries may be established and a 'memory' of past plate boundaries preserved, influencing the nucleation of future boundaries. Surface temperature is critical for this model—hotter surface conditions like those found today on Venus lead to rapid healing and grain-size increase, so that long-lived plate boundaries do not form and the stagnant lid prevails. Passive 'mid-ocean ridges' can also be generated, while horizontal variations in viscosity can generate transforms. In support of their theory, David and Yanick point to evidence for grain-size reduction caused by severe stress in rocks known as 'mylonites', commonly found in mountain belts and believed to have formed by plastic deformation deep within the crust.

A third possible mechanism involves weakening of a negatively buoyant lithosphere by impacts. It has been suggested that asteroids or comets could have ruptured the Earth's early stagnant lid, inducing subduction. The problem here, of course, is that such collisions were common on all rocky planets during the early days of the solar system, yet only one went on to develop sustained subduction (so far as we know). However, what if the 'impact' came from *beneath* the lithosphere rather than above? There is good evidence that impacting mantle plume heads have weakened and ruptured the Earth's lithosphere in the past, creating triple junctions such as the one that divided the Ontong Java, Manihiki, and Hikurangi plateaux (Chapter 7), or the East African plume that generated the triple junction beneath the Afar region. On Venus and Mars, both of which sport active convection beneath stagnant lids, heat is lost via large volcanic eruptions, believed to be the surface expression of rising mantle plumes. If Venus' stagnant

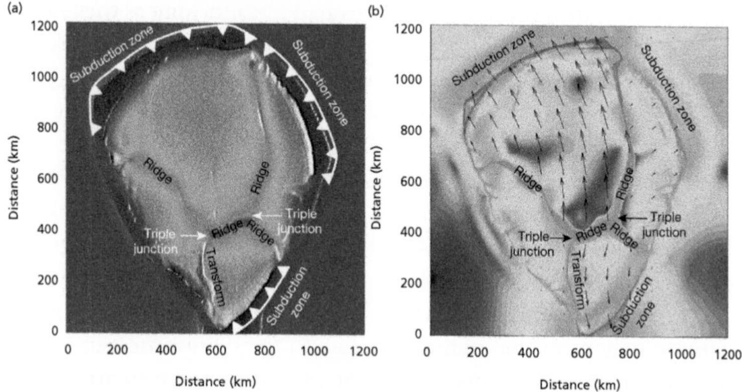

Figure 11.6 Plan view of miniature plate tectonic system generated by model plume impact. The surface is domed by the impact, resulting in the formation of radial rifts, while the dome is circumscribed by incipient subduction zones. [From Gerya et al. (2015).]

lid today is an analogue of the early Earth, then heat loss on the young planet was very likely dominated by 'plume tectonics'.

Taras Gerya (of ETH Zürich) and Stephan Sobolev (of the University of Potsdam) have carried out a series of computer simulations to see whether large plume heads impacting on the base of the lithosphere could, in fact, lead to subduction.[39] In their models, a strong, negatively buoyant, stagnant lid is weakened from beneath by magma generated by the rising plume head, allowing the plume to penetrate the lithosphere and produce voluminous volcanism, creating a large oceanic plateau. As the plateau cools, it spreads outward over the lithosphere, forcing it down into the mantle, resulting in a ring of subduction zones surrounding the plateau (Figure 11.6). Tears develop in this subducting ring of lithosphere, forming a set of discrete, curved, subduction zones, which may or may not develop into self-sustaining subduction systems, depending on the values chosen for the various model parameters. As the newly formed subduction zones retreat, rifts are generated within the plateau, evolving into a network of mid-ocean ridges and transforms separating a set of new plates. Hence, if the early Earth was dominated by plume tectonics, as some believe, then this would offer a plausible mechanism for getting plates moving.

[39] Ueda et al. (2008); Gerya et al. (2015); Baes et al. (2016).

A crucial factor in this model is the degree of lithosphere weakening induced by magma rising from the plume. If this is insufficient, the plume will spread *beneath* the stagnant lid, creating thickened lithosphere by underplating, and plate tectonics will not be initiated. Secondly, the stagnant lid needs to be old enough that it has become negatively buoyant, but not so old that it is too strong to penetrate. Where the lithosphere is young and buoyant, the subducting 'proto-slabs' may break off, ending subduction. Thirdly, a weak, hydrated, crust (like today's oceanic crust) has to be present to lubricate the downgoing slabs, allowing deep subduction to develop; otherwise slabs may simply freeze at shallow depth. In such cases, the circular plateau remains, but plate tectonics is not initiated. As Stephan and Taras pointed out, this may offer an explanation for the existence of 'coronae' on the surface of Venus,[40] a suggestion that appears to be supported by recent laboratory modelling involving nanospheres in fluids heated from below.[41] In their own model of early mantle convection, Takashi Nakagawa and Paul Tackley also find that mantle plumes instigate subduction within just 60 million years of the Earth's formation, giving rise to mobile-lid or episodic-lid convection.[42] They note that the arrival of plumes increases stress on the lithosphere and produces a thickened basaltic crust, the lower part of which may be altered by pressure to eclogite, thereby encouraging subduction.

It may seem odd that plate tectonics, an efficient mode of heat loss, was only instigated after the planet had cooled appreciably. Heat flow in the Archean is estimated at three to five times that of today (global mean 65 mW m^{-2}), and mantle melting was far greater, producing abundant volcanism and a much thicker crust. Volcanism, involving the transfer of hot magma from the interior to the surface, where it cools and solidifies, is itself a highly efficient mode of heat loss, which must have played a major part in cooling the stagnant-lid Earth. It has been suggested that the mode of cooling prior to plate tectonics may have resembled that observed on Jupiter's moon Io, which loses the enormous amounts of heat generated by tidal forces from the giant planet by a mechanism known as 'heat pipes'.[43] In this mode, heat is transported upwards in

[40] In fact, Don Turcotte of Cornell University suggested that the coronae might be sites of incipient subduction as long ago as 1993. See Turcotte (1993).

[41] Davaille et al. (2017).

[42] Nakagawa and Tackley (2015).

[43] O'Reilly and Davies (1981).

narrow pipes by magma, perhaps generated by rising mantle plumes, and erupted to form a new surface. William Moore and Alexander Webb, at Hampton University, Virginia, and Louisiana State University, respectively, suggested a few years ago that a similar mechanism might have operated on the early Earth. They believed that the predictions of the heat-pipe model, including a rapid transition to plate tectonics, matched observations in ancient cratons, such as the Pilbara in western Australia and the Barberton in South Africa (of which more anon). William has recently carried out computer simulations[44] showing how the heat-pipe mode of heat loss might have evolved into plate tectonics. In the model, heat loss via heat pipes thickens the stagnant lid and suppresses plate tectonics until the Earth has cooled sufficiently for volcanism to begin to wane. At some point, conduction of heat through the lithosphere becomes dominant, producing variations in lid thickness that focus stress on thinner regions that evolve into plate boundaries.

For subduction to begin, a broken plate must bend. Weakening must therefore extend beyond the narrow plate-boundary zone, allowing the plate to deform at a hinge line and descend into the mantle. Water penetrating surface faults could perform the weakening required to permit bending—hence a surface ocean may be necessary for subduction initiation. Alternatively, the injection of heat and magma from a mantle plume could conceivably produce the required weakening. Putative mantle plumes on Venus and Mars are presently unable to provide sufficient weakening to escape the stagnant-lid mode, although they may have done in the past. Grain-size reduction, however, while it may create narrow plate boundaries, would require some other mechanism to weaken the entire lithosphere in the past and allow bending. As mentioned above, computer modelling by Fabio Crameri, Paul Tackley, and others[45] found that allowing a free upper surface on the lithosphere (in which the surface may rise and fall, as on the real Earth) led to the spontaneous initiation of asymmetrical, sloping, subduction zones, which were difficult or impossible to reproduce in free-slip models (in which the level of the surface was fixed, allowing only horizontal movement). It appeared that the free surface assisted with the plate bending required to get subduction underway.

[44] Moore and Lenardic (2015); Kankanamge and Moore (2016).
[45] Crameri et al. (2012).

Other Earths

Hubertus Strughold may sound like a character from the Goon Show (for those old enough to remember it), but was, in fact, a German physiologist who became Chief Scientist at NASA's Aerospace Medical Division in the 1960s, where he had a hand in designing spacesuits for the Apollo astronauts. His achievements were, however, overshadowed by reports concerning his involvement in Nazi war crimes involving experiments on children. Our interest in Herr Strughold arises from a paper he published in 1955, entitled 'The Ecosphere of the Sun', in which he estimated the surface temperatures of Mars and Venus, concluding that they, like Earth, orbited the Sun within the range in which liquid water could exist on their surfaces. His calculations were wrong insofar as Venus was concerned, but his 'ecosphere' concept evolved into the widely adopted modern notion of a 'circumstellar habitable zone'—the range of planetary orbits in which liquid surface water could exist and potentially support life. As Michael Hart of NASA later pointed out, for life to evolve along the lines seen on Earth, a planet needs to reside within a much narrower 'continuously habitable zone' over a period of billions of years, adapting itself as the output from its star changes with time.

As we have seen, plate tectonics provides the thermostat that maintains conditions at the surface within narrow limits over the billions of years necessary to retain liquid water and to prevent a descent into 'frozen planet' mode or, worse, a runaway greenhouse, resulting in a Venus-like desert. It may be, therefore, that the theoretical 'plate tectonics zone' also approximates the continuously habitable zone of a star. Perhaps Venus passed out of the Sun's continuously habitable zone 600 million years ago and Mars billions of years earlier, while Mercury never entered it. The term 'habitable zone', as generally defined, is thus a misnomer, since it takes a lot more than simply being in the liquid-water zone for life to get a foothold on a planet—a long list of other requirements has to be fulfilled, too. For a start, any self-respecting planet intending to host animal and plant life needs to be rocky, that is, composed largely of silicate rock and iron, with a decent supply of radioactive fuel on board. Secondly, it needs to start off hot and be large enough to retain much of its internal heat for billions of years, which effectively means being about the size of the Earth (but see the section 'Size Matters, or Maybe Not' below). Thirdly, it needs to have an atmosphere, which implies that it requires a convecting molten iron core capable of generating a strong surface magnetic field to deflect the stream of charged particles from its star that would otherwise erode atmospheric gases.[46] Fourthly, there needs to be constant replenishment of essential elements—nutrients—to enable life to evolve. In addition, our ideal planet must have the right kind of reducing chemistry at the right time for life to get started, which probably means it must have warm (rather than hot) hydrothermal vents in which an RNA world can become established. On Earth, most of these services are provided by plate tectonics, and we know of no alternative mechanisms for many of them.

But this is not all. For, as the Marquis of Salisbury observed so presciently (Chapter 5), it takes billions of years to turn jellyfish into Man, or, more accurately, microbes into jellyfish, highlighting the requirement, not just for vast spans of time, but for stability of the conditions for life over periods representing a significant slice of the age of the universe. This requires that such a planet exists within what has been described as the 'galactic habitable zone'—too near the galactic centre, and there is a severe risk of being hit by a large piece of flying debris in the form of planetismals, comets, or meteors, not to mention radiation from supernovae, that could finish off life before it has got a foothold. On the other hand, being too far out could result in a lack of 'metals',[47] with which to form rocky planets. Thus, while it appears that microbes can evolve very early in a planet's evolution, and should therefore be common throughout the universe, animals, plants, and fungi are a different matter altogether, requiring several billion years of incubation on worlds that experience a high level of stability. No wonder, then, that some have taken to referring to Earth as the 'Goldilocks planet'.[48]

Which brings us to the 'astrobiologists', who, as the name implies, study life beyond Earth. Fate has cruelly thwarted these folk in their noble aim by restricting their sample size to just the one planet, thus rendering the distinction between astrobiologists and their run-of-the-mill biologist colleagues somewhat delusory, if not faintly ridiculous,

[46] But, I hear you say, Venus has no magnetic field yet it sports a very thick atmosphere. The explanation is that Venus does, in fact, have a magnetosphere, produced by interaction between the upper atmosphere and the solar wind. Apparently, this provides partial protection for its dense CO_2 atmosphere.

[47] Another example of loose terminology, this time by astronomers. It just means elements heavier than hydrogen and helium.

[48] Zalasiewicz and Williams (2012).

just as if a group of self-styled astrosociologists planned to study homelessness in the Milky Way. George Gaylord Simpson, Professor of Vertebrate Paleontology at Harvard University in Massachusetts, was one of the first spoilsports to pour cold water on the subject (then known as 'exobiology'). In a 1964 *Science* article entitled, 'The Nonprevalence of Humanoids', he famously pointed out that 'this "science" has yet to demonstrate that its subject matter exists!' More recently, in a 2005 review of a book on the development of astrobiology, Jeffrey Bada, Professor of Marine Chemistry at Scripps Institution of Oceanography, came to the conclusion that 'one of the field's greatest attractions was money—and lots of it'. A cynic might thus conclude that the invention of astrobiology was little more than an attempt (successful, as it turned out) by sundry biologists and biochemists to tap the abundant funds of national and international space programmes. Good luck to them!

Nevertheless, the discovery of rocky 'exoplanets'—planets orbiting stars other than our Sun—provides potentially valuable opportunities to study the role of different modes of mantle convection in planetary evolution, particularly where planets both older and younger than the Earth are observed (assuming they could be dated), and allowing ideas concerning transitions such as that from stagnant-lid to mobile-lid to be tested (assuming there was a way to identify these modes). To their credit, the astrobiologists have not been deterred by the impossibility of their task. On the contrary, their discoveries of numerous new exoplanets have become a standard fallback for journalists needing to fill the inside pages of newspapers (or the front page of *The Sun*) on slow news days. On one such day in February 2017, NASA put out a press release announcing the discovery of seven new exoplanets orbiting an 'ultra-cool dwarf star' in the constellation of Aquarius, known as TRAPPIST-1, a story that was taken up by press news agencies around the world.[49] Based on density estimates, at least six of these planets were rocky and more or less Earth-sized, and all appeared to orbit within the

[49] The planets were identified using NASA's *Spitzer Space Telescope*, along with the ESO's ground-based Very Large Telescope. According to Wikipedia (yes, I know), this story was even less newsworthy, since three of the exoplanets had been discovered the year before by Michaël Gillon and fourteen colleagues, using the Transiting Planets and Planetesimals Small Telescope in the Chilean Andes, hence the star's name, 'TRAPPIST-1'.

estimated habitable zone of the star. On their website, NASA helpfully provided an exciting 360° view of the surface of one of these worlds, showing large, rocky, outcrops amongst what appeared to be snow and small pools of water: 'a world swimming in water in perpetual twilight', all backed by an attractive red sky. The viewer's excitement dimmed somewhat, however, on reading the words 'This is an artist's interpretation' beneath the graphic.

Calculations by astronomers suggested that the outermost of the seven TRAPPIST-1 satellites might have surface water and atmospheres, making them prime targets for further research and perhaps future space missions. Sadly, given the vast distances and the unfeasibility of ever sending probes (at least, according to our present understanding of the laws of physics), neither we nor our descendents are ever likely to view the surfaces of planets such as these, far less carry out research on their interior structures and modes of cooling. At the expense of spoiling a wonderful illusion, it should be pointed out that, assuming life exists on one or other of these 'Earth-like' worlds, and given that they are 40 light years distant, even a telephone call is out of the question—you would need to wait for (literally) a lifetime for an answer to any questions you might wish to put to any of its techno-savvy inhabitants. More recent work[50] has served only to deepen the disappointment, with preliminary climate modelling suggesting that all bar one of the 'new' exoplanets are likely to reside outside the habitable zone after all.

The race to spot the greatest number of exoplanets around the galaxy was begun by NASA with the launch of its *Kepler* space telescope in 2009. Within a short time, careful measurements of tiny fluctuations in the amount of light arriving from stars between the constellations Cygnus and Lyra revealed the existence of five previously unknown exoplanets,[51] none of which were like the Earth. The first Earth-sized exoplanet, Kepler-186f, was discovered in 2014, orbiting within the habitable zone of a main-sequence M1-type dwarf star, some 500 light years distant. From the diminution of light intensity during occultation, the radius of Kepler-186f was estimated to be a little larger than Earth's, but, apart from its orbital period of 130 days, nothing more could be

[50] Wolf (2017).

[51] Dozens of exoplanets had, however, been known previously, since the first discovery in 1992.

observed. This discovery marked the beginning of a free-for-all in which thousands of suspected exoplanets would be discovered, providing instant kudos for the astronomers involved.[52] By April 1st, 2017, 3607 exoplanets had been discovered, of which around 10 per cent were probably rocky.

For any of them to be truly Earth-like, rather than, say, Venus-like (i.e. lifeless and searingly hot) or Mars-like (lifeless and bitterly cold), they would need to be cooled by plate tectonics. Hence, an Earth-sized planet exhibiting plate tectonics and orbiting within the habitable zone was, and remains, a prime target in the search for life, intelligent or otherwise (most probably otherwise), begging the question: 'How do you tell if an exoplanet has plate tectonics?' I have to tell you now that the answer, both today and in the foreseeable future, is 'You cannot'. As others have pointed out, it took a bloody long time for us to detect plate tectonics on Earth, even though it operates right beneath our feet, generating thousands of volcanoes and earthquakes. In order to spot plate tectonics with any confidence on a distant planet, the minimum requirement is imagery of its surface, which means sending a probe. To reach Pluto, a dwarf planet within our own solar system, took the *New Horizons* spacecraft nine years, travelling at 36,000 mph. To reach the nearest star, Proxima Centauri, where an exoplanet slightly larger than Earth was deemed to be 'a viable candidate habitable planet',[53] even at twice that speed, would take roughly 40,000 years, so don't hold your breath.[54] Still, we can no doubt look forward to announcements of discoveries of extraterrestrial life (or, as the *Daily Mail* puts it, 'aliens'), based on flimsy evidence such as the spectrum of light from dim and

[52] Michaël Gillon, along with two other astronomers, was rewarded by *Time* magazine by being named amongst the 100 most influential people on Earth in 2016.

[53] Despite being tidally-locked to its star and receiving 30 times more extreme-UV radiation and 250 times more X-rays than the surface of the Earth.

[54] There is, perhaps, hope. The *Daily Mirror* announced in December 2016 that Stephen Hawking is working with Russian billionaire Yuri Milner on a $100 million project to construct a fleet of tiny 'nanocraft', fitted with even tinier cameras and driven by powerful ground-based lasers, that could reach exoplanets before humans finish wrecking the Earth. Travelling at 20 per cent of the speed of light, these miniature starships could arrive at Alpha Centauri in twenty years. However, since there is no means of slowing the missiles, any pictures of the exoplanets will have to be taken with a very fast shutter speed as the craft flash by at 135 million mph, and will take a further four and a half years to reach Earth.

distant planets as they transit their stars, and soon on direct observations using ever bigger and more expensive kit, such as NASA's $8.8 billion James Webb Telescope and ESA's €1 billion Extremely Large Telescope and equally expensive PLATO mission.[55] Yet, in reality, the best we will be able to do is predict the *likelihood*, attended by large uncertainties, of plate tectonics and life on any of these planets.

Size Matters, or Maybe Not

A decade ago, a small group at Harvard University published a paper[56] confidently announcing the 'Inevitability of Plate Tectonics on Super-Earths'. 'Super-Earths' were defined as exoplanets with a mass of between one and ten times that that of the Earth, a few of which had recently been detected orbiting stars within our galaxy.[57] If not preposterous, then this assertion was, at least, somewhat presumptuous, since there is more to the Earth than mere mass. Even if we include density in the definition, specifying, say, a mean density of 5000 kg m^{-3} or greater, this still does not guarantee anything like our Earth, since factors like surface temperature and the existence of oceans also need to be taken into account. Such planets might just as well be called 'super-Venuses'.

Leaving aside the questionable terminology, we might ask just how inevitable plate tectonics really is on rocky planets larger than the Earth. The Harvard group's claim was based on calculations of plate thickness and convective stress, suggesting that, since the former declined with increasing mass, while the latter increased, the lithospheres of all planets larger than Earth were bound to be broken into plates, even on planets without surface water. Their calculations made a number of simplifying assumptions, for example, that viscosity within these large planets is unaffected by increasing pressure. Their conclusions were quickly challenged by rival modellers,[58] who concluded on the basis of their own simulations that, on the contrary, large rocky

[55] Although it seems inevitable that, once the media and public tire of discoveries that don't involve aliens, the budgets for exoplanet research will be cut.

[56] Valencia et al. (2007).

[57] Models of planet formation predict that, while planets in the range 1–10 times the Earth's mass are common, larger bodies in the range 10–100 times the Earth's mass will be rare.

[58] O'Neill and Lenardic (2007).

planets would be *less* likely to have plate tectonics, since an increase in size would lead to an increase in the strength of the lithosphere. A third group[59] found that the increased temperature within 'super-Earths' would lead to more melting and a much thicker crust, making the lithosphere more buoyant and less likely to subduct. In 2010, Jun Korenaga, at Yale University, presented 'scaling laws' for mantle convection that he believed showed that the size of a planet 'plays a rather minor role',[60] directly contradicting the Harvard group's findings. He suggested that other factors, in particular, the presence of surface water, determine whether plate tectonics operates or whether a stagnant lid is formed. The waters, already turbid, were muddied further the following year by more modelling results[61] suggesting that, while plate tectonics is theoretically more likely on larger planets in which the mantle is heated from below (i.e. by the core), when heat is produced internally (i.e. by radioactive decay), as is believed to be the case universally, the likelihood of plate tectonics is the same for planets of all sizes. If larger planets are denser, however, then this study concluded that plate tectonics becomes more likely, although other factors, such as surface temperature and abundance of water, are more important than size.

This was, however, by no means the end of the argument—the 'plate tectonic wars', as one journalist called them,[62] continued. More results were reported in 2012,[63] invoking surface temperature (determined by the incidence of solar radiation and atmospheric composition), in addition to planetary mass, as a major control on whether stagnant-lid or plate tectonics prevailed. According to these results, plate tectonics would be favoured by larger, cooler, planets. But, by 2013, the tide was turning against the Harvard team. A German–Canadian group[64] produced results of computer simulations, incorporating heating both from below and within the mantle, combined with the use of slightly different parameters in their convection model, suggesting that, as rocky planets increase in size, the temperature effect on viscosity dominates,

[59] Kite et al. (2009).
[60] Korenaga (2010).
[61] Van Heck and Tackley (2011).
[62] Chorost (2013).
[63] Foley et al. (2012).
[64] Stein et al. (2013).

producing a stagnant-lid mode of convection. Other models[65] have tended to support this view, but emphasize the importance of a planet's initial temperature in determining which road is taken—the 'standard' value assumed by most researchers leading to a stagnant lid rather than plate tectonics. Recent work[66] suggests that the reason for the conflicting results of the competing groups lies in their assumptions concerning the internal heating of the mantle by radioactive decay. Young planets with a full load of radioactive elements will tend to experience a hot stagnant-lid regime, while old planets, in which 'bottom heating' from the core becomes more important as the contribution from radioactive decay declines, tend to enter a cold stagnant-lid mode. Somewhere between the two extremes, conditions for mobile plates exist, depending on surface temperature, which, in turn, depends on climate. Ultimately, it is becoming clear that the emergence of plate tectonics depends heavily on initial conditions and, perhaps, random occurrences, such that computer models are unable to predict when a planet might make the transition from stagnant-lid to mobile-lid tectonics.

By now, you may be thinking that much of plate tectonics research has been reduced (or, indeed, elevated) to computer modelling. Such models exploit the formidable and growing power of computers to simulate highly complex processes operating within the Earth and also (since they are driven by physical principles) on other planets of different sizes and compositions. A difficulty with this approach, as with climate modelling, is the very complexity that requires such simulations. Computer models can only tell you whether a particular set of assumptions and parameters is able to reproduce something like plate tectonics. There may be numerous combinations of parameters that could produce realistic results, hence such 'forward modelling', as it is called, in inherently non-unique. Moreover, mantle convection, leading to rising plumes and sinking plates, is a chaotic phenomenon, such that very small changes in initial conditions such as temperature or viscosity can lead to very different results. There may thus be a limit to what computer modelling alone can tell us about processes on the early Earth. John Dewey, who spent much of his career in the field, shares this view and bemoans that 'Research in the earth sciences is moving away, too fast and too far from the reality of the rocks. Ultimately, the

[65] Noack and Breuer (2014).

[66] Weller et al. (2015); Weller and Lenardic (2016, 2017).

current, blinkered, obsession with process must be tempered by the historical perspective of the record preserved in the rocks.'[67] What the emergence of plate tectonics showed was that observation and theory are like two wheels on the same axle: together, they keep things moving in the right direction, but if one comes off, you veer off into the mire. So, what do the rocks tell us about the initiation of plate tectonics on Earth?

In the Beginning

In their 2014 paper in *Nature*, David Bercovici and Yanick Ricard opined that 'The emergence of plate tectonics is arguably Earth's defining moment'.[68] But exactly when was that moment? Was it during the planet's salad days, now represented by the oldest rocks in existence, or did it, like the emergence of animals and plants, require billions of years of gestation? Since the computer modellers cannot answer this question, we have no choice but to turn to the geologists for help in deciphering the record of conditions on the young Earth, which, owing to the thoroughness of subduction in eliminating oceanic crust from the planet's surface, is limited to the most ancient rocks preserved on the continents. For their part, geologists have been only too happy to oblige, providing estimates that range from the Hadean, 4.5 billion years ago, to the Ordovician, around 485 million years ago—a span representing 89 per cent of Earth history.

At one extreme, there are those who suggest that plate tectonics was up and running soon after the Moon-forming impact, around 4.5 billion years ago. This group includes Mark Harrison (at the University of California, Los Angeles), who presented evidence from zircons[69] that he believed showed that conditions like the present Earth, with liquid water, felsic (continental) crust, and even plate boundaries, were in existence during the Hadean.[70] Most of the arguments for a very early onset of plate tectonics are likewise based on the age of zircons. If the oldest zircons (from the Canadian Acasta gneiss) were formed in rocks as old as 4.4 billion years, and if these rocks were felsic, as is generally the case with zircons,

[67] Dewey (2007).
[68] Bercovici and Ricard (2014).
[69] See Chapter 7 for a discussion of their use.
[70] Harrison (2009).

and if these felsic rocks formed continents created at volcanic arcs like those existing today, then it follows that plate tectonics must have been operating in the Hadean. However, there are rather too many 'ifs' in there for most people's liking. More solid evidence from actual rocks leads many others to believe that plate tectonics had commenced by the early Archean, around 4 billion years ago. One such is Kevin Burke, who, on the basis of his extensive field experience, believes that Archean greenstone[71] belts are a clear analogue of modern oceanic processes, representing deformed crust and volcanic arcs from former oceans. Kevin sees no fundamental differences between plate tectonic processes on the early Earth and those operating today, while offering the caveat that many of his previous suggestions have been proved wrong.[72]

There are problems with interpretations such as Kevin's and Mark's. One is that Archean rocks sport a very different structure to later rock outcrops, consisting of domed granite and gneiss bodies, surrounded by downward-folded greenstone belts,[73] implying formation by processes unlike those we see operating today. These ancient outcrops lack the diagnostic markers of plate tectonics, such as ophiolites, blueschists, and ultra-high-pressure terranes. Ophiolites—chunks of oceanic plate stranded on continental crust, like those in Cyprus and Oman that record the death throes of the Tethys Ocean—typically (but not universally) occur in the sequence pillow lavas, sheeted dykes, gabbros, and peridotites, the latter frequently altered to serpentinites. These days, most researchers believe that the crust preserved in ophiolites formed, not at mid-ocean ridges, but by sea-floor spreading in the fore-arc region above subduction zones, or maybe behind volcanic arcs, in back-arc basins. Hence, ophiolites may at once represent evidence both for spreading *and* subduction. Many geologists have searched for ancient ophiolites that would be accepted as convincing evidence of Archean plate tectonics. The prize for the world's oldest ophiolite was claimed in 2007 for an outcrop of sheeted dykes in the Isua region of Greenland, dated at 3.8 billion years.[74] This interpretation has, however, been challenged on various grounds, in particular, that the sequence is by no

[71] Greenstone belts were discussed in Chapter 6. They comprise a number of rock types, but are dominated by metamorphosed basalts, containing green amphibole minerals.

[72] Burke (2011).

[73] Or 'synclines', as they are known to geologists.

[74] Furnes et al. (2007).

means complete in the same way as younger ophiolites, and that sheeted dykes may in any case form in other ways besides sea-floor spreading. The oldest *complete* ophiolite, the Dongwanzi ophiolite, was discovered in China, around 250 km northeast of Beijing.[75] This has a well-defined age of 2.5 billion years and must therefore qualify as the world's oldest undisputed ophiolite.

Blueschists are metamorphosed basalts containing characteristic sodium-rich blue amphibole minerals, interpreted as oceanic crust that has been altered in the low-temperature, but high-pressure, conditions of subduction zones at depths of 40–60 km. It appears that the oldest blueschists are also found in China, where the Aksi blueschists have been dated at around 750 million years. Ultra-high-pressure terranes are outcrops of metamorphosed continental rocks containing coesite and diamonds that have also been subducted, perhaps to over 100 km depth, following which, like blueschists, the altered rocks have somehow managed to find their way back to the surface to testify to their experience. The oldest ultra-high-pressure terranes so far identified are located in Africa, and are dated at 620 million years.

Despite the absence of these markers in Archean rocks, a number of geologists have attempted to fit the geological and geochemical evidence from Archean cratons into standard plate tectonics theory, identifying greenstone belts as oceanic remnants accreted to the margins of embryonic granite and gneiss continents. Ali Polat (at the University of Windsor, Canada) and his colleagues presented, in 2015, an interpretation of the geology of the West Greenland craton,[76] formed during the period from 3.85 to 2.55 billion years ago, entirely in terms of 'uniformitarian'[77] plate tectonics, invoking deep subduction much like that observed today. Fault-bounded outcrops within the craton were identified as relict island arcs accreted to Archean continents at convergent plate margins, following closure of ocean basins beginning around 3.7 billion years ago. Gneisses, representing crust metamorphosed by heat and pressure, form the majority of outcrops, and are interpreted by Ali and

[75] Kusky et al. (2001).

[76] Polat et al. (2015).

[77] Uniformitarianism, in addition to earning a potentially hefty score at Scrabble (actually, Super Scrabble, since it is too long to fit on a standard Scrabble board), is the term applied to the principle that geological evolution may be explained, in the words of the famous nineteenth-century geologist Charles Lyell, 'by reference to causes now in operation'.

colleagues as having originated as plutons like those existing beneath the modern Andean volcanic arc. More recently, Ali has joined with Kevin Burke and several others to present an interpretation of a 3 billion-year-old greenstone belt, also in West Greenland, as an ophiolite, representing crust produced by spreading in a back-arc basin.[78] Allen Nutman (at the University of Wollongong, Australia) and colleagues have likewise interpreted rocks in the Isua belt of southwest Greenland as an assemblage developed around 3.7 billion years ago at an intra-oceanic subduction zone.[79] The same authors also published a proposal that the Wilson Cycle was already underway at this time,[80] with all ten of the oldest surviving gneiss complexes forming a supercontinent by 3.66 billion years ago. Break-up of the supercontinent, they suggested, was prompted by the arrival of a mantle plume head beneath the Pilbara craton (now in western Australia), producing a LIP that incorporated felsic rocks produced by crustal melting.

At the opposite extreme to these researchers, we find a trio of well-respected mavericks: Bob Stern, John Dewey, and Warren Hamilton, all of whom believe that plate tectonics began within the past billion years (i.e. recently). In his 2008 article, entitled 'Modern-Style Plate Tectonics Began in Neoproterozoic Time: An Alternative Interpretation of Earth's Tectonic History', Bob accepts the lack of evidence for Archean blueschists, ultra-high-pressure terranes, and ophiolites as evidence of absence (Figure 11.7). He suggests a date of around 1 billion years for the onset of 'modern-style' plate tectonics, whilst allowing the probability of a primitive form of plate tectonics—'proto-plate tectonics', he calls it—as early as 2.7 billion years ago, before which unstable stagnant-lid tectonics was, he believes, dominant. John Dewey concurs with much of this,[81] but believes that plate tectonics began even later, just 600 million years ago, based mainly on the same reasoning, namely that blueschists, ophiolites, and ultra-high-pressure terranes become common after this time.[82] Prior to this, John believes that plate tectonics underwent a long gestation, beginning around 3 billion years ago (following the early permobile regime) with what he also calls proto-plate tectonics,

[78] Polat et al. (2016).

[79] Nutman et al. (2015a).

[80] Nutman et al. (2015b).

[81] Dewey (2007).

[82] However, recent calculations (Palin and White, 2015) suggest that the absence of older blueschists may be a result, not of the absence of deep subduction, but of the higher magnesium content of subducted Precambrian basalt crust.

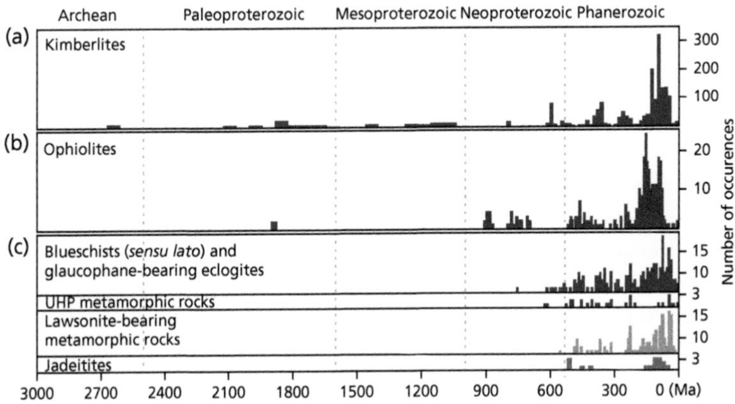

Figure 11.7 Distinctive plate tectonic and subduction indicators for the past 3 Ga of Earth history. [From Stern et al. (2016).]

which lasted only 300 million years or so, and was followed by a long period of roughly 2 billion years of 'evolving' plate tectonics (plate tectonics *sensu lato*), prior to the final transition to plate tectonics (plate tectonics *sensu stricto*) in the late Precambrian.

The most extreme views were expressed by Warren Hamilton. In a defiant article published in 2011,[83] Warren declared that 'Plate tectonics began in Neoproterozoic time, and plumes from deep mantle have never operated'. Leaving aside the second claim, which we dealt with in Chapter 10, his assertion that plate tectonics began late in Earth history was based, once again, on the absence of direct evidence for subduction, such as blueschists, ophiolites, and ultra-high-pressure terranes. In apparent contradiction of the title of his article, Warren suggested that fully modern plate tectonics processes did not begin until the Ordovician, just 485 million years ago, conceding that subduction may have begun somewhat earlier, around 850 million years ago. He also suggested that a thick, mafic, 'protocrust' formed over the entire globe as early as 4.45 billion years ago, from which all later continental crust was derived.

Consensus, of Sorts

Despite all the controversy, it is fair to say that the consensus view today is that plate tectonics began somewhere around 3 billion years ago

[83] Hamilton (2011).

(although one should always bear in mind that consensus, as well illustrated by the continental drift controversy, may ultimately count for little). An important study in 2011, by Steven Shirey and Stephen Richardson[84] of the Carnegie Institution of Washington and the University of Cape Town, respectively, examined a large suite of minerals that were trapped inside diamonds brought up in kimberlite pipes from depths between 125 and 175 km, corresponding to the mantle part of the lithosphere that underlies the crust of ancient cratons (often referred to as 'deep continental keels'). Such minerals are believed to be the oldest and most pristine samples of mantle compositions available, and should therefore record any major changes in mantle melting through Earth history. Precise dating of trapped silicate and sulphide minerals revealed an abrupt change, 3 billion years ago, from purely peridotite compositions, indicating partial melting of typical mantle, to compositions incorporating eclogite, which, as we know, is formed by high-pressure alteration of (subducted) basalt crust, implying transport from the surface to depths of over 125 km beneath continents. Hence, Steve and Steve concluded, this date marked the onset of deep subduction of oceanic crust, and hence of modern-style plate tectonics. They envisaged a change from a regime of crustal formation by mantle melting above rising hot plumes, during the period from 4 to 3 billion years ago, to one of crustal formation above steeply dipping slabs of cold oceanic crust after 3 billion years ago.

As you may recall from Chapter 7, a profound change in the net rate of continental crust production is believed to have occurred at this time. In 2012, a Bristol group, including Bruno Dhuime and Chris Hawkesworth, estimated that this rate had declined from around 3 km^3 per year before 3 billion years ago, to just 0.8 km^3 per year afterwards,[85] which they attributed to the onset of subduction that has since removed continental crust almost as fast as it was created.[86] In 2015, the group used the ratio of rubidium to strontium in a collection of over 13,000 igneous rock samples from crust ranging in age from the Hadean to the Phanerozoic to argue that the crust prior to 3 billion years ago was largely mafic (silica content in the range 45–52 per cent), as might be

[84] Shirey and Richardson (2011).

[85] Dhuime et al. (2012). Within errors, this agrees with the conclusion of Bob Stern and David Scholl that the yang of continent removal more-or-less balances the yin of its creation (Chapter 6).

[86] Mainly by subduction erosion at the base of the overriding plate.

found in oceanic plateaux, while higher ratios after that time showed that the crust became more felsic (silica content around 57 per cent, making it, strictly, intermediate) and probably thicker.[87] Like the two Steves, they ascribed this transition to the onset of plate tectonics and the founding of the modern subduction factory. High-pressure laboratory experiments[88] suggested that the earlier Archean (i.e. prior to 3 billion years ago) was dominated by limited shallow subduction of thick mafic crust like that forming oceanic plateaux today. Such shallow subduction would not produce a transition of basaltic crust to eclogite, and would not, therefore, provide a driving traction for plate motions. In addition, seismic studies show that the crust of most cratons older than 2.5 billion years is thinner and more felsic than later crust, with a sharp Moho.[89] These characteristics are generally explained by delamination of the more mafic lower crust in the hotter mantle of the early Earth, as discussed in Chapter 6. Younger cratons tend to have thicker, denser, intermediate crust, and frequently a less sharp Moho. Geochemical evidence also points to a major shift 2.5 billion years ago,[90] suggesting a decrease in the degree and depth of mantle melting at that time.

Two of the oldest (3.65 to 3.22 billion years) and best-preserved Archean cratons are the Kaapvaal craton of southern Africa and the Pilbara craton of western Australia (mentioned previously). As an illustration of the subjectivity involved in geological interpretations, Martin Kranendonk, at the University of New South Wales, and his colleagues[91] point out that, despite the very close similarities between the two, published explanations for their formation are diametrically opposed, with the Barberton terrane in the Kaapvaal craton commonly explained by modern-style subduction–accretion, while the East Pilbara terrane in western Australia is attributed to an origin involving a thick volcanic plateau formed above a mantle plume. As they demonstrate, both can be explained by a tectonic regime very different to modern plate tectonics. In this form of tectonics, thick volcanic plateaux, rather like the Ontong Java Plateau (Chapter 7), formed upon older continental crust as a consequence of extensive mantle melting above 'upwelling mantle' (they cautiously avoid using the word 'plume'), rather than by moving

[87] Dhuime et al. (2015).
[88] Hastie et al. (2016).
[89] Yuan (2015).
[90] Keller and Schoene (2012); Condie et al. (2016).
[91] For example, Smithies et al. (2007) and Van Kranendonk et al. (2015).

plates. This type of crust is, Martin and colleagues note, distinct from Archean gneiss terranes, which they believe are formed as a result of shallow subduction and melting of the thickened and more buoyant plateau lithosphere, a process dubbed 'sagduction'. After 3.2 billion years ago, both the Barberton and Pilbara cratons were affected by 'modern-style' plate tectonics, including rifting, steep subduction, and terrane accretion, that together heralded the commencement of the Wilson Cycle. The thick, buoyant, continental keels created previously ensured that these cratons avoided subduction in the new plate tectonics regime, and so formed the nucleus of our modern continents.

Incorporating what is perhaps one of the world's worst puns into the title of their paper, Jean Bédard, at the Geological Survey of Canada, and his Canadian colleagues have likened the search for evidence of Archean plate tectonics to the hunting of the Snark.[92] In particular, they conclude that there is no convincing evidence for the existence of Archean oceanic crust created by spreading, and thus Archean arcs, such as those proposed by Ali Polat and Allen Nutman, are really Snarks. Based on their interpretation of cratons such as the Pilbara, Jean and his friends propose that Archean tectonics involved the drifting of cratonic cores, now represented by granitic domes, driven by viscous drag of the hot, convecting, mantle on their deep keels, a mechanism that you will instantly recognize as virtually identical to that suggested by Arthur Holmes as the main driving force for continental drift. During their travels, the cratons would sweep up basaltic plateaux and other cratonic fragments, resulting in compression at their leading edges and, perhaps, sea-floor spreading in their wakes. Oceanic plateaux might be overridden and forced down into the hot mantle, where they melted to produce the tonalite magmas characteristic of Archean cratons, but deep subduction did not occur. Hence, they propose, continental drift began in the early Archean, a billion years before modern-style subduction and plate tectonics. In a more recent paper,[93] Jean suggests that Archean basalts and komatiites[94] (high-magnesium

[92] Bédard et al. (2013). For anyone unfamiliar with Lewis Carroll, this is a futile search for a non-existent animal.

[93] Bédard (2017).

[94] Named after the Komati river in the Barberton region of South Africa. The Barberton greenstone belt was the place where the first finds of gold were found in South Africa.

lava types produced by high-temperature mantle melting, common in Archean rocks but only rarely found in younger outcrops) formed above giant, long-lived, mantle plumes during stagnant-lid intervals lasting hundreds of millions of years. During these intervals, convection in the hot upper mantle led to periodic instability and overturn of the stagnant lid like that suggested for Venus. Cooling of the mantle and lithosphere gradually increased the density and rigidity of the oceanic lithosphere, so that, by 2.5 billion years ago, modern-style deep subduction zones began to form, taking slabs all the way to the core–mantle boundary and initiating whole-mantle convection that cooled the mantle further. The Archean was thus, according to Jean, a time of mantle plume volcanism and episodic continental drift, with periodic overturn of stagnant lids. Archean plumes differed from modern plumes, since, while the latter mainly originate in slab graveyards and hence carry the geochemical imprint of subducted plates, Archean plumes ascended from a deep mantle in which no subducted plates were present.

Summing Up

In a slightly abstruse, but nonetheless fascinating, review of the subject, Gary Ernst, Norman (Norm) Sleep (both at Stanford University), and Tatsuki Tsujimori (at Tohoku University) have set out a likely sequence of events leading to modern plate tectonics.[95] Following the impact that resulted in formation of the Moon around 4.5 million years ago, a 300 km-deep magma ocean existed for tens of millions of years, before a solid lid formed, beneath which vigorous convection continued, driven by decay of long-lived (e.g. ^{238}U) and short-lived (e.g. ^{26}Al) radioisotopes, together with core formation. They refer to this period (approximately 4.5 to 4.4 billion years ago) as the 'Hadean stage', during which small-scale mantle convection cells exerted strong tractions on thin surface 'platelets' and eliminated all the original 'magma ocean' crust by shallow subduction (which they call 'underflow'). This was followed by the 'Hadean–Archean' stage (4.4 to 2.7 billion years ago), at the beginning of which the planet cooled sufficiently for steam to have rained out of the atmosphere, creating a global ocean. By 4 billion years ago, the

95 Ernst et al. (2016).

Earth had entered Burke and Dewey's 'permobile' or 'proto-plate tectonics' mode, during which 'the Earth's surface was covered by small, soft, hot, weakly subductable platelets'. Thick oceanic crust was produced by decompression melting of numerous giant plumes rising rapidly through the hotter, lower-viscosity, mantle. Partial melting of this crust produced the granite–greenstone–tonalite complexes typical of Archean continental crust. The mantle, being hotter by several hundred degrees, was able to melt the thinner, sinking, slabs, producing komatiite magmas that are common in Archean terranes. Continued growth of these complexes led, eventually, to the creation of the first supercontinent around 2.7 billion years ago, near the end of the Archean, marking the onset of the Wilson cycle. This, in turn, resulted in limited stagnant-lid tectonics (i.e. thermal blanketing) beneath the supercontinent, raising the temperature of the upper mantle, leading to break-up.

As the mantle cooled during the later Archean, there was a shift from plume-dominated (or 'bottom-up' tectonics) to plate-dominated (or 'top-down') tectonics. As Gary and Norm put it, 'plume ascent was more important attending the high-temperature early Earth, whereas by degrees, plate sinking came to dominate later, cooler, stages of mantle flow'. Continued mantle cooling led to the creation of thicker, cooler, and larger plates with thinner crust, which became denser than the underlying mantle and thus unstable, initiating subduction. By 3 billion years ago, the scale of plate tectonics had increased to the point that mantle convection encompassed the entire mantle, and rigid plates with horizontal dimensions of thousands of kilometres began to form. The contrast between Archean komatiites, greenstones, and grey gneisses, and post-Archean alkaline igneous rocks, blueschists, and eclogites can therefore be attributed to the reduction in the internal temperature of the Earth.

It is becoming apparent from recent research that the transition from 'proto-plate tectonics' to true 'modern-style' plate tectonics took place over a period of hundreds of millions of years, with deep subduction developing in some places even as plume-lid tectonics continued elsewhere. This interpretation is supported by computer modelling of subduction on a hotter Earth by Ria Fischer and Taras Gerya,[96] suggesting that the transition from plume-lid tectonics to plate tectonics

[96] Fischer and Gerya (2016a; 2016b).

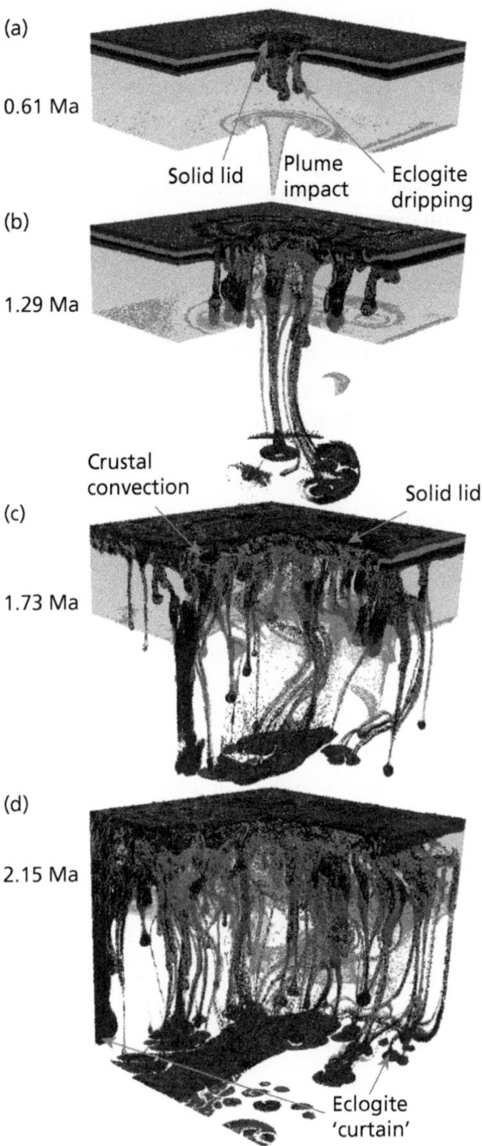

Figure 11.8 Computer model of Archean plume-lid tectonics by Fischer and Gerya (2016a). (a) a mantle plume impacts the lithosphere causing partial melting and volcanism, resulting in a thickened mafic crust above the plume. The additional pressure causes the lower crust to turn to dense, unstable, eclogite which begins to 'drip' into the mantle. (b) small buoyant 'plumes' or 'diapirs' are produced by the sinking eclogite, which rise to the surface. (c/d) a thin lid forms on top of a vigorously convecting crust and a 'curtain' of eclogite drippings is created.

occurred gradually over an extended period as the mantle cooled. At the hottest mantle conditions, corresponding to the early Archean, their model produced 'small-scale mantle convection, unstable dripping lithosphere, thick basaltic crust and small plates'. At lower temperatures, intermediate characteristics, such as shallow subduction, were observed, becoming, by degrees, more like modern plate tectonics, with deep subduction replacing dripping slabs and lithosphere (Figure 11.8). In addition, as others have noted, it may be the case that periods of deep subduction alternated, or co-existed, with plume-lid tectonics for more than a billion of years. Hence, somewhat anticlimactically, it is beginning to seem likely that there was no 'defining moment' in Earth history.

What Has Plate Tectonics Ever Done for Us?

As geophysicists and astrobiologists now acknowledge, plate tectonics plays a central role in preparing a planet for life, and also in maintaining that life once sparked. Where plates are created at mid-ocean ridges, hydrothermal vents provide a likely setting for the origin of microbial life and the subsequent evolution of ecosystems based on reduced chemicals emanating from the newly formed crust. Where plates are recycled into the mantle, the subduction factory generates continental crust and releases water and carbon dioxide back into the atmosphere, creating a world in which continents composed of thick, buoyant, crust, are surrounded by thinner, denser, oceanic crust. The basins thus formed are, in the case of the Earth, filled with just the right volume of surface water to produce oceans (in which life could originate) and land (on which salamanders could evolve into *Homo sapiens*). Without plate tectonics, there would be hardly any land, making the evolution of, say, mammals, impossible, since even marine mammals evolved from land-dwelling ancestors.[97]

Plate tectonics also controls long-term climate, primarily through the carbon cycle, as we discovered in Chapter 9. Carbon dioxide is delivered to the oceans and atmosphere at mid-ocean ridges and volcanic arcs, and removed by weathering of exposed rocks, primarily on continents but also on the ocean floor. Supply and removal are tightly linked, forming a negative-feedback mechanism that keeps CO_2 concentration at trace

[97] The same goes for insects and reptiles.

levels, presently 0.04 per cent of the atmosphere. In a nutshell, an increase in supply of CO_2 through volcanism leads to an increase in surface temperature by means of the greenhouse effect, which leads to faster weathering, which leads to faster removal of CO_2 from the atmosphere, followed by transport of carbonates in rivers and deposition in the oceans, which leads to a reduction in atmospheric CO_2, which leads to a decrease in surface temperature.[98] Likewise, a reduction in volcanism results in a reduced rate of weathering and a decrease in the rate of CO_2 removal, reversing the initial cooling effect. Hence, these processes constitute a thermostat, keeping the Earth's surface temperature within the range of liquid water through geological time (apart from the odd ice age), and making the place habitable.

Yet, just as there is little point in having a posh kitchen if you don't have the ingredients to cook anything, so there is not much value in a planet possessing the right physical conditions for life if the raw materials are not available. For life as we know it, the recipe includes a shopping list of elements needed to synthesize large organic molecules such as nucleic acids and proteins. On the early Earth, it seems that hydrothermal vents provided reduced chemicals—such as methane, hydrogen sulphide, and hydrogen—that powered the first microbes. Seawater circulating through the hot crust leached out these chemicals and delivered them conveniently, if unreliably, at hot springs that very likely became the cradle of microbial life. Animals and plants have more exacting requirements, including the nutrients potassium, calcium, sulphur, iron, nitrogen, and, in particular, phosphorus. Without phosphorus, you cannot make RNA, DNA, or ATP (not to mention teeth and bones), and, without these, you can forget it. Once again, plate tectonics steps in, concentrating these elements in the magma produced by partial melting of the mantle beneath mid-ocean ridges—magma that ultimately forms oceanic crust. Re-melting of this mafic crust, as we saw in Chapter 7, produces granitic magma, concentrating the vital nutrients even further. Granite rock, however, while very useful for kitchen worktops, is not very handy if all you want are the trace elements within it to cook up some life. Yet again, plate tectonics comes to the rescue by creating mountain ranges, exposing granitic rock to rapid

[98] Given enough time, the same processes will remove the CO_2 that we have thoughtlessly contributed to the atmosphere. The devil, however, is in the words 'enough time'.

weathering: the same process that removes carbon dioxide and operates the global thermostat also transfers phosphorus, potassium, and iron to rivers that then transport them to the oceans, constantly replenishing nutrients for marine life.

Of course, animal life would have got nowhere without oxygen. Oxygen is the second most abundant element forming the planet as a whole, but, owing to its penchant for Austin Metros, it appears to have been confined to the interior of the Earth for the first 2 billion years of the planet's history, with little or none in the early atmosphere. No wonder, then, that life was restricted to rather boring microbes dependent on fickle hot springs for much of that time. Around 2.4 billion years ago, however, things began to improve, and oxygen concentration increased to something like 0.02 per cent of the atmosphere,[99] a development referred to grandly as 'the Great Oxygenation Event'. Oxygen in the atmosphere and oceans is contributed by photosynthesizing plants and phytoplankton, using sunlight to split water into its two constituent elements and combine them with carbon dioxide to produce living tissue. As every schoolchild knows,[100] photosynthesis releases oxygen gas directly into the atmosphere and oceans, some of which is recombined with carbon or hydrogen during respiration and organic decay. A proportion of organic matter, however, escapes decay and becomes buried in organic sediments, with the result that oxygen accumulates in the atmosphere. The more organic carbon that is buried, the more oxygen that is liberated, so an increase in carbon burial will produce a corresponding increase in the concentration of oxygen in the atmosphere and oceans, other things being equal.

Meanwhile, as we discussed in Chapter 10, deep subduction drives mantle convection that cools the core and regulates the geodynamo, creating a protecting veil over the planet that helps to conserve the atmosphere and reduce genetic damage to living cells caused by the solar wind. Another threat to living cells is ultraviolet radiation, especially at wavelengths between 200 and 300 nm. As you will no doubt remember, the appearance of a large hole in the stratospheric ozone layer above Antarctica caused consternation during the 1980s, because of the potential for damage to the DNA of living cells caused by the

[99] Planavsky et al. (2014).
[100] This may be wishful thinking.

increased intensity of ultraviolet light reaching the surface.[101] Without oxygen (O_2) in the atmosphere, there can be no ozone (O_3) layer, so that, in the absence of continents and continental weathering, terrestrial Archean microbes would have been exposed to genetic damage, perhaps inhibiting (or maybe promoting) biological evolution. Studies of the ratios of sulphur isotopes preserved in Precambrian sediments[102] have revealed an abrupt change in the fractionation[103] of the rare stable isotopes ^{33}S and ^{34}S, synchronous with the Great Oxidation Event, a change widely attributed to the appearance of oxygen in the atmosphere and the creation of an ozone shield around 2.4 billion years ago.

However, life was slow to exploit these changes until almost a billion years after the Great Oxidation Event, almost certainly because the levels of oxygen in the atmosphere and shallow seas were still very low compared with the concentrations that multicellular animals like us demand. All this changed in what geologists now call the 'Neoproterozoic' (see Figure 7.10), when oxygen levels rose a hundred-fold to more than 2 per cent of the atmosphere. Once again, this has been linked to an increase in productivity and deposition of organic sediments following a period of enhanced nutrient supply during mountain building, this time associated with the assembly of Gondwana, during which extensive mountain ranges were formed.[104] Now the oxygen concentration was sufficient to allow animals to evolve rapidly, producing, first, the Ediacaran forms between 610 and 541 million years ago, followed by the Cambrian explosion between 541 and 500 million years ago, which saw the appearance of the first vertebrates.

As you probably noticed, the Great Oxidation Event occurred close to the time that many believe 'true' plate tectonics began, reinforcing the theory that oxidation of the atmosphere and oceans, like the regulation of climate, was ultimately facilitated by the formation of plates

[101] Leading to the Montreal Protocol to limit CFCs. You may be surprised, and perhaps dismayed, to learn that the hole is still with us, and is even larger now than when discovered in 1985.

[102] Farquar et al. (2000).

[103] Today, fractionation of sulphur isotopes is dependent on their masses (i.e. the number of neutrons in the nucleus). However, before 2.4 billion years ago, this was not the case, and fractionation was mass-independent, a phenomenon explained by photochemical reactions between sulphur compounds released by volcanoes and ultraviolet light during the Archean.

[104] Santosh et al. (2014).

and the onset of deep subduction.[105] One explanation for this is that the manufacture of more felsic continental crust at subduction zones reduced the potential for the removal of oxygen by weathering, since iron- and magnesium-rich mafic crust is more readily oxidized than the andesitic and felsic rocks produced at volcanic arcs.[106] Once the newly created continents were subjected to collisions, the first supercontinents could form, the collision zones being marked by mountain belts that quickly eroded, releasing nutrients such as iron and phosphorus into rivers and oceans, promoting a global bloom of photosynthesising microbes.[107] Burial of organic carbon produced by these microbes allowed some of the oxygen gas so generated to remain in the atmosphere, leading to a sustained increase in concentration. Cin-Ty Lee at Rice University, and colleagues,[108] have followed the same line of reasoning, suggesting that the production of large volumes of felsic continental crust was the result of the transition from stagnant-lid to mobile-lid tectonics and the initiation of deep subduction. The latter serves to return water to the upper mantle, where it enables the production of hydrous magmas necessary for the operation of the subduction factory and the manufacture of felsic crust. The same folk suggest that the second, Neoproterozoic, rise in atmospheric oxygen concentration was triggered by the accumulation of carbon on the continents, leading to an increase in CO_2 in the oceans and atmosphere, and hence increased organic productivity and burial, thereby liberating more oxygen. They conclude that the two-step rise in oxygen may thus be 'a natural consequence of plate tectonics, continent formation, and the growth of a carbon reservoir'.

Given the still-shaky level of current understanding of the processes controlling biological evolution, it may seem somewhat premature, or even presumptuous, to start contemplating the factors that could lead, not just to single-celled microbial life, and not just to complex life such as jellyfish, but to the kind of life you could befriend on Facebook, or, in Bob Stern's words, 'technological species'.[109] Yet Bob has recently taken

[105] For a more balanced discussion of the Great Oxidation Event and alternative explanations (of which there are many) for the oxidation of the atmosphere and oceans, see Lenton and Watson (2011) or Langmuir and Broecker (2012).

[106] The primary sinks for atmospheric oxygen are Fe^{2+} and S^{2-} ions, both of which are less abundant in felsic rocks than in mafic rocks.

[107] Campbell and Allen (2008).

[108] Lee et al. (2016).

[109] Stern (2016).

the bold step of speculating on this subject, adopting the parsimonious view that conditions similar to those here on Earth would be most conducive to the descent of something like Man elsewhere in the Cosmos. He conducts a thought experiment, imagining evolution on two planets, one with plate tectonics, and one without, noting that, with regard to the appearance of technological species, environmental pressures that result in adaptation, innovation, and speciation are necessary, and concluding that such pressures are far more likely to be experienced on a plate tectonics planet. Successful evolution resulting from what Bob calls the 'plate tectonics pump' creates 'bioassets', such as appendages for grasping and manipulating, efficient eyesight, and a complex central nervous system, allowing intelligent creatures (as he somewhat fancifully puts it) 'to examine the night sky and wonder'.

Of course, like the Romans, plate tectonics has its drawbacks, as the folks in Indonesia, Japan, and New Zealand know only too well. In the geological past, widespread volcanism, heralding the rifting of Pangea and the birth of the North Atlantic,[110] resulted in the emplacement of the LIP known as the Central Atlantic Magmatic Province within North America, Europe, and Africa, an event that was coeval with a mass extinction that ended the Triassic period, 201 million years ago. LIPs formed by the arrival of mantle plumes have caused even greater chaos, notably the Siberian Traps and Deccan Traps that ended the Permian and Cretaceous periods, respectively. On the other hand, it might be argued that, by clearing the planet of unwanted reptiles, these catastrophes opened the way for the evolution and spread of mammals.

Around ten million years after the last of these mass extinctions, a group of small tree-dwelling Asian monkeys found their way across the shrinking Tethys Ocean into North Africa (which then included Arabia), where they found the extensive tropical forests much to their liking. They proceeded to evolve into a range of new species that thrived in their adopted continent, as did later arrivals, such as rodents and cats. About 30 million years ago, things started to change as Africa's northward motion slowed and a hot mantle plume impacted the north-eastern part of the continent (beneath present-day Ethiopia), ripping great gashes in the crust that subsequently evolved to become the Red Sea and Gulf of Aden. A third gash penetrated southward into

[110] Thought to be related to thermal blanketing rather than a mantle plume.

East Africa without opening a new ocean, transforming the landscape from tropical forested lowland to a volcanic rift valley flanked by jagged mountain peaks. Within the newly created rift, shielded from moist winds by the adjacent uplands, the climate became much drier, and what rainfall there was collected in large lakes formed in depressions created by rifting. Tropical forest was replaced by a fragmented patchwork of vegetation, including newly-evolved C_4 grasses. Surprisingly, the descendents of the Asian primates adapted well to this new, unstable, landscape, using their rapidly developing brains and dexterous limbs to survive. By two million years ago, a bipedal species emerged that was much larger than any that had gone before, with a brain roughly twice the size of its predecessors, giving it the ability to create simple tools, with which it hunted in organized groups, eventually returning to Asia and expanding its range far to the east. Many anthropologists now believe that fragmentation of the East African landscape and the development of ephemeral lakes, resulting from the formation of a new plate boundary, were the major drivers of the emergence of the genus *Homo*.[111] More controversially, a major pulse of volcanism within the Rift between 370,000 and 170,000 years ago could have been the trigger for further evolution associated with the appearance of *Homo sapiens*.[112] As a species, therefore, we have an awful lot to thank plate tectonics for.

Afterthoughts

Recall that, prior to 1965, continental drift was little more than a somewhat esoteric academic theory. The rise of first-generation plate tectonics in the late 1960s provided the necessary physical and mathematical framework that eventually convinced even Americans of its veracity, yet it remained largely confined to academia. In schools and colleges, students were (and frequently still are) given garbled accounts of drift theory, incorporating ideas that were superseded long ago. And, along with the frequent misuse of 'tectonic plates' as a political metaphor, there remains widespread misunderstanding of plate tectonics and its central role in the evolution of this planet.

[111] Trauth et al. (2015).

[112] For a more expert, and no doubt more coherent, account of human evolution, see Maslin (2017).

Today, second-generation plate tectonics is maturing into a dynamic theory of planetary cooling that involves the entire globe, from the surface to the inner core. Drifting continents are recognized as the visible expression of moving plates, and plates, in turn, constitute the surface boundary layer of a great engine that delivers heat energy from the interior to the surface by mantle convection. At the core–mantle boundary, this engine is linked to another great engine that transfers heat produced by inner-core formation and radioactive decay in the outer core up into the mantle. Plates, cooled at the surface, regulate slow mantle convection by means of subduction, and mantle convection, in turn, regulates the much faster convection in the core, incidentally generating the reversing geomagnetic field that protects the atmosphere and records plate motions. At the surface, the plates link mantle convection to a third great heat engine driving the circulation of the atmosphere and oceans. Plate tectonics regulates this engine by operating the global thermostat, supplying greenhouse gases at mid-ocean ridges and volcanic arcs, and removing them by creating mountain ranges where carbon dioxide is extracted during rock weathering and then transported in solution to the oceans, where it is locked away in sediments. At the same time, vital nutrients are weathered out and transported to support life in the oceans. Furthermore, plate tectonics alters the geography of continents and ocean basins, leading to long-term changes in ocean currents and global climate. By making and breaking connections between adjacent continents and oceans, the movement of plates guides the course of biological evolution, encouraging speciation on isolated continents, and merging formerly independent ecosystems on colliding continents. Hence, and quite remarkably, plate tectonics has evolved from a geometrical description of moving paving stones into the crucial factor that permits the emergence of 'technological species' here, and perhaps elsewhere.

In their violent youth, all planets are hotheads, starting out with so much surplus energy that their surfaces melt to form magma oceans. Such oceans lose internal heat to space rapidly—so rapidly that, within a few tens of millions of years, the surface of a young planet cools and solidifies to create a hot stagnant lid above a vigorously convecting mantle. Plumes rising through the mantle may generate a new oceanic crust in a 'bottom-up' form of tectonics, the high degree of melting producing a thick, light, crust above a thin layer of rigid mantle. Such lithosphere is buoyant and remains at the surface, forming a stagnant lid that inhibits subduction—a condition that Bob Stern calls 'trench lock'. Heating from below makes the base of the lid unstable, causing it to delaminate and sink, or, in extreme cases, the entire lid to collapse into the interior, as may have occurred beneath the northern hemisphere of Mars early in its history, creating the Martian Dichotomy, or on Venus around 600 million years ago, completely renewing its surface.

Sooner or later, every planet arrives at a major fork in the road to heat death. Most will continue along the stagnant-lid highway, perhaps with occasional overturn of their surfaces, but a few will take the road less travelled, the road that leads to plate tectonics. This was the road chosen by the Earth back at the end of the Archean. For us, and all multicellular life, this turned out to be a very good thing, but, like all good things, it will come to an end. For plate tectonics is its own worst enemy, being such an efficient mode of shedding heat[113] that, in a billion years or so, it will have cooled the Earth sufficiently that the upper mantle will no longer be able to melt as plates are ripped apart at mid-ocean ridges. At about the same time, the oceans will have evaporated owing to the gradual increase in the Sun's luminosity,[114] finally exposing the tennis-ball seam just as sea-floor spreading comes to an end. The consequence will be what Bob Stern calls 'ridge lock': opposing plates will become welded together and the planet will pass into its cold stagnant-lid phase. The Earth, therefore, currently occupies a narrow space between two fatal extremes, and has done so for several billion years, long enough for microbes to evolve into Man. The demise of plate-driven convection will, however, spell the end of the world as we know it: without the life-support services provided by plate tectonics, the Earth will once again be inhabited only by microbes, eventually becoming as lifeless as its neighbours. If the inhabitants of a planet orbiting a star in the elliptical galaxy known affectionately as NGC 4709, 215 million light-years distant in the constellation Centaurus, launched a spacecraft today, travelling, like Stephen Hawking's laser-propelled nanoprobes, at a fifth of the speed of light, it would arrive to find a planet rather like Venus: hot and dead.

At the turn of the century, a highly influential book appeared, entitled *Rare Earth*. Written by palaeontologist Peter Ward and astronomer

[113] The present ratio of heat production within the Earth to heat loss at the surface is around 0.4.

[114] Laconte et al. (2013).

Donald Brownlee,[115] it marshalled all the evidence then available concerning the conditions necessary for the evolution of animal life on a rocky planet, creating a long list of requirements that included a circular orbit at the right distance from a star in the right region of the galaxy, the stabilizing effect of a large moon, and protection from asteroid impacts by gas giant planets orbiting further out. At the top of their list was plate tectonics, required to manufacture continents and subsequently regulate atmospheric greenhouse gases via the global thermostat. Pete and Don remarked that 'It may seem odd to think that plate tectonics could be not only the cause of mountain chains and ocean basins but also, and most enigmatically, a key to the evolution and preservation of complex metazoans on Earth', concluding: 'Of all of the attributes that make Earth rare, plate tectonics may be one of the most profound and—in terms of the evolution and maintenance of animal life—one of the most important.' They noted that, while surface water in the liquid state was necessary, too much would produce oceans so deep that no land existed and therefore no rock weathering could occur to draw down carbon dioxide and prevent a runaway greenhouse. Too little would very likely prevent continents from being created in the first place, and, even if continents did exist, there would be insufficient shallow seas in which carbonates could be sequestered. Once again, the Earth finds itself in a very narrow space between two fatal extremes, with sufficient exposed continental crust to remove atmospheric carbon dioxide and supply nutrients, and sufficient shallow seas to sequester carbonates and provide a home for marine life.

All in all, it is becoming apparent that a combination of rather unlikely circumstances is required in order for a planet to become habitable for animal life, chief of which is the operation of plate tectonics. And even in the very rare instances in which such conditions conspire to produce a world just right for multicellular life to flourish, biological evolution may be interrupted or terminated by unfortunate events such as giant volcanic eruptions, unwonted arrivals from space, or the appearance of 'technological species'.

[115] Ward and Brownlee (2000).

References

Allaby, M. (Ed.) (2013) *A Dictionary of Geology and Earth Sciences*, 4th edition. Oxford University Press, Oxford.

Allwardt, A. O. (1990) The roles of Arthur Holmes and Harry Hess in the development of modem global tectonics. PhD thesis, University of California, Santa Cruz.

Anderson, D. L. (1982) Hotspots, polar wander, Mesozoic convection and the geoid. *Nature*, 297, 391–3.

Arculus, R. J. (2003) Use and abuse of the terms calcalkaline and calcalkalic. *Journal of Petrology*, 44, 929–35.

Arculus, R. J., Ishizuka, O., Bogus, K. A., et al. (2015) A record of spontaneous subduction initiation in the Izu–Bonin–Mariana arc. *Nature Geoscience*, 8, 728–33.

Argus, D. F., Gordon, R. G. & DeMets, C. (2011) Geologically current motion of 56 plates relative to the no-net-rotation reference frame. *Geochemistry Geophysics Geosystems*, 12, 1–13.

Ariel, A. & Berger, N. A. (2006) *Plotting the Globe: Stories of Meridians, Parallels, and the International Date Line*. Praeger, Westport, CT.

Armstrong, R. L. (1981) Radiogenic isotopes: The case for crustal recycling on a near-steady-state no-continental-growth Earth. *Philosophical Transactions of the Royal Society of London A: Mathematical, Physical and Engineering Sciences*, 301, 443–72.

Arndt, N. & Davaille, A. (2013) Episodic Earth evolution. *Tectonophysics*, 609, 661–74.

Arndt, N. T. & Nisbet, E. G. (2012) Processes on the young Earth and the habitats of early life. *Annual Review of Earth and Planetary Sciences*, 40, 521–49.

Austermann, J., Kaye, B. T., Mitrovica, J. X. & Huybers, P. (2014) A statistical analysis of the correlation between large igneous provinces and lower mantle seismic structure. *Geophysical Journal International*, 197, 1–9.

Badro, J., Fiquet, G., Guyot, F., et al. (2003) Iron partitioning in Earth's mantle: Toward a deep lower mantle discontinuity. *Science*, 300, 789–91.

Baes, M., Gerya, T. & Sobolev, S. V. (2016) 3-D thermo-mechanical modeling of plume-induced subduction initiation. *Earth and Planetary Science Letters*, 453, 193–203.

Baes, M., Govers, R. & Wortel, R. (2011) Subduction initiation along the inherited weakness zone at the edge of a slab: Insights from numerical models. *Geophysical Journal International*, 184, 991–1008.

Baker, E. T. (2009) Relationships between hydrothermal activity and axial magma chamber distribution, depth, and melt content. *Geochemistry Geophysics Geosystems*, 10, 1–15.

Baker, E. T., German, C. R. & Elderfield, H. (1995) Hydrothermal plumes over spreading-center axes: Global distributions and geological inferences. In Humphris, S. E., Zierenberg, R. A., Mullineaux, L. S. & Thomson, R. E. (Eds.) *Seafloor Hydrothermal Systems: Physical, Chemical, Biological, and Geological Interactions*. Geophysical Monograph 91, American Geophysical Union, Washington, DC, 47–71.

Baker, E. T., Massoth, G. J. & Feely, R. A. (1987) Cataclysmic hydrothermal venting on the Juan de Fuca Ridge. *Nature*, 329, 149–51.

Baker, J. A. (1967) *The Peregrine*. Collins, London.

Ballard, R. D. (2000) *The Eternal Darkness*. Princeton University Press, Princeton, NJ.

Barazangi, M. & Dorman, J. (1969) World seismicity maps compiled from ESSA, Coast and Geodetic Survey, epicenter data, 1961–1967. *Bulletin of the Seismological Society of America*, 59, 369–80.

Barrett, P. J., Fielding, C. R. & Wise, S. W. (1998) Initial Report on CRP-1, Cape Roberts Project, Antarctica. *Terra Antarctica*, 5, 1–187.

Barron, E. J., Fawcett, P. J., Pollard, D. & Thompson, S. (1993) Model simulations of Cretaceous climates: The role of geography and carbon dioxide. *Philosophical Transactions of the Royal Society of London B: Biological Sciences*, 341, 307–15.

Barron, E. J., Sloan, J. L. & Harrison, C. G. A. (1980) Potential significance of land–sea distribution and surface albedo variations as a climatic forcing factor; 180 m.y. to the present. *Palaeogeography Palaeoclimatology Palaeoecology*, 30, 17–40.

Barron, E. J. & Washington, W. M. (1984) The role of geographic variables in explaining paleoclimates: Results from Cretaceous climate model sensitivity studies. *Journal of Geophysical Research: Atmospheres*, 89, 1267–79.

Beaulieu, S. E., Baker, E. T. & German, C. R. (2015) Where are the undiscovered hydrothermal vents on oceanic spreading ridges? *Deep-Sea Research, Part Ii: Topical Studies in Oceanography*, 121, 202–12.

Bédard, J. H. (2017) Stagnant lids and mantle overturns: Implications for Archaean tectonics, magmagenesis, crustal growth, mantle evolution, and the start of plate tectonics. *Geoscience Frontiers*, http://dx.doi.org/10.1016/j.gsf.2017.01.005.

Bédard, J. H., Harris, L. B. & Thurston, P. C. (2013) The hunting of the snArc. *Precambrian Research*, 229, 20–48.

Bell, D. B., Jung, S. J. A., Kroon, D., et al. (2015) Atlantic deep-water response to the Early Pliocene shoaling of the Central American Seaway. *Scientific Reports*, 5, 1–12.

Bell, R. T. & Jefferson, C. W. (1987) An hypothesis for an Australian–Canadian connection in the Late Proterozoic and the birth of the Pacific Ocean. *Proceedings, Pacific Rim Congress '87: Parkville, Victoria, Australian Institute of Mining and Metallurgy*.

Belliver, A. (1956) Henri Poincaré ou la vocation souveraine. Gallimard, Paris.

Benfield, A. E. (1950) The Earth's magnetism. *Scientific American*, 182(6), 20–4.

Bercovici, D. & Ricard, Y. (2014) Plate tectonics, damage and inheritance. *Nature*, 508, 513–16.

Bercovici, D., Tackley, P. J. & Ricard, Y. (2015) The generation of plate tectonics from mantle dynamics. In Schubert, G. (Ed.) *Treatise on Geophysics*,2nd edition. Elsevier, Oxford, 271–318.

Berner, R. A. (2006) GEOCARBSULF: A combined model for Phanerozoic atmospheric O_2 and CO_2. *Geochimica et Cosmochimica Acta*, 70, 5653–64.

Berner, R. A. & Lasaga, A. C. (1989) Modeling the geochemical carbon cycle. *Scientific American*, 260(3), 74–81.

Berner, R. A., Lasaga, A. C. & Garrels, R. M. (1983) The carbonate–silicate geochemical cycle and its effect on atmospheric carbon dioxide over the past 100 million years. *American Journal of Science*, 283, 641–83.

Berner, R. A. & Maasch, K. A. (1996) Chemical weathering and controls on atmospheric O_2 and CO_2: Fundamental principles were enunciated by J. J. Ebelmen in 1845. *Geochimica et Cosmochimica Acta*, 60, 1633–37.

Biggin, A. J., Piispa, E. J., Pesonen, L. J., et al. (2015) Palaeomagnetic field intensity variations suggest Mesoproterozoic inner-core nucleation. *Nature*, 526, 245–8.

Birch, F. (1954) The Earth's mantle elasticity and constitution. *Transactions of the American Geophysical Union*, 35, 79–98.

Bird, P. (2003) An updated digital model of plate boundaries. *Geochemistry Geophysics Geosystems*, 4, 1–52.

Blackett, P. M. S. (1952) A Negative experiment relating to magnetism and the Earth's rotation. *Philosophical Transactions of the Royal Society of London A: Mathematical and Physical Sciences*, 245, 309–70.

Blackett, P. M. S., Clegg, J. A. & Stubbs, P. H. S. (1960) An analysis of rock magnetic data. *Proceedings of the Royal Society of London A: Mathematical, Physical and Engineering Sciences*, 256, 291–322.

Brass, G. W., Saltmann, E., Sloan, J. L., et al. (1982) Ocean circulation, plate tectonics, and climate. In *Climate in Earth History: Studies in Geophysics*. National Academies Press, Washington, DC, 83–9.

Bruckshaw, J. M. & Robertson, E. I. (1948) The measurement of magnetic properties of rocks. *Journal of Scientific Instruments*, 25, 444.

Bryson, B. (2003) *A Short History of Nearly Everything*. Doubleday, New York.

Buck, W. R., Lavier, L. L. & Poliakov, A. N. B. (2005) Modes of faulting at mid-ocean ridges. *Nature*, 434, 719–23.

Buffett, B. (2012) Geomagnetism under scrutiny. *Nature*, 485, 319–20.

Buffett, B. A. (2000) Earth's core and the geodynamo. *Science*, 288, 2007–12.

Buffett, B. A. (2015) Core–mantle interactions. In Schubert, G. (Ed.) *Treatise on Geophysics* 2nd edition. Elsevier, Oxford, 213–24.

Bull, A. L., Domeier, M. & Torsvik, T. H. (2014) The effect of plate motion history on the longevity of deep mantle heterogeneities. *Earth and Planetary Science Letters*, 401, 172–82.

Bullard, E. (1969) The origin of the oceans. *Scientific American*, 221(3), 66–75.

Bullard, E. (1975a) The emergence of plate tectonics: A personal view. *Annual Review of Earth and Planetary Sciences*, 3, 1–30.

Bullard, E. (1975b) William Maurice Ewing. 12 May 1906–4 May 1974. *Biographical Memoirs of Fellows of the Royal Society of London*, 21, 268–311.

Bullard, E. C. & Day, A. (1961) The flow of heat through the floor of the Atlantic Ocean. *Geophysical Journal of the Royal Astronomical Society*, 4, 282–92.

Bullard, E. C., Everett, J. E. & Smith, A. G. (1965) The fit of the continents around the Atlantic. *Philosophical Transactions of the Royal Society of London A: Mathematical and Physical Sciences*, 258, 41–51.

Bullen, K. E. (1938) Composition of the Earth at a depth of 500–700 km. *Nature*, 142, 671–72.

Bullen, K. E. (1963) *An Introduction to the Theory of Seismology*, 3rd edition. Cambridge University Press, Cambridge.

Burchfiel, B. C. & Davis, G. A. (1972) Structural framework and evolution of the southern part of the Cordilleran Orogen, Western United States. *American Journal of Science*, 272, 97–118.

Bürgmann, R. & Dresen, G. (2008) Rheology of the lower crust and upper mantle: Evidence from rock mechanics, geodesy, and field observations. *Annual Review of Earth and Planetary Sciences*, 36, 531–67.

Burke, K. (2007). Dancing continents (review of *Supercontinent*, by Ted Nield). *Science*, 318, 1385.

Burke, K. (2011) Plate tectonics, the Wilson Cycle, and mantle plumes: Geodynamics from the top. *Annual Review of Earth and Planetary Sciences*, 39, 1–29.

Burke, K. & Torsvik, T. H. (2004) Derivation of large igneous provinces of the past 200 million years from long-term heterogeneities in the deep mantle. *Earth and Planetary Science Letters*, 227, 531–8.

Burke, K. & Wilson, J. T. (1972) Is the African plate stationary? *Nature*, 239, 387–90.

Burke, K. C. & Wilson, J. T. (1976) Hot spots on the Earth's surface. *Scientific American*, 235(2), 46–57.

Burke, W. H., Denison, R. E., Hetherington, E. A., et al. (1982) Variation of seawater $^{87}Sr/^{86}Sr$ throughout Phanerozoic time *Geology*, 10, 516–19.

Burov, E. B. & Watts, A. B. (2006) The long-term strength of continental lithosphere: 'Jelly sandwich' or 'crème brûlée'? *GSA Today*, 16, 4–10.

Byrne, P. K., Klimczak, C., Sengor, A. M. C., et al. (2014) Mercury's global contraction much greater than earlier estimates. *Nature Geoscience*, 7, 301–7.

Campbell, I. H. & Allen, C. M. (2008) Formation of supercontinents linked to increases in atmospheric oxygen. *Nature Geoscience*, 1, 554–8.

Canales, J. P., Sohn, R. A. & Demartin, B. J. (2007) Crustal structure of the Trans-Atlantic Geotraverse (TAG) segment (Mid-Atlantic Ridge, 26°10'N): Implications for the nature of hydrothermal circulation and detachment faulting at slow spreading ridges. *Geochemistry Geophysics Geosystems*, 8, 1–18.

Cane, M. A. & Molnar, P. (2001) Closing of the Indonesian seaway as a precursor to east African aridification around 3-4 million years ago. *Nature*, 411, 157–62.

Cann, J. (1991) Onions and leaks: magma at mid-ocean ridges. *Oceanus*, 34, 36–41.

Cann, J. R. (1974) A model for oceanic crustal structure developed. *Geophysical Journal of the Royal Astronomical Society*, 39, 169–87.

Cann, J. R., Blackman, D. K., Smith, D. K., et al. (1997) Corrugated slip surfaces formed at ridge–transform intersections on the Mid-Atlantic Ridge. *Nature*, 385, 329–32.

Cann, J. R. & Strens, M. R. (1982) Black smokers fuelled by freezing magma. *Nature*, 298, 147–9.

Cann, J. R. & Vine, F. J. (1966) An area on the crest of the Carlsberg Ridge: Petrology and magnetic survey. *Philosophical Transactions of the Royal Society of London A: Mathematical and Physical Sciences*, 259, 198–217.

Carey, S. W. (1988) *Theories of the Earth and Universe: A History of Dogma in the Earth Sciences*, Stanford University Press, Stanford, CA.

Castle, J. C., Creager, K. C., Winchester, J. P. & van der Hilst, R. D. (2000) Shear wave speeds at the base of the mantle. *Journal of Geophysical Research: Solid Earth*, 105, 21543–57.

Cawood, P. A. & Buchan, C. (2007) Linking accretionary orogenesis with super-continent assembly. *Earth-Science Reviews*, 82, 217–56.

Cawood, P. A. & Hawkesworth, C. J. (2014) Earth's middle age. *Geology*, 42, 503–6.

Cawood, P. A., Hawkesworth, C. J. & Dhuime, B. (2013) The continental record and the generation of continental crust. *Geological Society of America Bulletin*, 125, 14–32.

Cawood, P. A., Strachan, R. A., Pisarevsky, S. A., et al. (2016) Linking collisional and accretionary orogens during Rodinia assembly and breakup: Implications for models of supercontinent cycles. *Earth and Planetary Science Letters*, 449, 118–26.

Chamberlin, T. C. (1899) Lord Kelvin's address on the age of the earth as an abode fitted for life. *Science*, 9, 889–901.

Chandler, M. T., Wessel, P., Taylor, B., et al. (2012) Reconstructing Ontong Java Nui: Implications for Pacific absolute plate motion, hotspot drift and true polar wander. *Earth and Planetary Science Letters*, 331, 140–51.

Chase, C. G. (1972) The *N* plate problem of plate tectonics. *Geophysical Journal of the Royal Astronomical Society*, 29, 117–22.

Chase, C. G. (1978) Plate kinematics: The Americas, East Africa, and the rest of the world. *Earth and Planetary Science Letters*, 37, 355–68.

Chase, C. G. (1979) Subduction, the geoid, and lower mantle convection. *Nature*, 282, 464–8.

Chase, C. G. (1981) Oceanic island Pb: Two-stage histories and mantle convection. *Earth and Planetary Science Letters*, 52, 277–84.

Chase, C. G. & Sprowl, D. R. (1983) The modern geoid and ancient plate boundaries. *Earth and Planetary Science Letters*, 62, 314–20.

Chen, J. H. (2016) Lower-mantle materials under pressure. *Science*, 351, 122–3.

Chen, W. P. & Molnar, P. (1983) Focal depths of intracontinental and intraplate earthquakes and their implications for the thermal and mechanical properties of the lithosphere. *Journal of Geophysical Research*, 88, 4183–214.

Choi, C. Q. (2017) Inner workings: diamond anvils probe the origins of Earth's magnetic field. *Proceedings of the National Academy of Sciences of the USA*, 114, 1215–16.

Chopin, C. (1984) Coesite and pure pyrope in high-grade blueschists of the western Alps: A first record and some consequences. *Contributions to Mineralogy and Petrology*, 86, 107–18.

Chorost, M. (2013) The plate tectonic wars. *Astronomy Now*, June, 18–20.

Chroston, P. N. & Simmons, G. (1989) Seismic velocities from the Kohistan Volcanic Arc, Northern Pakistan. *Journal of the Geological Society*, 146, 971–9.

Chung, S. L., Chu, M. F., Zhang, Y. Q., et al. (2005) Tibetan tectonic evolution inferred from spatial and temporal variations in post-collisional magmatism. *Earth-Science Reviews*, 68, 173–96.

Clegg, J. A., Almond, M. & Stubbs, P. H. S. (1954) The remanent magnetism of some sedimentary rocks in Britain. *Philosophical Magazine*, 45, 583–98.

Clegg, J. A., Deutsch, E. R. & Griffiths, D. H. (1956) Rock magnetism in India. *Philosophical Magazine*, 1, 419–31.

Coakley, B. J. (1998) Forty days in the belly of the beast. In 'The Oceans'. *Scientific American Presents*, 9(3), 36–7.

Coltice, N., Phillips, B. R., Bertrand, H., et al. (2007) Global warming of the mantle at the origin of flood basalts over supercontinents. *Geology*, 35, 391–4.

Condie, K. C., Aster, R. C. & van Hunen, J. (2016) A great thermal divergence in the mantle beginning 2.5 Ga: Geochemical constraints from greenstone basalts and komatiites. *Geoscience Frontiers*, 7, 543–53.

Coney, P. J., Jones, D. L. & Monger, J. W. H. (1980) Cordilleran suspect terranes. *Nature*, 288, 329–33.

Connerney, J. E. P., Acuna, M. H., Ness, N. F., et al. (2005) Tectonic implications of Mars crustal magnetism. *Proceedings of the National Academy of Sciences of the USA*, 102, 14970–5.

Contreras, L., Pross, J., Bijl, P. K., et al. (2013) Early to Middle Eocene vegetation dynamics at the Wilkes Land Margin (Antarctica). *Review of Palaeobotany and Palynology*, 197, 119–42.

Coode, A. M. (2011) Part of the history of the origin of transform faults. *Earth Sciences History*, 30, 58–62.

Corfield, R. (2003) *The Silent Landscape*. Joseph Henry Press, Washington, DC.

Costa, F. & Chakraborty, S. (2008) The effect of water on Si and O diffusion rates in olivine and implications for transport properties and processes in the upper mantle. *Physics of the Earth and Planetary Interiors*, 166, 11–29.

Cox, A. (1973) *Plate Tectonics and Geomagnetic Reversals.* W H Freeman, New York.

Cox, A., Dalrymple, G. B. & Doell, R. R. (1964) Geomagnetic polarity epochs. *Science*, 143, 351–2.

Cox, A. & Doell, R. R. (1960) Review of paleomagnetism. *Geological Society of America Bulletin*, 71, 645–768.

Cox, A., Doell, R. R. & Dalrymple, G. B. (1964) Reversals of the Earth's magnetic field. *Science*, 144, 1537–43.

Coxall, H. K., Wilson, P. A., Palike, H., et al. (2005) Rapid stepwise onset of Antarctic glaciation and deeper calcite compensation in the Pacific Ocean. *Nature*, 433, 53–7.

Crameri, F. & Tackley, P. J. (2016) Spontaneous development of arcuate single-sided subduction in global 3-D mantle convection models with a free surface. *Journal of Geophysical Research-Solid Earth*, 119, 5921–42.

Crameri, F., Tackley, P. J., Meilick, I., et al. (2012) A free plate surface and weak oceanic crust produce single-sided subduction on Earth. *Geophysical Research Letters*, 39, 1–7.

Creager, K. C. & Jordan, T. H. (1986) Aspherical structure of the core–mantle boundary from PKP travel times. *Geophysical Research Letters*, 13, 1497–500.

Creer, J. M., Irving, E. & Runcorn, S. K. (1954) The direction of the geomagnetic field in remote epochs in Great Britain. *Journal of Geomagnetism and Geoelectricity*, 6, 163–8.

Creer, K. M. & Irving, E. (2012) Testing continental drift: Constructing the first palaeomagnetic path of polar wander (1954). *Earth Sciences History*, 31, 111–45.

Croll, J. (1875) *Climate and Time in their Geological Relations: A Theory of Secular Changes of the Earth's Climate.* Adam and Charles Black, Edinburgh.

Crowell, J. C. & Frakes, L. A. (1970) Phanerozoic glaciation and the causes of ice ages. *American Journal of Science*, 268, 193–224.

Crowley, J. W., Katz, R. F., Huybers, P., et al. (2015) Glacial cycles drive variations in the production of oceanic crust. *Science*, 347, 1237–40.

Daly, R. A. (1923) The Earth's crust and its stability. *American Journal of Science*, Ser. 5, 5, 349–71.

Dalziel, I. W. D. (1991) Pacific margins of Laurentia and East Antarctica–Australia as a conjugate rift pair: Evidence and implications for an Eocambrian supercontinent. *Geology*, 19, 598–601.

Davaille, A., Smrekar, S. E. & Tomlinson, S. (2017) Experimental and observational evidence for plume-induced subduction on Venus. *Nature Geoscience*, 10, 349–55.

Davies, D. R., Goes, S., Davies, J. H., et al. (2012) Reconciling dynamic and seismic models of Earth's lower mantle: The dominant role of thermal heterogeneity. *Earth and Planetary Science Letters*, 353, 253–69.

Davies, D. R., Goes, S. & Lau, H. C. P. (2015) Thermally dominated deep mantle LLSVPs: A review. In Khan, A. & Deschamps, F. (Eds.) *Earth's Heterogeneous Mantle: A Geophysical, Geodynamical, and Geochemical Perspective.* Springer, Cham, Switzerland, 441–77.

Davies, G. F. (1977) Whole-mantle convection and plate tectonics. *Geophysical Journal of the Royal Astronomical Society*, 49, 459–86.

Davies, G. F. & Gurnis, M. (1986) Interaction of mantle dregs with convection: Lateral heterogeneity at the core–mantle boundary. *Geophysical Research Letters*, 13, 1517–20.

Davy, B. (2014) Rotation and offset of the Gondwana convergent margin in the New Zealand region following Cretaceous jamming of Hikurangi Plateau large igneous province subduction. *Tectonics*, 33, 1577–95.

DeConto, R. M. & Pollard, D. (2003) Rapid Cenozoic glaciation of Antarctica induced by declining atmospheric CO_2. *Nature*, 421, 245–49.

DeMets, C., Gordon, R. G., Argus, D. F. & Stein, S. (1990) Current plate motions. *Geophysical Journal International*, 101, 425–78.

Dewey, J. F. (2003) Plate tectonics and geology, 1965 to today. In Oreskes, N. (Ed.) *Plate Tectonics: An Insider's History of the Modern Theory of the Earth*. Westview Press, Boulder, CO, 227–42.

Dewey, J. F. (2007) The secular evolution of plate tectonics and the continental crust: An outline. In Hatcher, R. D., Jr, Carlson, M. P., Mcbride, J. H. & Martínez Catalán, J. R. (Eds.) *4-D Framework of Continental Crust*. Geological Society of America Memoir 200. Geological Society of America, Boulder, CO, 1–7.

Dewey, J. F. (2015) A harbinger of plate tectonics: a commentary on Bullard, Everett and Smith (1965) 'The fit of the continents around the Atlantic'. *Philosophical Transactions of the Royal Society of London A: Mathematical and Physical Sciences*, 373, 1–9.

Dewey, J. F. & Burke, K. (1974) Hot spots and continental break-up: Implications for collisional orogeny. *Geology*, 2, 57–60.

Dewey, J. F. & Burke, K. C. A. (1973) Tibetan, Variscan, and Precambrian basement reactivation: Products of continental collision. *Journal of Geology*, 81, 683–92.

Dewey, J. F., Cande, S. & Pitman, W. C. (1989) Tectonic evolution of the India/Eurasia collision zone. *Eclogae Geologicae Helvetiae*, 82, 717–34.

Dewey, J. F. & Horsfield, B. (1970) Plate tectonics, orogeny and continental growth. *Nature*, 225, 521–5.

Dhuime, B., Hawkesworth, C. & Cawood, P. (2011) When continents formed. *Science*, 331, 154–5.

Dhuime, B., Wuestefeld, A. & Hawkesworth, C. J. (2015) Emergence of modern continental crust about 3 billion years ago. *Nature Geoscience*, 8, 552–5.

Dickinson, W. R. (1970) Relations of andesites, granites, and derivative sandstones to arc-trench tectonics. *Reviews of Geophysics and Space Physics*, 8, 813–60.

Dickinson, W. R. & Hatherton, T. (1967) Andesitic volcanism and seismicity around the Pacific. *Science*, 157, 801–3.

Dietz, R. S. (1961) Continent and ocean basin evolution by spreading of the sea floor. *Nature*, 190, 854–7.

Dietz, R. S. (1977) Plate tectonics: A revolution in geology and geophysics. *Tectonophysics*, 38, 1–6.

Dietz, R. S. (1994) Earth, sea, and sky: Life and times of a journeyman geologist. *Annual Review of Earth and Planetary Sciences*, 22, 1–32.

Dietz, R. S. & Holden, J. C. (1970) Reconstruction of Pangaea: Breakup and dispersion of continents, Permian to Present. *Journal of Geophysical Research*, 75, 4939–56.

Dobson, D. (2016) Earth's core problem. *Nature*, 534, 45–5.

Doell, R. R. (1971) Memorial to John Warren Graham 1918–1971. *Geological Society of America Memorials*, III, 105–8.

Donn, W. L. & Shaw, D. M. (1977) Model of climate evolution based on continental drift and polar wandering. *Geological Society of America Bulletin*, 88, 390–6.

Dott, R. H. (1979) The geosyncline—First major geological concept 'made in America'. In Schneer, C. (Ed.) *Two Hundred Years of Geology in America: Proceedings of the New Hampshire Bicentennial Conference on the History of Geology*. University Press of New England, Hanover, NH, 238–67.

DSDP (2007) *Deep Sea Drilling Project Reports and Publications*, Volume 29. http://www.deepseadrilling.org/29/dsdp_toc.htm.

Du Bois, P. M., Irving, E., Opdyke, N. D., et al. (1957) The geomagnetic field in Upper Triassic times in the United States. *Nature*, 180, 1186–7.

du Toit, A. L. (1937) *Our Wandering Continents: An Hypothesis of Continental Drifting*, Oliver and Boyd, Edinburgh.

Duque-Caro, H. (1990) The Choco Block in the northwestern corner of South America: Structural, tectonostratigraphic, and paleogeographic implications. *Journal of South American Earth Sciences*, 3, 71–84.

Dziewonski, A. M., Hager, B. H. & Oconnell, R. J. (1977) Large-scale heterogeneities in the lower mantle. *Journal of Geophysical Research*, 82, 239–55.

Eakins, B. W. & Sharman, G. F. (2012) Hypsographic curve of Earth's surface from ETOPO1. *NOAA Technical Memorandum NESDIS NGDC-2*. NOAA National Geophysical Data Center, Boulder, CO.

Eaton, J. P. & Murata, K. J. (1960) How volcanoes grow. *Science*, 132, 925–38.

Elsasser, W. M. (1946) Induction effects in terrestrial magnetism, Part I. Theory. *Physical Review*, 69, 106–16.

Elsasser, W. M. (1968) The mechanics of continental drift. *Proceedings of the American Philosophical Society*, 112, 344–53.

Elsasser, W. M. (1971) Sea-floor spreading as thermal convection. *Journal of Geophysical Research*, 76, 1101–12.

Elsasser, W. M., Olson, P. & Marsh, B. D. (1979) The depth of mantle convection. *Journal of Geophysical Research*, 84, 147–55.

Elsworth, G., Galbraith, E., Halverson, G. & Yang, S. M. (2017) Enhanced weathering and CO_2 drawdown caused by latest Eocene strengthening of the Atlantic meridional overturning circulation. *Nature Geoscience*, 10, 213–16.

Emiliani, C. & Edwards, G. (1953) Tertiary ocean bottom temperatures. *Nature*, 171, 887–8.

England, P., Molnar, P., and Richter, F. (2017) John Perry's neglected critique of Kelvin's age for the Earth: A missed opportunity in geodynamics. *GSA Today*, 17(1), 4–9.

Ernst, W. G., Sleep, N. H. & Tsujimori, T. (2016) Plate-tectonic evolution of the Earth: Bottom-up and top-down mantle circulation. *Canadian Journal of Earth Sciences*, 53, 1103–20.

Escalona, A. & Mann, P. (2011) Tectonics, basin subsidence mechanisms, and paleogeography of the Caribbean–South American plate boundary zone. *Marine and Petroleum Geology*, 28, 8–39.

Exon, N. F., Kennett, J. P., Malone, M. J. (2004) Leg 189 Synthesis: Cretaceous–Holocene history of the Tasmanian Gateway. In Exon, N. F., Kennett, J. P. & Malone, M.J. (Eds.) *Proceedings of the Ocean Drilling Program, Scientific Results*, Volume 189. 1. http://www-odp.tamu.edu/publications/189_SR/synth/synth.htm.

Fei, H. Z., Wiedenbeck, M., Yamazaki, D. & Katsura, T. (2013) Small effect of water on upper-mantle rheology based on silicon self-diffusion coefficients. *Nature*, 498, 213–15.

Ferreiro, L. D. (2011) *Measure of the Earth: The Enlightenment Expedition that Reshaped our World*. Basic Books, New York.

Fischer, A. G. (1981) Climatic oscillations in the biosphere. In Nitecki, M. H. (Ed.) *Biotic Crises in Ecological and Evolutionary Time*. Academic Press, New York, 103–31.

Fischer, A. G. (1984) The two Phanerozoic supercycles. In Berggren, W. A. & Van Couvering, J. (Eds.) *Catastrophies and Earth History*. Princeton University Press, Princeton, NJ, 129–50.

Fischer, R. & Gerya, T. (2016a) Early Earth plume-lid tectonics: A high-resolution 3D numerical modelling approach. *Journal of Geodynamics*, 100, 198–214.

Fischer, R. & Gerya, T. (2016b) Regimes of subduction and lithospheric dynamics in the Precambrian: 3D thermomechanical modelling. *Gondwana Research*, 37, 53–70.

Foley, B. J., Bercovici, D. & Landuyt, W. (2012) The conditions for plate tectonics on super-Earths: Inferences from convection models with damage. *Earth and Planetary Science Letters*, 331, 281–90.

Forsyth, D. W. (1972) Mechanisms of earthquakes and plate motions in the East Pacific. *Earth and Planetary Science Letters*, 17, 189–93.

Fortey, R. (2004) *The Earth: An Intimate History*. HarperCollins, London.

Fowler, A. C. (1985) Fast thermoviscous convection. *Studies in Applied Mathematics*, 72, 189–219.

Fowler, C. M. R. (1978) Mid-Atlantic Ridge: Structure at 45°N. *Geophysical Journal of the Royal Astronomical Society*, 54, 167–83.

Frankel, H. (1987) Jan Hospers and the rise of paleomagnetism. *Eos, Transactions American Geophysical Union*, 68, 577–81.

Frankel, H. R. (2012a) *The Continental Drift Controversy*, Volume I: *Wegener and the Early Debate.* Cambridge University Press, Cambridge.

Frankel, H. R. (2012b) *The Continental Drift Controversy*, Volume II: *Paleomagnetism and Confirmation of Drift.* Cambridge University Press, Cambridge.

Frankel, H. R. (2012c) *The Continental Drift Controversy*, Volume III: *Introduction of Seafloor Spreading.* Cambridge University Press, Cambridge.

Frankel, H. R. (2012d) *The Continental Drift Controversy*, Volume IV: *Evolution into Plate Tectonics.* Cambridge University Press, Cambridge.

Frankel, H. R. (2014) Edward Irving's palaeomagnetic evidence for continental drift (1956). *Episodes*, 37, 59–62.

Frohlich, C. (2006) *Deep Earthquakes.* Cambridge University Press, Cambridge.

Fukao, Y. & Obayashi, M. (2015) Deep Earth structure—Subduction zone structure in the mantle transition zone. In Schubert, G. (Ed.) *Treatise on Geophysics*, 2nd edition. Elsevier, 641–54.

Furnes, H., de Wit, M., Staudigel, H., et al. (2007) A vestige of Earth's oldest ophiolite. *Science*, 315, 1704–7.

Fyfe, W. S. (1960) The possibility of *d*-electron coupling in olivine at high pressures. *Geochimica et Cosmochimica Acta*, 19, 141–3.

Gailler, L., Arcay, D., Münch, P., et al. (2017) Forearc structure in the Lesser Antilles inferred from depth to the Curie temperature and thermo-mechanical simulations. *Tectonophysics*, 706/707, 71–90.

Garcia, E. S., Sandwell, D. T. & Smith, W. H. F. (2014) Retracking CryoSat-2, Envisat and Jason-1 radar altimetry waveforms for improved gravity field recovery. *Geophysical Journal International*, 196, 1402–22.

Garnero, E. J. & Helmberger, D. V. (1995) A very slow basal layer underlying large-scale low-velocity anomalies in the lower mantle beneath the Pacific: Evidence from core phases. *Physics of the Earth and Planetary Interiors*, 91, 161–76.

Garnero, E. J. & McNamara, A. K. (2008) Structure and dynamics of Earth's lower mantle. *Science*, 320, 626–8.

Gass, I. G. & Masson-Smith, D. (1963) The geology and gravity anomalies of the Troodos Massif, Cyprus. *Philosophical Transactions of the Royal Society of London A: Mathematical and Physical Sciences*, 255, 417–67.

German, C. R., Livermore, R. A., Baker, E. T., et al. (2000) Hydrothermal plumes above the East Scotia Ridge: an isolated high-latitude back-arc spreading centre. *Earth and Planetary Science Letters*, 184, 241–50.

Gerya, T. V., Connolly, J. A. D. & Yuen, D. A. (2008) Why is terrestrial subduction one-sided? *Geology*, 36, 43–6.

Gerya, T. V., Stern, R. J., Baes, M., et al. (2015) Plate tectonics on the Earth triggered by plume-induced subduction initiation. *Nature*, 527, 221–5.

Gilbert, W. (1600) *De Magnete, magneticisque corporibus, et de magno magnete tellure.* Chiswick Press, London.

Girard, J., Amulele, G., Farla, R., et al. (2016) Shear deformation of bridgmanite and magnesiowüstite aggregates at lower mantle conditions. *Science*, 351, 144–7.

Glatzmaier, G. A. & Roberts, P. H. (1995) A three-dimensional self-consistent computer simulation of a geomagnetic-field reversal. *Nature*, 377, 203–9.

Glen, W. (1982) *The Road to Jaramillo: Critical Years of the Revolution in Earth Science.* Stanford University Press, Stanford, CA.

Goes, S., Capitanio, F. A., Morra, G., et al. (2011) Signatures of downgoing plate-buoyancy driven subduction in Cenozoic plate motions. *Physics of the Earth and Planetary Interiors*, 184, 1–13.

Goff, J. A. (2015) Comment on 'Glacial cycles drive variations in the production of oceanic crust'. *Science*, 349, 1065.

Graham, J. W. (1949) The stability and significance of magnetism in sedimentary rocks. *Journal of Geophysical Research*, 54, 131–67.

Grand, S. P. (1994) Mantle shear structure beneath the Americas and surrounding oceans. *Journal of Geophysical Research-Solid Earth*, 99, 11591–621.

Gutenberg, B. (Ed.) (1939) *The Internal Constitution of the Earth.* McGraw-Hill, New York.

Gutenberg, B. & Richter, C. F. (1941) *Seismicity of the Earth.* Geological Society of America, New York.

Hacker, B. R. & Gerya, T. V. (2013) Paradigms, new and old, for ultrahigh-pressure tectonism. *Tectonophysics*, 603, 79–88.

Hager, B. H., Clayton, R. W., Richards, M. A., et al. (1985) Lower mantle heterogeneity, dynamic topography and the geoid. *Nature*, 313, 541–6.

Haldane, J. B. S. (1963) Review of *The Truth About Death. Journal of Genetics*, 58, 463.

Halley, E. (1686) An Account of the Cause of the Change of the Variation of the Magnetical Needle; With an Hypothesis of the Structure of the Internal Parts of the Earth: As It Was Proposed to the Royal Society in One of Their Late Meetings. By Edm. Halley. *Philosophical Transactions*, 16, 563–78.

Hamilton, W. & Myers, W. B. (1966) Cenozoic tectonics of the Western United States. *Reviews of Geophysics*, 4, 509–49.

Hamilton, W. B. (1990) On terrane analysis. *Philosophical Transactions of the Royal Society of London A: Mathematical, Physical and Engineering Sciences*, 331, 511–22.

Hamilton, W. B. (2011) Plate tectonics began in Neoproterozoic time, and plumes from deep mantle have never operated. *Lithos*, 123, 1–20.

Hansen, U. & Yuen, D. A. (1988) Numerical simulations of thermal-chemical instabilities at the core–mantle boundary. *Nature*, 334, 237–40.

Harrison, C. G. A. (2016) The present-day number of tectonic plates. *Earth, Planets and Space*, 68, 1–14.

Harré, R. (1981) *Great Scientific Experiments: Twenty Experiments that Changed our View of the World.* Phaidon Press, Oxford.

Harrison, T. M. (2009) The Hadean crust: Evidence from >4 Ga zircons. *Annual Review of Earth and Planetary Sciences*, 37, 479–505.

Hastie, A. R., Fitton, J. G., Bromiley, G. D., et al. (2016) The origin of Earth's first continents and the onset of plate tectonics. *Geology*, 44, 855–8.

Hawkesworth, C., Cawood, P. & Dhuime, B. (2013) Continental growth and the crustal record. *Tectonophysics*, 609, 651–60.

Hawkesworth, C., Cawood, P., Kemp, T., et al. (2009) A matter of preservation. *Science*, 323, 49–50.

Hawkesworth, C. J., Cawood, P. A. & Dhuime, B. (2016) Tectonics and crustal evolution. *GSA Today*, 26, 4–11.

Hawkesworth, C. J. & Kemp, A. I. S. (2006) Evolution of the continental crust. *Nature*, 443, 811–17.

Hayden, L. A. & Watson, E. B. (2007) A diffusion mechanism for core–mantle interaction. *Nature*, 450, 709–11.

Haymon, R. M., Fornari, D. J., Edwards, M. H., et al. (1991) Hydrothermal vent distribution along the East Pacific Rise crest (9°09'–54'N) and its relationship to magmatic and tectonic processes on fast-spreading mid-ocean ridges. *Earth and Planetary Science Letters*, 104, 513–34.

Haymon, R. M., Fornari, D. J., Von Damm, K. L., et al. (1993) Volcanic eruption of the mid-ocean ridge along the East Pacific Rise crest at 9°45–52'N: Direct submersible observations of seafloor phenomena associated with an eruption event in April, 1991. *Earth and Planetary Science Letters*, 119, 85–101.

Hazen, R. M. (2012) *The Story of Earth: The First 4.5 Billion Years, from Stardust to Living Planet.* Viking, New York.

Herron, E. M. (1972) Two Small Crustal Plates in the South Pacific near Easter Island. *Nature*, 240, 35–7.

Herron, E. M. (1978) Structure of the East Pacific Rise crest from multichannel seismic reflection data. *Geological Society of America Memoirs*, 154, 683–701.

Hess, H. H. (1962) History of ocean basins. In Engel, A. E. G., James, H. L. & Leonard, B. F. (Eds.) *Petrologic Studies: A Volume To Honor A. F. Buddington.* Geological Society of America, New York, 599–620.

Hey, R. & Vogt, P. (1977) Spreading center jumps and sub-axial asthenosphere flow near the Galapagos hotspot. *Tectonophysics*, 37, 41–52.

Hey, R. N., Johnson, P. D., Martinez, F., et al. (1995) Plate boundary reorganization at a large-offset, rapidly propagating rift. *Nature*, 378, 167–70.

Hey, R. N., Kleinrock, M. C., Miller, S. P., et al. (1986) Sea beam/deep-tow investigation of an active oceanic propagating rift system, Galapagos 95.5°W. *Journal of Geophysical Research: Solid Earth and Planets*, 91, 3369–93.

Hide, R. (1996) Stanley Keith Runcorn, 4 November 1922–5 December 1995. *Quarterly Journal of the Royal Astronomical Society*, 37, 463–5.

Hill, D. J., Haywood, A. M., Valdes, P. J., et al. (2013) Paleogeographic controls on the onset of the Antarctic circumpolar current. *Geophysical Research Letters*, 40, 5199–204.

Hoffman, P. F. (1991) Did the breakout of Laurentia turn Gondwanaland inside-out? *Science*, 252, 1409–12.

Hofmann, A. W. & White, W. M. (1982) Mantle plumes from ancient oceanic crust. *Earth and Planetary Science Letters*, 57, 421–36.

Holden, J. C. & Vogt, P. R. (1977) Graphic solutions to problems of plumacy. *Eos, Transactions American Geophysical Union*, 56, 573–80.

Holmes, A. (1913) *The Age of the Earth*. Harper & Brothers, London.

Holmes, A. (1931) Radioactivity and earth movements. *Nature*, 128, 496–6.

Holmes, A. (1944) *Principles of Physical Geology*. Thomas Nelson, London.

Holmes, A. (1965) *Principles of Physical Geology*, 2nd edition. Thomas Nelson, London.

Hore, P. (2002) *Patrick Blackett: Sailor, Scientist, Socialist*. Frank Cass, London.

Hospers, J. (1954) Rock magnetism and polar wandering. *Nature*, 173, 1183–4.

Houser, C. (2016) Global seismic data reveal little water in the mantle transition zone. *Earth and Planetary Science Letters*, 448, 94–101.

Hu, X. M., Garzanti, E., Wang, J. G., et al. (2015) The timing of India–Asia collision onset—Facts, theories, controversies. *Earth-Science Reviews*, 160, 264–99.

Hurley, P. M. & Rand, J. R. (1969) Pre-drift continental nuclei. *Science*, 164, 1229–42.

Ichikawa, H., Yamamoto, S., Kawai, K. & Kameyama, M. (2016) Estimate of subduction rate of island arcs to the deep mantle. *Journal of Geophysical Research: Solid Earth*, 121, 5447–60.

IPCC (2003) *Climate Change 2013: The Physical Science Basis. Contribution of Working Group I to the Fifth Assessment Report of the Intergovernmental Panel on Climate Change*, Cambridge University Press, Cambridge.

Irving, E. (1956) Palaeomagnetic and palaeoclimatological aspects of polar wandering. *Geofisica pura e applicata*, 33, 63–70.

Irving, E. (1988) The paleomagnetic confirmation of continental drift. *Eos, Transactions American Geophysical Union*, 69, 994–1014.

Irving, E. & Green, R. (1958) Polar movement relative to Australia. *Geophysical Journal of the Royal Astronomical Society*, 1, 64–72.

Isacks, B., Oliver, J. & Sykes, L. R. (1968) Seismology and the new global tectonics. *Journal of Geophysical Research*, 73, 5855–99.

Ishii, M. & Tromp, J. (1999) Normal-mode and free-air gravity constraints on lateral variations in velocity and density of Earth's mantle. *Science*, 285, 1231–6.

Jackson, J. (2002) Strength of the continental lithosphere: Time to abandon the jelly sandwich? *GSA Today*, 12, 4–9.

Jacobs, J. A., Russell, R. D. & Wilson, J. T. (1974) *Physics and Geology*, 2nd edition. McGraw-Hill, New York.

Jagoutz, O. & Behn, M. D. (2013) Foundering of lower island-arc crust as an explanation for the origin of the continental Moho. *Nature*, 504, 131–4.

Jagoutz, O. & Kelemen, P. B. (2015) Role of arc processes in the formation of continental crust. *Annual Review of Earth and Planetary Sciences*, 43, 363–404.

Jakeš, P. & White, A. J. R. (1969) Structure of the Melanesian arcs and correlation with distribution of magma types. *Tectonophysics*, 8, 223–36.

Jeffreys, H. (1928) The times of transmission and focal depths of large earthquakes. *Monthly Notices of the Royal Astronomical Society, Geophysical Supplement*, 1, 500–21.

Johnson, E. A., Murphy, T. & Torreson, O. W. (1948) Pre-history of the Earth's magnetic field. *Terrestrial Magnetism and Atmospheric Electricity*, 53, 349–72.

Joly, J. (1925) *The Surface-History of the Earth.* Clarendon Press, Oxford.

Jones, G. M. (1977) Thermal interaction of the core and the mantle and long-term behavior of the geomagnetic field. *Journal of Geophysical Research*, 82, 1703–9.

Jones, M. T., Jerram, D. A., Svensen, H. H. & Grove, C. (2016) The effects of large igneous provinces on the global carbon and sulphur cycles. *Palaeogeography Palaeoclimatology Palaeoecology*, 441, 4–21.

Jordan, T. H. & Lynn, W. S. (1974) A velocity anomaly in the lower mantle. *Journal of Geophysical Research*, 79, 2679–85.

Julian, B. R., Foulger, G. R., Hatfield, O., et al. (2015) Hotspots in hindsight. *Geological Society of America Special Papers*, 514, 105–21.

Kankanamge, D. G. J. & Moore, W. B. (2016) Heat transport in the Hadean mantle: From heat pipes to plates. *Geophysical Research Letters*, 43, 3208–14.

Karson, J. A., Kelley, D. S., Fornari, D. J., et al. (2015) *Discovering the Deep: A Photographic Atlas of the Seafloor and Ocean Crust*, Cambridge University Press, Cambridge.

Kaufman, S. B. (2015) The difference between extrAversion and extrOversion. https://blogs.scientificamerican.com/beautiful-minds/the-difference-between-extraversion-and-extroversion/.

Keigwin, L. (1982) Isotopic paleoceanography of the Caribbean and East Pacific: Role of Panama uplift in Late Neogene time. *Science*, 217, 350–2.

Keigwin, L. D. (1978) Pliocene closing of the Isthmus of Panama, based on biostratigraphic evidence from nearby Pacific Ocean and Caribbean Sea cores. *Geology*, 6, 630–4.

Kelemen, P. B. & Behn, M. D. (2016) Formation of lower continental crust by relamination of buoyant arc lavas and plutons. *Nature Geoscience*, 9, 197–205.

Keller, C. B. & Schoene, B. (2012) Statistical geochemistry reveals disruption in secular lithospheric evolution about 2.5 Gyr ago. *Nature*, 485, 490–3.

Kelley, D. S., Karson, J. A., Blackman, D. K., et al. (2001) An off-axis hydrothermal vent field near the Mid-Atlantic Ridge at 30°N. *Nature*, 412, 145–9.

Kennett, J. P. (1977) Cenozoic evolution of Antarctic glaciation, the Circum-Antarctic Ocean, and their impact on global paleoceanography. *journal of Geophysical Research-Oceans and Atmospheres*, 82, 3843–60.

Kennett, J. P., Houtz, R. E., Andrews, P. B., et al. (1974) Development of the Circum-Antarctic Current. *Science*, 186, 144–7.

Kennett, J. P. & Shackleton, N. J. (1976) Oxygen isotopic evidence for the development of the psychrosphere 38 Myr ago. *Nature*, 260, 513–15.

Kent, G. M., Harding, A. J., Orcutt, J. A., et al. (1994) Uniform accretion of oceanic crust south of the Garrett transform at 14°15′S on the East Pacific Rise. *Journal of Geophysical Research: -Solid Earth*, 99, 9097–116.

Kerrick, D. M. (2001) Present and past nonanthropogenic CO_2 degassing from the solid Earth. *Reviews of Geophysics*, 39, 565–85.

Kite, E. S., Manga, M. & Gaidos, E. (2009) Geodynamics and rate of volcanism on massive Earth-like planets. *Astrophysical Journal*, 700, 1732–49.

Knesel, K. M., Cohen, B. E., Vasconcelos, P. M. & Thiede, D. S. (2008) Rapid change in drift of the Australian plate records collision with Ontong Java Plateau. *Nature*, 454, 754–7.

Knittle, E. & Jeanloz, R. (1991) Earth's core–mantle boundary: Results of experiments at high pressures and temperatures. *Science*, 251, 1438–43.

Knopf, A. (1948) The geosynclinal theory. *Geological Society of America Bulletin*, 59, 649–69.

Koelemeijer, P., Ritsema, J., Deuss, A. & van Heijst, H. J. (2016) SP12RTS: a degree-12 model of shear- and compressional-wave velocity for Earth's mantle. *Geophysical Journal International*, 204, 1024–39.

Konôpková, Z., McWilliams, R. S., Gomez-Perez, N. & Goncharov, A. F. (2016) Direct measurement of thermal conductivity in solid iron at planetary core conditions. *Nature*, 534, 99–101.

Köppen, W. & Wegener, A. (1924) *Die Klimate der geologischen Vorzeit*. Gebrüder Borntraeger, Berlin. [English translation by B. Oelkers, edited by J. Thiede, K. Lochte, and A. Dummermuth (2015) *The Climates of the Geological Past*. Schweizerbart Science Publishers, Stuttgart.]

Korenaga, J. (2010) Scaling of plate tectonic convection with pseudoplastic rheology. *Journal of Geophysical Research: Solid Earth*, 115, 1–24.

Kusky, T. M., Li, J. H. & Tucker, R. D. (2001) The Archean Dongwanzi ophiolite complex, North China craton: 2.505-billion-year-old oceanic crust and mantle. *Science*, 292, 1142–45.

Labrosse, S., Hernlund, J. W. & Coltice, N. (2007) A crystallizing dense magma ocean at the base of the Earth's mantle. *Nature*, 450, 866–9.

Laj, C., Kissel, C., Garnier, F. & HerreroBervera, E. (1996) Relative geomagnetic field intensity and reversals for the last 1.8 My from a central equatorial Pacific core. *Geophysical Research Letters*, 23, 3393–6.

Laj, C., Kissel, C. & Guillou, H. (2002) Brunhes' research revisited: Magnetization of volcanic flows and baked clays. *Eos, Transactions American Geophysical Union*, 83, 381–7.

Langmuir, C. H. & Broecker, W. (2012) *How to Build a Habitable Planet: The Story of Earth from the Big Bang to Humankind*, revised and expanded edition. Princeton University Press, Princeton, NJ.

Larmor, J. (1919) How could a rotating body such as the Sun become a magnet? *Report of the British Association for the Advancement of Science 87th Meeting.*

Lay, T. (2015) Deep Earth structure: Lower mantle and D″. In Schubert, G. (Ed.) *Treatise on Geophysics*, 2nd edition. Elsevier, Oxford, 683–723.

Lay, T. & Helmberger, D. V. (1983) The shear-wave velocity gradient at the base of the mantle. *Journal of Geophysical Research*, 88, 8160–70.

Le Grand, H. E. (2002) Plate tectonics, terranes and continental geology. *Geological Society, London, Special Publications*, 192, 199–213.

Leconte, J., Forget, F., Charnay, B., et al. (2013) Increased insolation threshold for runaway greenhouse processes on Earth-like planets. *Nature*, 504, 268–71.

Lee, C. T. A., Thurner, S., Paterson, S. & Cao, W. R. (2015) The rise and fall of continental arcs: Interplays between magmatism, uplift, weathering, and climate. *Earth and Planetary Science Letters*, 425, 105–19.

Lekic, V., Cottaar, S., Dziewonski, A. & Romanowicz, B. (2012) Cluster analysis of global lower mantle tomography: A new class of structure and implications for chemical heterogeneity. *Earth and Planetary Science Letters*, 357, 68–77.

Lenton, T. & Watson, A. (2011) *Revolutions that Made the Earth*. Oxford University Press, Oxford.

Lessios, H. A. (2008) The Great American Schism: Divergence of marine organisms after the rise of the Central American Isthmus. *Annual Review of Ecology Evolution and Systematics*, 39, 63–91.

Li, Z. X., Bogdanova, S. V., Collins, A. S., et al. (2008) Assembly, configuration, and break-up history of Rodinia: A synthesis. *Precambrian Research*, 160, 179–210.

Livermore, R., Cunningham, A., Vanneste, L. & Larter, R. (1997) Subduction influence on magma supply at the East Scotia Ridge. *Earth and Planetary Science Letters*, 150, 261–75.

Lonsdale, P. (1983) Overlapping rift zones at the 5.5°S offset of the East Pacific Rise. *Journal of Geophysical Research*, 88, 9393–406.

Lonsdale, P. (1989) Segmentation of the Pacific-Nazca spreading center, 1°N–20°S. *Journal of Geophysical Research: Solid Earth and Planets*, 94, 12197–225.

Lund, D. C. & Asimow, P. D. (2011) Does sea level influence mid-ocean ridge magmatism on Milankovitch timescales? *Geochemistry Geophysics Geosystems*, 12, 1–26.

Lunt, D. J., Valdes, P. J., Haywood, A. & Rutt, I. C. (2008) Closure of the Panama Seaway during the Pliocene: Implications for climate and Northern Hemisphere glaciation. *Climate Dynamics*, 30, 1–18.

Luyendyk, B. P. (1970) Origin and history of abyssal hills in the Northeast Pacific Ocean. *Geological Society of America Bulletin*, 81, 2237–60.

Lyell, C. (1837) *Principles of Geology: Being an Inquiry How Far the Former Changes of the Earth's Surface are Referable to Causes Now in Operation*, 9th edition. John Murray, London.

Macdonald, K. C., Becker, K., Spiess, F. N. & Ballard, R. D. (1980) Hydrothermal heat flux of the 'black smoker' vents on the East Pacific Rise. *Earth and Planetary Science Letters*, 48, 1–7.

Macdonald, K. C. & Fox, P. J. (1983) Overlapping spreading centres: New accretion geometry on the East Pacific Rise. *Nature*, 302, 55–8.

Macdonald, K. C., Fox, P. J., Alexander, R. T., et al. (1996) Volcanic growth faults and the origin of Pacific abyssal hills. *Nature*, 380, 125–9.

Macdonald, K. C., Fox, P. J., Perram, L. J., et al. (1988) A new view of the mid-ocean ridge from the behaviour of ridge-axis discontinuities. *Nature*, 335, 217–25.

Macdonald, K. C., Sempere, J. C. & Fox, P. J. (1986) Reply: The debate concerning overlapping spreading centers and mid-ocean ridge processes. *Journal of Geophysical Research: Solid Earth and Planets*, 91, 501–11.

Malin, S. R. C. & Bullard, E. (1981) The direction of the Earth's magnetic field at London, 1570–1975. *Philosophical Transactions of the Royal Society of London A: Mathematical, Physical and Engineering Sciences*, 299, 357–423.

Marshall, L. G., Webb, S. D., Sepkoski, J. J. & Raup, D. M. (1982) Mammalian Evolution and the Great American Interchange. *Science*, 215, 1351–357.

Maslin, M. (2013) *Climate: A Very Short Introduction*. Oxford University Press, Oxford.

Maslin, M. (2017) *The Cradle of Humanity: How the Changing Landscape of Africa Made Us So Smart*. Oxford University Press, Oxford.

Mason, R. (2003) Stripes on the sea floor. In Oreskes, N. (Ed.) *Plate Tectonics: An Insider's History of the Modern Theory of the Earth*. Westview Press, Boulder, CO, 31–45.

Mason, W. G., Moresi, L., Betts, P. G. & Miller, M. S. (2010) Three-dimensional numerical models of the influence of a buoyant oceanic plateau on subduction zones. *Tectonophysics*, 483, 71–9.

Masters, G., Laske, G., Bolton, H. & Dziewonski, A. (2000) The Relative behavior of shear velocity, bulk sound speed, and compressional velocity in the mantle: Implications for chemical and thermal structure. In Karato, S.-I., Forte, A., Liebermann, R., Masters, G. & Stixrude, L. (Eds.) *Earth's Deep Interior: Mineral Physics and Tomography From the Atomic to the Global Scale*. American Geophysical Union, Washington, DC, 63–87.

Matthews, R. K. & Poore, R. Z. (1980) Tertiary $\delta^{18}O$ record and glacio-eustatic sea-level fluctuations. *Geology*, 8, 501–4.

McAdoo, D. C. & Marks, K. M. (1992) Gravity fields of the Southern Ocean from Geosat data. *Journal of Geophysical Research: Solid Earth*, 97, 3247–60.

McKenzie, D. (1972) Active tectonics of the Mediterranean region. *Geophysical Journal of the Royal Astronomical Society*, 30, 109–85.

McKenzie, D. (1994) The relationship between topography and gravity on Earth and Venus. *Icarus*, 112, 55–88.

McKenzie, D. (2003) Plate tectonics: A surprising way to start a scientific career. In Oreskes, N. (Ed.) *Plate Tectonics: An Insider's History of the Modern Theory of the Earth*. Westview Press, Boulder, CO, 169–90.

McKenzie, D., Ford, P. G., Johnson, C., et al. (1992) Features on Venus generated by plate boundary processes. *Journal of Geophysical Research: Planets*, 97, 13533–44.

McKenzie, D. & Nimmo, F. (1997) Elastic thickness estimates for Venus from line of sight accelerations. *Icarus*, 130, 198–216.

McKenzie, D. & Sclater, J. G. (1971) The evolution of the Indian Ocean since the Late Cretaceous. *Geophysical Journal of the Royal Astronomical Society*, 24, 437–528.

McKenzie, D. P. (1970) Plate tectonics of the Mediterranean region. *Nature*, 226, 239–43.

McKenzie, D. P. & Morgan, W. J. (1969) Evolution of triple junctions. *Nature*, 224, 125–33.

McKenzie, D. P. & Parker, R. L. (1967) North Pacific: An example of tectonics on a sphere. *Nature*, 216, 1276–80.

McKenzie, D. P. & Sclater, J. G. (1973) The evolution of the Indian Ocean. *Scientific American*, 228(5), 62–72.

McMenamin, M. A. S. & McMenamin, D. L. S. (1990) *The Emergence of Animals: The Cambrian Breakthrough*. Columbia University Press, New York.

McNamara, A. K. & Zhong, S. J. (2005) Thermochemical structures beneath Africa and the Pacific Ocean. *Nature*, 437, 1136–9.

Meert, J. G. & Torsvik, T. H. (2003) The making and unmaking of a supercontinent: Rodinia revisited. *Tectonophysics*, 375, 261–88.

Menard, H. W. (1986) *The Ocean of Truth: A Personal History of Global Tectonics*. Princeton University Press, Princeton, NJ.

Menard, H. W. & Mammeric. J (1967) Abyssal hills, magnetic anomalies and the East Pacific Rise. *Earth and Planetary Science Letters*, 2, 465–72.

Merrill, R. T. (2010) *Our Magnetic Earth, The Science of Geomagnetism*. University of Chicago Press, Chicago.

Meyerhoff, A. A. (1968) Arthur Holmes: Originator of spreading ocean floor hypothesis. *Journal of Geophysical Research*, 73, 6563–5.

Mian, Z. U. & Tozer, D. C. (1990) No water, no plate tectonics: convective heat transfer and the planetary surfaces of Venus and Earth. *Terra Nova*, 2, 455–9.

Miller, K. G., Browning, J. V., Aubry, M. P., et al. (2008) Eocene–Oligocene global climate and sea-level changes: St. Stephens Quarry, Alabama. *Geological Society of America Bulletin*, 120, 34–53.

Mills, B., Daines, S. J. & Lenton, T. M. (2014) Changing tectonic controls on the long-term carbon cycle from Mesozoic to present. *Geochemistry Geophysics Geosystems*, 15, 4866–84.

Minster, J. B. & Jordan, T. H. (1978) Present-day plate motions. *Journal of Geophysical Research*, 83, 5331–54.

Minster, J. B., Jordan, T. H., Molnar, P. & Haines, E. (1974) Numerical modelling of instantaneous plate tectonics. *Geophysical Journal International*, 36, 541–76.

Miyashiro, A. (1961) Evolution of metamorphic belts. *Journal of Petrology*, 2, 277–311.

Molnar, P. (2011) Jack Oliver (1923–2011). *Nature*, 470, 176.

Molnar, P., Fitch, T. J. & Wu, F. T. (1973) Fault plane solutions of shallow earthquakes and contemporary tectonics in Asia. *Earth and Planetary Science Letters*, 19, 101–12.

Molnar, P. & Tapponnier, P. (1975) Cenozoic tectonics of Asia: Effects of a continental collision. *Science*, 189, 419–26.

Moore, W. B. & Lenardic, A. (2015) The efficiency of plate tectonics and non-equilibrium dynamical evolution of planetary mantles. *Geophysical Research Letters*, 42, 9255–60.

Moore, W. B. & Webb, A. A. G. (2013) Heat-pipe Earth. *Nature*, 501, 501–5.

Moores, E. M. (1991) Southwest U.S.–East Antarctic (SWEAT) connection: A hypothesis. *Geology*, 19, 425–8.

Moores, E. M. & Vine, F. J. (1971) The Troodos Massif, Cyprus and other ophiolites as oceanic crust: Evaluation and implications. *Philosophical Transactions of the Royal Society of London A: Mathematical and Physical Sciences*, 268, 443–66.

Moresi, L. & Solomatov, V. (1998) Mantle convection with a brittle lithosphere: Thoughts on the global tectonic styles of the Earth and Venus. *Geophysical Journal International*, 133, 669–82.

Morgan, W. J. (1968) Rises, trenches, great faults, and crustal blocks. *Journal of Geophysical Research*, 73, 1959–82.

Morgan, W. J. (1971) Convection plumes in the lower mantle. *Nature*, 230, 42–3.

Morgan, W. J. (1972) Deep mantle convection plumes and plate motions. *American Association of Petroleum Geologists Bulletin*, 56, 203–13.

Morris, S. & Canright, D. (1984) A boundary-layer analysis of Benard convection in a fluid of strongly temperature-dependent viscosity. *Physics of the Earth and Planetary Interiors*, 36, 355–73.

Murakami, M., Ohishi, Y., Hirao, N. & Hirose, K. (2012) A perovskitic lower mantle inferred from high-pressure, high-temperature sound velocity data. *Nature*, 485, 90–4.

Murphy, J. B. & Nance, R. D. (2003) Do supercontinents introvert or extrovert? Sm–Nd isotope evidence. *Geology*, 31, 873–6.

Murphy, J. B. & Nance, R. D. (2005) Do supercontinents turn inside-in or inside-out? *International Geology Review*, 47, 591–619.

Nagata, T. & Uyeda, S. (1956) Production of self-reversal of thermoremanent magnetism by heat treatment of ferromagnetic minerals. *Nature*, 177, 179–80.

Nairn, A. E. M. (1961) *Descriptive Palaeoclimatology*. Interscience, New York.

Nakagawa, T. & Tackley, P. J. (2015) Influence of plate tectonic mode on the coupled thermochemical evolution of Earth's mantle and core. *Geochemistry Geophysics Geosystems*, 16, 3400–13.

Nance, D. R. & Murphy, B. J. (2013) Origins of the supercontinent cycle. *Geoscience Frontiers*, 4, 439–48.

Nelson, K. D., Zhao, W. J., Brown, L. D., et al. (1996) Partially molten middle crust beneath southern Tibet: Synthesis of project INDEPTH results. *Science*, 274, 1684–8.

Ni, J. & Barazangi, M. (1984) Seismotectonics of the Himalayan collision zone: Geometry of the UNDERTHRUSTING Indian plate beneath the Himalaya. *Journal of Geophysical Research*, 89, 1147–63.

Ni, S. D., Tan, E., Gurnis, M. & Helmberger, D. (2002) Sharp sides to the African superplume. *Science*, 296, 1850–2.

Nield, T. (2007) *Supercontinent: Ten Billion Years in the Life of Our Planet*. Granta, London.

Nisbet, E. G. & Fowler, C. M. R. (1978) The Mid-Atlantic Ridge at 37 and 45°N: Some geophysical and pertrological constraints. *Geophysical Journal of the Royal Astronomical Society*, 54, 631–60.

Nishi, M., Irifune, T., Tsuchiya, J., et al. (2014) Stability of hydrous silicate at high pressures and water transport to the deep lower mantle. *Nature Geoscience*, 7, 224–7.

Noack, L. & Breuer, D. (2014) Plate tectonics on rocky exoplanets: Influence of initial conditions and mantle rheology. *Planetary and Space Science*, 98, 41–9.

Nutman, A. P., Bennett, V. C. & Friend, C. R. L. (2015a) The emergence of the Eoarchaean proto-arc: Evolution of a c. 3700 Ma convergent plate boundary at Isua, southern West Greenland. *Geological Society, London, Special Publications*, 389, 113–33.

Nutman, A. P., Bennett, V. C. & Friend, C. R. L. (2015b) Proposal for a continent 'Itsaqia' amalgamated at 3.66 Ga and rifted apart from 3.53 Ga: Initiation of a Wilson Cycle near the start of the rock record. *American Journal of Science*, 315, 509–36.

Nye, M. J. (1999) Temptations of theory, strategies of evidence: P. M. S. Blackett and the Earth's magnetism, 1947–52. *British Journal for the History of Science*, 32, 69–92.

Ocean Studies Board (1999) *Global Ocean Science: Toward an Integrated Approach* (Ocean Studies Board/Commission on Geosciences, Environment and Resources/ National Research Council). National Academy Press, Washington, DC.

Oganov, A. R. & Ono, S. (2004) Theoretical and experimental evidence for a post-perovskite phase of $MgSiO_3$ in Earth's D″ layer. *Nature*, 430, 445–8.

Ohta, K., Kuwayama, Y., Hirose, K., et al. (2016) Experimental determination of the electrical resistivity of iron at Earth's core conditions. *Nature*, 534, 95–8.

Olive, J. A., Behn, M. D., Ito, G., et al. (2015) Sensitivity of seafloor bathymetry to climate-driven fluctuations in mid-ocean ridge magma supply. *Science*, 350, 310–13.

Olive, J. A., Behn, M. D., Ito, G., et al. (2016) Response to Comment on 'Sensitivity of seafloor bathymetry to climate-driven fluctuations in mid-ocean ridge magma supply'. *Science*, 353, 1405.

Oliver, J. (2003) Earthquake seismology in the plate tectonics revolution. In Oreskes, N. (Ed.) *Plate Tectonics: An Insider's History of the Modern Theory of the Earth*. Westview Press, Boulder, CO, 155–66.

Oliver, J. & Isacks, B. (1967) Deep earthquake zones, anomalous structures in the upper mantle, and the lithosphere. *Journal of Geophysical Research*, 72, 4259–75.

Oliver, J. E. (1996) *Shocks and Rocks*, American Geophysical Union, Washington, DC.

Olson, P. (2016) Mantle control of the geodynamo: Consequences of top-down regulation. *Geochemistry Geophysics Geosystems*, 17, 1935–56.

Olson, P., Deguen, R., Hinnov, L. A. & Zhong, S. J. (2013) Controls on geomagnetic reversals and core evolution by mantle convection in the Phanerozoic. *Physics of the Earth and Planetary Interiors*, 214, 87–103.

Olson, P., Deguen, R., Rudolph, M. L. & Zhong, S. J. (2015) Core evolution driven by mantle global circulation. *Physics of the Earth and Planetary Interiors*, 243, 44–55.

Olson, P. & Kincaid, C. (1991) Experiments on the interaction of thermal convection and compositional layering at the base of the mantle. *Journal of Geophysical Research: Solid Earth and Planets*, 96, 4347–54.

O'Neill, C. & Lenardic, A. (2007) Geological consequences of super-sized Earths. *Geophysical Research Letters*, 34, L19204.

O'Neill, C., Lenardic, A., Weller, M., et al. (2016) A window for plate tectonics in terrestrial planet evolution? *Physics of the Earth and Planetary Interiors*, 255, 80–92.

O'Neil, J. & Carlson, R. W. (2017) Building Archean cratons from Hadean mafic crust. *Science*, 355, 1199–202.

O'Reilly, T. C. & Davies, G. F. (1981) Magma transport of heat on Io: A mechanism allowing a thick lithosphere. *Geophysical Research Letters*, 8, 313–16.

Oreskes, N. (1999) *The Rejection of Continental Drift: Theory and Method in American Earth Science*, Oxford University Press, Oxford.

Otsuka, K. & Karato, S. (2012) Deep penetration of molten iron into the mantle caused by a morphological instability. *Nature*, 492, 243–6.

Palin, R. M. & White, R. W. (2015) Emergence of blueschists on Earth linked to secular changes in oceanic crust composition. *Nature Geoscience*, 9, 60–4.

Pallister, J. S. & Hopson, C. A. (1981) Samail ophiolite plutonic suite: Field relations, phase variation, cryptic variation and layering, and a model of a spreading ridge magma chamber. *Journal of Geophysical Research*, 86, 2593–644.

Palmer, M. R. & Elderfield, H. (1985) Sr isotope composition of sea water over the past 75 Myr. *Nature*, 314, 526–8.

Pearson, D. G., Brenker, F. E., Nestola, F., et al. (2014) Hydrous mantle transition zone indicated by ringwoodite included within diamond. *Nature*, 507, 221–4.

Pearson, P. N. (2012) Oxygen isotopes in foraminifera: Overview and historical review. In Ivany, L. C. & Huber, B. T. (Eds.) *Reconstructing Earth's Deep-Time Climate: The State of the Art in 2012. Paleontological Society Short Course, November 3, 2012. The Paleontological Society Papers*, Volume 18. Paleontological Society, Boulder, CO, 1–38.

Peregrinus de Maricourt, P. (1269) *Epistola Petri Peregrini de Maricourt ad Sygerum de Foucaucourt, militem, de magnete*.

Perry, J. (1895) On the age of the Earth. *Nature*, 51, 224–7.

Peterman, Z. E., Hedge, C. E. & Tourtelo, H. A. (1970) Isotopic composition of strontium in sea water throughout Phanerozoic time. *Geochimica et Cosmochimica Acta*, 34, 105–20.

Petitgirard, S., Malfait, W. J., Sinmyo, R., et al. (2015) Fate of $MgSiO_3$ melts at core–mantle boundary conditions. *Proceedings of the National Academy of Sciences of the USA*, 112, 14186–90.

Phillips, B. R. & Bunge, H. P. (2007) Supercontinent cycles disrupted by strong mantle plumes. *Geology*, 35, 847–50.

Phillips, R. J. & Hansen, V. L. (1994) Tectonic and magmatic evolution of Venus. *Annual Review of Earth and Planetary Sciences*, 22, 597–654.

Phillips, R. J. & Hansen, V. L. (1998) Geological evolution of Venus: Rises, plains, plumes, and plateaus. *Science*, 279, 1492–7.

Pitman, W. C. & Heirtzler, J. R. (1966) Magnetic anomalies over the Pacific–Antarctic Ridge. *Science*, 154, 1164–71.

Planavsky, N. J., Reinhard, C. T., Wang, X. L., et al. (2014) Low Mid-Proterozoic atmospheric oxygen levels and the delayed rise of animals. *Science*, 346, 635–8.

Plattner, C., Malservisi, R., Dixon, T. H., et al. (2007) New constraints on relative motion between the Pacific Plate and Baja California microplate (Mexico) from GPS measurements. *Geophysical Journal International*, 170, 1373–80.

Platz, T., Byrne, P. K., Massironi, M. & Hiesinger, H. (2015) Volcanism and tectonism across the inner solar system: An overview. *Geological Society, London, Special Publications*, 401, 1–56.

Polat, A., Wang, L. & Appel, P. W. U. (2015) A review of structural patterns and melting processes in the Archean craton of West Greenland: Evidence for crustal growth at convergent plate margins as opposed to non-uniformitarian models. *Tectonophysics*, 662, 67–94.

Powell, C. M. A. & Conaghan, P. J. (1973) Plate tectonics and the Himalayas. *Earth and Planetary Science Letters*, 20, 1–12.

Pozzo, M., Davies, C., Gubbins, D. & Alfe, D. (2012) Thermal and electrical conductivity of iron at Earth's core conditions. *Nature*, 485, 355–U99.

Priestley, K., Jackson, J. & McKenzie, D. (2008) Lithospheric structure and deep earthquakes beneath India, the Himalaya and southern Tibet. *Geophysical Journal International*, 172, 345–62.

Pross, J., Contreras, L., Bijl, P. K., et al. (2012) Persistent near-tropical warmth on the Antarctic continent during the early Eocene epoch. *Nature*, 488, 73–7.

Raff, A. D. & Mason, R. G. (1961) Magnetic survey off the west coast of North America, 40°N latitude to 52°N latitude *Geological Society of America Bulletin*, 72, 1267–70.

Raymo, M. E. (1991) Geochemical evidence supporting T. C. Chamberlin's theory of glaciation. *Geology*, 19, 344–7.

Raymo, M. E. & Ruddiman, W. F. (1992) Tectonic forcing of Late Cenozoic climate. *Nature*, 359, 117–22.

Regenauer-Lieb, K. (2006) Water and geodynamics. *Reviews in Mineralogy and Geochemistry*, 62, 451–73.

Reimink, J. R., Chacko, T., Stern, R. A. & Heaman, L. M. (2014) Earth's earliest evolved crust generated in an Iceland-like setting. *Nature Geoscience*, 7, 529–33.

Reimink, J. R., Davies, J. H. F. L., Chacko, T., Stern, R. A., Heaman, L. M., et al. (2016) No evidence for Hadean continental crust within Earth's oldest evolved rock unit. *Nature Geoscience*, 9, 777–80.

Richard, G., Bercovici, D. & Karato, S. I. (2006) Slab dehydration in the Earth's mantle transition zone. *Earth and Planetary Science Letters*, 251, 156–67.

Richter, F. M., Rowley, D. B. & DePaolo, D. J. (1992) Sr isotope evolution of seawater: the role of tectonics. *Earth and Planetary Science Letters*, 109, 11–23.

Rickers, F., Fichtner, A. & Trampert, J. (2013) The Iceland–Jan Mayen plume system and its impact on mantle dynamics in the North Atlantic region: Evidence from full-waveform inversion. *Earth and Planetary Science Letters*, 367, 39–51.

Ringwood, A. E. (1974) The petrological evolution of island arc systems: Twenty-seventh William Smith Lecture. *Journal of the Geological Society*, 130, 183–204.

Ringwood, A. E. & Major, A. (1967) High pressure reconnaissance investigations in the system Mg_2SiO_4–MgO–H_2O. *Earth and Planetary Science Letters*, 2, 130–3.

Rintoul, S. R., Hughes, C. W. & Olbers, D. (2001) The Antarctic Circumpolar Current system. In Siedler, G., Church, J. & Gould, J. (Eds.) *Ocean Circulation and Climate: Observing and Modelling the Global Ocean*, Academic Press, San Diego, 271–302.

Ritsema, J., Deuss, A., van Heijst, H. J. & Woodhouse, J. H. (2011) S40RTS: A degree-40 shear-velocity model for the mantle from new Rayleigh wave dispersion, teleseismic traveltime and normal-mode splitting function measurements. *Geophysical Journal International*, 184, 1223–36.

Ritsema, J., van Heijst, H. J. & Woodhouse, J. H. (1999) Complex shear wave velocity structure imaged beneath Africa and Iceland. *Science*, 286, 1925–8.

Robbins, J. W., Smith, D. E. & Ma, C. (1993) Horizontal crustal deformation and large scale plate motions inferred from space geodetic techniques. In Smith, D. E. & Turcotte, D. L. (Eds.) *Contributions of Space Geodesy to Geodynamics: Crustal Dynamics.* American Geophysical Union, Washington, DC, 21–36.

Roberts, N. M. W., Van Kranendonk, M. J., Parman, S. & Clift, P. D. (2015) Continent formation through time. *Geological Society, London, Special Publications*, 389, 1–16.

Rogers, A. D., Tyler, P. A., Connelly, D. P., et al. (2012) The discovery of new deep-sea hydrothermal vent communities in the Southern Ocean and implications for biogeography. *PLOS Biology*, 10, 17.

Rogers, J. J. W. & Santosh, M. (2003) Supercontinents in earth history. *Gondwana Research*, 6, 357–68.

Rogers, J. J. W. & Santosh, M. (2004) *Continents and Supercontinents*, Oxford University Press, Oxford.

Rolf, T., Coltice, N. & Tackley, P. J. (2014) Statistical cyclicity of the supercontinent cycle. *Geophysical Research Letters*, 41, 2351–8.

Rudwick, M. J. S. (2014) *Earth's Deep History*, University of Chicago Press, Chicago.

Runcorn, S. K. (1956a) Paleomagnetic comparisons between Europe and North America. *Proceedings of the Geological Association of Canada*, 8, 77–85.

Runcorn, S. K. (1956b) Paleomagnetism, polar wandering and continental drift. *Geologie en Mijnbouw*, 18, 253–8.

Rusby, R. I. (1992) GLORIA and other geophysical studies of the tectonic pattern and history of the Easter Microplate, southeast Pacific. *Geological Society, London, Special Publications*, 60, 81–106.

Sager, W. W., Sano, T. & Geldmacher, J. (2016) Formation and evolution of Shatsky Rise oceanic plateau: Insights from IODP Expedition 324 and recent geophysical cruises. *Earth-Science Reviews*, 159, 306–36.

Sandwell, D. T. (1992) Antarctic marine gravity field from high-density satellite altimetry. *Geophysical Journal International*, 109, 437–48.

Sandwell, D. T., Muller, R. D., Smith, W. H. F., et al. (2014) New global marine gravity model from CryoSat-2 and Jason-1 reveals buried tectonic structure. *Science*, 346, 65–7.

Sandwell, D. T. & Schubert, G. (1992) Evidence for retrograde lithospheric subduction on Venus. *Science*, 257, 766–70.

Santosh, M., Maruyama, S., Sawaki, Y. & Meert, J. G. (2014) The Cambrian Explosion: Plume-driven birth of the second ecosystem on Earth. *Gondwana Research*, 25, 945–65.

Scher, H. (2017) Carbon–ocean gateway links. *Nature Geoscience*, 10, 164–5.

Scher, H. D., Whittaker, J. M., Williams, S. E., et al. (2015) Onset of Antarctic Circumpolar Current 30 million years ago as Tasmanian Gateway aligned with westerlies. *Nature*, 523, 580–3.

Schouten, H., Klitgord, K. D. & Gallo, D. G. (1993) Edge-driven microplate kinematics. *Journal of Geophysical Research: Solid Earth*, 98, 6689–701.

Schubert, G. & Sandwell, D. (1989) Crustal volumes of the continents and of oceanic and continental submarine plateaus. *Earth and Planetary Science Letters*, 92, 234–46.

Schubert, G., Turcotte, D. L. & Olson, P. (2001) *Mantle Convection in the Earth and Planets*, Cambridge University Press, Cambridge.

Schuchert, C. (1925) *Biographical Memoir of Joseph Barrell, 1869–1919*. National Academy of Sciences, Washington, DC.

Searle, M. (2003) *Colliding Continents*, Oxford University Press.

Searle, R. (1984) GLORIA Survey of the East Pacific Rise near 3.5°S: Tectonic and volcanic characteristics of a fast spreading mid-ocean rise. *Tectonophysics*, 101, 319–44.

Searle, R. C., Rusby, R. I., Engeln, J., et al. (1989) Comprehensive sonar imaging of the Easter microplate. *Nature*, 341, 701–5.

Şengör, A. M. C. & Dewey, J. F. (1990) Terranology: Vice or Virtue? *Philosophical Transactions of the Royal Society of London A: Mathematical, Physical and Engineering Sciences*, 331, 457–77.

Seton, M., Muller, R. D., Zahirovic, S., et al. (2012) Global continental and ocean basin reconstructions since 200 Ma. *Earth-Science Reviews*, 113, 212–70.

Shackleton, N. (1967) Oxygen isotope analyses and Pleistocene temperatures re-assessed. *Nature*, 215, 15–17.

Shackleton, N. & Kennett, B. L. N. (1975) Paleotemperature history of the Cenozoic and the initiation of Antarctic glaciation: Oxygen and carbon isotope analyses in DSDP Sites 277, 279, and 281. Initial Reports of the Deep Sea Drilling Project. http://deepseadrilling.org/29/volume/dsdp29_17.pdf.

Shirey, S. B. & Richardson, S. H. (2011) Start of the Wilson Cycle at 3 Ga Shown by diamonds from subcontinental mantle. *Science*, 333, 434–6.

Sidorin, I., Gurnis, M. & Helmberger, D. V. (1999) Evidence for a ubiquitous seismic discontinuity at the base of the mantle. *Science*, 286, 1326–31.

Simmons, N. A., Myers, S. C., Johannesson, G., et al. (2015) Evidence for long-lived subduction of an ancient tectonic plate beneath the southern Indian Ocean. *Geophysical Research Letters*, 42, 9270–8.

Singh, S. C., Crawford, W. C., Carton, H., et al. (2006) Discovery of a magma chamber and faults beneath a Mid-Atlantic Ridge hydrothermal field. *Nature*, 442, 1029–32.

Skilbeck, J. N. & Whitehead, J. A. (1978) Formation of discrete islands in linear island chains. *Nature*, 272, 499–501.

Sleep, N. H. (1994) Martian plate tectonics. *Journal of Geophysical Research: Planets*, 99, 5639–55.

Smith, D. C. (1984) Coesite in clinopyroxene in the Caledonides and its implications for geodynamics. *Nature*, 310, 641–4.

Smithies, R. H., Van Kranendonk, M. J. & Champion, D. C. (2007) The Mesoarchean emergence of modern-style subduction. *Gondwana Research*, 11, 50–68.

Sobolev, N. V. & Shatsky, V. S. (1990) Diamond inclusions in garnets from metamorphic rocks: A new environment for diamond formation. *Nature*, 343, 742–6.

Solomatov, V. S. (1995) Scaling of temperature- and stress-dependent viscosity convection. *Physics of Fluids*, 7, 266–74.

Solomatov, V. S. & Moresi, L. N. (1996) Stagnant lid convection on Venus. *Journal of Geophysical Research: Planets*, 101, 4737–53.

Spiess, F. N., Macdonald, K. C., Atwater, T., et al. (1980) East Pacific Rise: Hot springs and geophysical experiments. *Science*, 207, 1421–33.

Stein, C., Lowman, J. P. & Hansen, U. (2013) The influence of mantle internal heating on lithospheric mobility: Implications for super-Earths. *Earth and Planetary Science Letters*, 361, 448–59.

Stern, R. J. (2004) Subduction initiation: spontaneous and induced. *Earth and Planetary Science Letters*, 226, 275–92.

Stern, R. J. (2008) Modern-style plate tectonics began in Neoproterozoic time: An alternative interpretation of Earth's tectonic history. *Geological Society of America Special Paper*, 440, 265–80.

Stern, R. J. (2010) The anatomy and ontogeny of modern intra-oceanic arc systems. *Geological Society, London, Special Publications*, 338, pp. 7–34.

Stern, R. J. (2016) Is plate tectonics needed to evolve technological species on exoplanets? *Geoscience Frontiers*, 7, 573–80.

Stern, R. J., Leybourne, M. I. & Tsujimori, T. (2016) Kimberlites and the start of plate tectonics. *Geology*, 44, 799–802.

Stern, R. J. & Scholl, D. W. (2010) Yin and yang of continental crust creation and destruction by plate tectonic processes. *International Geology Review*, 52, 1–31.

Stille, H. (1941) *Einführung in den bau Amerikas*. Gebrüder Borntraeger, Berlin.

Stow, D. (2010) *Vanished Ocean: How Tethys Reshaped the World*. Oxford University Press, Oxford.

Strutt, R. J. (1904) *The Becquerel Rays and the Properties of Radium*. Edward Arnold, London.

Strutt, R. J. (1909) The accumulation of helium in geological time. III. (Zircons). *Proceedings of the Royal Society of London A*, 83, 298–301.

Summerhayes, C. P. (2015) *Earth's Climate Evolution*. Wiley-Blackwell, Oxford.

Sykes, L. R. (1967) Mechanism of earthquakes and nature of faulting on the mid-ocean ridges *Journal of Geophysical Research*, 72, 2131–53.

Tackley, P. J. (2000a) The quest for self-consistent generation of plate tectonics in mantle convection models. In Richards, M. A., Gordon, R. G. & Van Der Hilst, R. D. (Eds.) *The History and Dynamics of Global Plate Motions*. American Geophysical Union, Washington, DC,47–72.

Tackley, P. J. (2000b) Mantle convection and plate tectonics: Toward an integrated physical and chemical theory. *Science*, 288, 2002–7.

Tackley, P. J. (2011) Living dead slabs in 3-D: The dynamics of compositionally-stratified slabs entering a 'slab graveyard' above the core–mantle boundary. *Physics of the Earth and Planetary Interiors*, 188, 150–62.

Talwani, M. (1964) A review of marine geophysics. *Marine Geology*, 2, 29–80.

Talwani, M., Le Pichon, X. & Heirtzler, J. R. (1965) East Pacific Rise: The magnetic pattern and the fracture zones. *Science*, 150, 1109–15.

Tan, E. & Gurnis, M. (2005) Metastable superplumes and mantle compressibility. *Geophysical Research Letters*, 32, 1–4.

Tan, E., Leng, W., Zhong, S. J. & Gurnis, M. (2011) On the location of plumes and lateral movement of thermochemical structures with high bulk modulus in the 3-D compressible mantle. *Geochemistry Geophysics Geosystems*, 12, 1–13.

Tapponnier, P., Peltzer, G., Le Dain, A. Y., et al. (1982) Propagating extrusion tectonics in Asia: New insights from simple experiments with Plasticine. *Geology*, 10, 611–16.

Thatcher, W. & Hill, D. P. (1995) A simple model for the fault-generated morphology of slow-spreading mid-oceanic ridges. *Journal of Geophysical Research: Solid Earth*, 100, 561–70.

Thatje, S., Marsh, L., Roterman, C. N., et al. (2015) Adaptations to hydrothermal vent life in *Kiwa tyleri*, a new species of yeti crab from the East Scotia Ridge, Antarctica. *PLOS ONE*, 10, e0127621.

Thomson, J. (1871) On the stratified rocks of Islay. *Report of the 41st Meeting of the British Association for the Advancement of Science, Edinburgh*, 110–11.

Thorne, M. S., Garnero, E. J. & Grand, S. P. (2004) Geographic correlation between hot spots and deep mantle lateral shear-wave velocity gradients. *Physics of the Earth and Planetary Interiors*, 146, 47–63.

Thorne, M. S., Garnero, E. J., Jahnke, G., et al. (2013) Mega ultra low velocity zone and mantle flow. *Earth and Planetary Science Letters*, 364, 59–67.

Tolstoy, M. (2015) Mid-ocean ridge eruptions as a climate valve. *Geophysical Research Letters*, 42, 1346–51.

Torsvik, T. H., Burke, K., Steinberger, B., et al. (2010) Diamonds sampled by plumes from the core–mantle boundary. *Nature*, 466, 352–U100.

Torsvik, T. H., Steinberger, B., Ashwal, L. D., et al. (2016) Earth evolution and dynamics—A tribute to Kevin Burke. *Canadian Journal of Earth Sciences*, 53, 1073–87.

Tozer, D. C. (1967) Some aspects of thermal convection theory for the Earth's mantle. *Geophysical Journal of the Royal Astronomical Society*, 14, 395–402.

Tucholke, B. E. & Lin, J. (1994) A geological model for the structure of ridge segments in slow spreading ocean crust. *Journal of Geophysical Research: Solid Earth*, 99, 11937–58.

Tucholke, B. E., Lin, J. & Kleinrock, M. C. (1998) Megamullions and mullion structure defining oceanic metamorphic core complexes on the mid-Atlantic ridge. *Journal of Geophysical Research: Solid Earth*, 103, 9857–66.

Turcotte, D. L. (1993) An episodic hypothesis for Venusian tectonics. *Journal of Geophysical Research: Planets*, 98, 17061–18.

Turcotte, D. L. & Oxburgh, E. R. (1972) Mantle convection and the new global tectonics. *Annual Review of Fluid Mechanics*, 4, 33–68.

Turnbull, M. J. M., Whitehouse, M. J. & Moorbath, S. (1996) New isotopic age determinations for the Torridonian, NW Scotland. *Journal of the Geological Society*, 153, 955–64.

Turner, H. H. (1922) On the arrival of earthquake waves at the Antipodes, and on the measurement of the focal depth of an earthquake. *Geophysical Journal International*, 1, 1–13.

Twain, M. (1872) *The Innocents at Home*. George Routledge and Sons, London.

Ueda, K., Gerya, T. & Sobolev, S. V. (2008) Subduction initiation by thermal–chemical plumes: Numerical studies. *Physics of the Earth and Planetary Interiors*, 171, 296–312.

Umbgrove, J. H. F. (1947) *The Pulse of the Earth*, 2nd edition. Martinus Nijhoff, Leiden.

Umemoto, K. & Hirose, K. C. G. L. (2015) Liquid iron–hydrogen alloys at outer core conditions by first-principles calculations. *Geophysical Research Letters*, 42, 7513–20.

United Nations (2017) UN Department of Economic and Social Affairs, Population Division. http://www.un.org/en/development/desa/population/index.shtml.

Vacquier, V. (1965) Transcurrent faulting in the ocean floor. *Philosophical Transactions of the Royal Society of London A, Mathematical and Physical Sciences*, 258, 77–81.

Vacquier, V. & Von Herzen, R. (1964) Evidence for connection between heat flow and the Mid-Atlantic Ridge magnetic anomaly. *Journal of Geophysical Research*, 69, 1093–101.

Vail, P. R., Mitchum, R. M., Todd, R. G., et al. (1977) Seismic stratigraphy and global changes of sea level. *American Association of Petroleum Geologists Memoir*, 26, 49–212.

Valencia, D., O'Connell, R. J. & Sasselov, D. D. (2007) Inevitability of plate tectonics on super-Earths. *Astrophysical Journal*, 670, L45–L48.

Valentine, J. W. & Moores, E. M. (1970) Plate-tectonic regulation of faunal diversity and sea level: A model. *Nature*, 228, 657–9.

Valentine, J. W. & Moores, E. M. (1972) Global tectonics and the fossil record. *Journal of Geology*, 80, 167–84.

Valley, J. W., Cavosie, A. J., Ushikubo, T., et al. (2014) Hadean age for a post-magma-ocean zircon confirmed by atom-probe tomography. *Nature Geoscience*, 7, 219–23.

Van Der Meer, D. G., Zeebe, R. E., van Hinsbergen, D. J. J., et al. (2014) Plate tectonic controls on atmospheric CO_2 levels since the Triassic. *Proceedings of the National Academy of Sciences of the USA*, 111, 4380–5.

van Heck, H. J. & Tackley, P. J. (2011) Plate tectonics on super-Earths: Equally or more likely than on Earth. *Earth and Planetary Science Letters*, 310, 252–61.

Van Kranendonk, M. J., Altermann, W., Beard, B. L., et al. (2012) A chronostratigraphic division of the Precambrian. In Gradstein, F. M., Ogg, J. G., Schmitz, M. & Ogg, G. (Eds.) *Geologic Time Scale 2012*. Elsevier, Amsterdam, 299–392.

Van Kranendonk, M. J., Kroner, A., Hoffmann, J. E., et al. (2014) Just another drip: Re-analysis of a proposed Mesoarchean suture from the Barberton Mountain Land, South Africa. *Precambrian Research*, 254, 19–35.

Van Kranendonk, M. J., Smithies, R. H., Griffin, W. L., et al. (2015) Making it thick: a volcanic plateau origin of Palaeoarchean continental lithosphere of the Pilbara and Kaapvaal cratons. *Geological Society, London, Special Publications*, 389, 83–111.

Vanacore, E. A., Rost, S. & Thorne, M. S. (2016) Ultralow-velocity zone geometries resolved by multidimensional waveform modelling. *Geophysical Journal International*, 206, 659–74.

Vening Meinesz, F. A. (1932) *Gravity Expeditions at Sea 1923–1930*, Volume I: *The Expeditions, the Computations and the Results*. Delft: NV Technische Boekhandel en Drukkerij J. Waltman Jr.

Vening Meinesz, F. A. & Poldervaart, A. (1955) Plastic buckling of the Earth's crust: The origin of geosynclines. *Crust of the Earth: A Symposium, Geological Society of America*.

Vine, F. J. (1966) Spreading of the ocean floor: new evidence. *Science*, 154, 1405–15.

Vine, F. J. (2010) Fred Vine: Plate tectonics revealed by a new discipline—marine geophysics. British Library interview, 20 August 2010. http://www.bl.uk/voices-of-science/interviewees/fred-vine/audio/fred-vine-plate-tectonics-revealed-by-a-new-discipline-marine-geophysics

Vine, F. J. & Matthews, D. H. (1963) Magnetic anomalies over ocean ridges. *Nature*, 199, 947–9.

Vine, F. J. & Wilson, J. T. (1965) Magnetic anomalies over a young ocean ridge off Vancouver Island. *Science*, 150, 485–9.

Vlaar, N. J. (1989) Vening Meinesz—A student of the Earth. *Eos, Transactions American Geophysical Union*, 70, 129–40.

von Huene, R., Ranero, C. R. & Vannucchi, P. (2004) Generic model of subduction erosion. *Geology*, 32, 913–16.

Wadati, K. (1928) Shallow and deep earthquakes. *Geophysical Magazine*, 1, 162–202.

Wadati, K. (1935) On the activity of deep-focus earthquakes in the Japan islands and neighbourhoods. *Geophysical Magazine*, 8, 305–25.

Wang, Q., Zhang, P. Z., Freymueller, J. T., et al. (2001) Present-day crustal deformation in China constrained by global positioning system measurements. *Science*, 294, 574–7.

Wang, W. T., Zheng, W. J., Zhang, P. Z., et al. (2017) Expansion of the Tibetan Plateau during the Neogene. *Nature Communications*, 8, 12.

Ward, P. D. & Brownlee, D. (2000) *Rare Earth: Why Complex Life is Uncommon in the Universe*. Copernicus, New York.

Watters, T. R., Daud, K., Banks, M. E., et al. (2016) Recent tectonic activity on Mercury revealed by small thrust fault scarps. *Nature Geoscience*, 9, 743–7.

Watts, A. B. & Burov, E. B. (2003) Lithospheric strength and its relationship to the elastic and seismogenic layer thickness. *Earth and Planetary Science Letters*, 213, 113–31.

Wegener, A. (1920) *Die Entstehung der Kontinente und Ozeane*, 2nd edition. Vieweg, Braunschweig.

Wegener, A. (1929) *Die Entstehung der Kontinente und Ozeane*, 4th edition. Vieweg, Braunschweig [*The Origin of Continents and Oceans*, translated by J. Biram (1966), Dover Publications, New York].

Weller, M. B. & Lenardic, A. (2016) The energetics and convective vigor of mixed-mode heating: Velocity scalings and implications for the tectonics of exoplanets. *Geophysical Research Letters*, 43, 9469–74.

Weller, M. B. & Lenardic, A. (2017) On the evolution of terrestrial planets: Bi-stability, stochastic effects, and the non-uniqueness of tectonic states. *Geoscience Frontiers*, http://dx.doi.org/10.1016/j.gsf.2017.03.001.

Weller, M. B., Lenardic, A. & O'Neill, C. (2015) The effects of internal heating and large scale climate variations on tectonic bi-stability in terrestrial planets. *Earth and Planetary Science Letters*, 420, 85–94.

Wen, L. X. & Anderson, D. L. (1997) Layered mantle convection: A model for geoid and topography. *Earth and Planetary Science Letters*, 146, 367–77.

White, D. A., Roeder, D. H., Nelson, T. H. & Crowell, J. C. (1970) Subduction. *Geological Society of America Bulletin*, 81, 3431–2.

Wilson, D. S. (1996) Fastest known spreading on the Miocene Cocos–Pacific plate boundary. *Geophysical Research Letters*, 23, 3003–6.

Wilson, J. T. (1951) On the growth of continents. *Papers and Proceedings of the Royal Society of Tasmania*, 85–111.

Wilson, J. T. (1963a) A possible origin of the Hawaiian islands. *Canadian Journal of Earth Sciences*, 51, 863–70.

Wilson, J. T. (1963b) Pattern of uplifted islands in the main ocean basins. *Science*, 139, 592–4.

Wilson, J. T. (1965) A new class of faults and their bearing on continental drift. *Nature*, 207, 343–7.

Wilson, J. T. (1966) Did the Atlantic close and then re-open? *Nature*, 211, 676–81.

Wilson, J. T. (1968) Static or mobile Earth: The current scientific revolution. *Proceedings of the American Philosophical Society*, 112, 309–20.

Wilson, J. T. (1982) Early days in university geophysics. *Annual Review of Earth and Planetary Sciences*, 10, 1–14.

Wingate, M. T. D., Pisarevsky, S. A. & Evans, D. A. D. (2002) Rodinia connections between Australia and Laurentia: No SWEAT, no AUSWUS? *Terra Nova*, 14, 121–8.

Wirth, R., Vollmer, C., Brenker, F., et al. (2007) Inclusions of nanocrystalline hydrous aluminium silicate 'Phase Egg' in superdeep diamonds from Juina (Mato Grosso State, Brazil). *Earth and Planetary Science Letters*, 259, 384–99.

Wolf, E. T. (2017) Assessing the habitability of the TRAPPIST-1 system using a 3D climate model. *Astrophysical Journal Letters*, 839, 6.

Wong, T. & Solomatov, V. S. (2016) Constraints on plate tectonics initiation from scaling laws for single-cell convection. *Physics of the Earth and Planetary Interiors*, 257, 128–36.

Worsley, T. R., Nance, D. & Moody, J. B. (1984) Global tectonics and eustasy for the past 2 billion years. *Marine Geology*, 58, 373–400.

Wright, C., Muirhead, K. J. & Dixon, A. E. (1985) The P wave velocity structure near the base of the mantle. *Journal of Geophysical Research: Solid Earth and Planets*, 90, 623–34.

Yin, A. (2012) An episodic slab-rollback model for the origin of the Tharsis rise on Mars: Implications for initiation of local plate subduction and final unification of a kinematically linked global plate-tectonic network on Earth. *Lithosphere*, 4, 553–93.

Yuan, H. Y. (2015) Secular change in Archaean crust formation recorded in Western Australia. *Nature Geoscience*, 8, 808–13.

Zachos, J., Pagani, M., Sloan, L., et al. (2001) Trends, rhythms, and aberrations in global climate 65 Ma to present. *Science*, 292, 686–93.

Zalasiewicz, J. & Williams, M. (2012) *The Goldilocks Planet*, Oxford University Press, Oxford.

Zhang, K.-J., Xia, B., Zhang, Y.-X., et al. (2014) Central Tibetan Meso-Tethyan oceanic plateau. *Lithos*, 210, 278–88.

Zhang, N. & Zhong, S. J. (2011) Heat fluxes at the Earth's surface and core–mantle boundary since Pangea formation and their implications for the geomagnetic superchrons. *Earth and Planetary Science Letters*, 306, 205–16.

Zhang, P. Z., Shen, Z., Wang, M., et al. (2004) Continuous deformation of the Tibetan Plateau from Global Positioning System data. *Geology*, 32, 809–12.

Zhang, Y. F., Wu, Y., Wang, C., et al. (2016) Experimental constraints on the fate of subducted upper continental crust beyond the 'depth of no return'. *Geochimica et Cosmochimica Acta*, 186, 207–25.

Zhao, W. L. & Morgan, W. J. (1987) Injection of Indian crust into Tibetan lower crust: A two-dimensional finite-element model study. *Tectonics*, 6, 489–504.

Zhong, S. & Rudolph, M. L. (2015) On the temporal evolution of long-wavelength mantle structure of the Earth since the early Paleozoic. *Geochemistry Geophysics Geosystems*, 16, 1599–615.

Index